Next Stop Mars

The Why, How, and When of Human Missions

Giancarlo Genta

Next Stop Mars

The Why, How, and When of Human Missions

 Springer

Published in association with
 Praxis Publishing
Chichester, UK

Giancarlo Genta
Torino, Italy

SPRINGER-PRAXIS BOOKS IN SPACE EXPLORATION

Springer Praxis Books
ISBN 978-3-319-44310-2   ISBN 978-3-319-44311-9   (eBook)
DOI 10.1007/978-3-319-44311-9

Library of Congress Control Number: 2016954425

Cover design: Jim Wilkie
Project editor: David M. Harland

Printed on acid-free paper

This Springer imprint is published by Springer Nature
The registered company is Springer International Publishing AG
The registered company address is: Gewerbestrasse 11, 6330 Cham, Switzerland

To Franca and Alessandro

Contents

Foreword

Every space enthusiast knows of the challenge made by President John F. Kennedy during an address to Congress on 25 May 1961, "I believe that this nation should commit itself to achieving the goal, before this decade is out, of landing a man on the Moon and returning him safely to the Earth. " This took the world by surprise, coming as it did at a time when the United States had yet to place a man into Earth orbit and NASA, less than 3 years old, manifestly did not have the faintest idea of how to reach the Moon.

Only a few cognoscenti know the genesis of Kennedy's speech. Deeply struck by the double blow to the US space program (and to national pride) delivered first in 1957 by Sputnik and then again by the pioneering orbital mission of Yuri Gagarin on 12 April 1961, the young President wanted to answer the Soviet achievements with a crushing blow that would clearly establish his nation's leadership in space.

According to the story told by Buzz Aldrin in his last book, Kennedy started by asking NASA to place an American crew on Mars. Seriously! The President tried to order NASA to make a Mars mission right away. Panicking NASA officials had the guts to say that Mars was out of the question. They then added hesitatingly, "However, Mr. President, we could maybe do the Moon, if you give us at least 15 years… "

In making his historic speech the next day, the daring President (who could hardly have imagined he would be assassinated a couple of years later) reduced the time estimate for a mission to the Moon to just under 9 years. And to their eternal credit, the space agency was able to fly its proud MISSION ACCOMPLISHED banner in July 1969, after a little over 8 years.

It is fascinating to speculate what might have happened had Kennedy accepted the requested 15 years. In all probability, we would have never been to the Moon. The financial, political and, above all, human and psychological burden of the Vietnam war, whose folly was finally acknowledged in 1975, would have stopped the Apollo program well before it could achieve its goal. As a matter of fact, 1969 was itself a marginal schedule and, as is well known, the program was cut short in 1972.

An indirect proof of this scenario comes from another little known event that concerns Wernher von Braun, who was not only the heart and soul of the Apollo program but also a visionary in the case for Mars. In August 1969, right after the successful Moon landing, von Braun, the hero of the day, told the Congressional Committee for Space, "The Moon is done. Now, on to Mars."

Von Braun proceeded to give a short presentation (available in the NASA archive) which outlined a detailed and very concrete plan to get to Mars by 1981. He compared the funding requirement to "the cost of a limited operation in a minor theater of war." His estimate wasn't a joke, it was true. Impressed by his presentation, the Committee voted on the project, possibly without realizing the historical importance of such a vote for future generations. In view of the pressure of the war, which was then at its peak in terms of US engagement and losses, the proposal was rejected by just a handful of votes. The Mars project was abandoned, Apollo was prematurely terminated in 1972, and in that year Von Braun left NASA. Since then, no human being has gone beyond a low Earth orbit.

Giancarlo Genta has now given us a book that represents a small step in the right direction for a giant leap to the Red Planet. This truly remarkable book has been badly needed. It contains not only Giancarlo's wisdom, experience and endurance, but also the work of the global IAA working group that he coordinated. As such, the book is even more precious. It addresses all aspects of the exploration of the Red Planet, from its early history to the upcoming plans for human missions of a variety of types. More than a textbook, it is a veritable pocket guide, and a must-read reference for the beginner and for the expert alike.

In addition, its publication comes at a time where new bold plans for landing people on Mars are under serious consideration all across the world. The clock is ticking.

Imagine you find yourself, all of a sudden, thanks to some magic space-time machine, in the US, in rural Ohio, in the dusty summer of 1930, only one year after the start of the Great Depression. You are actually near the small town of Wakaponeta. It is the end of August and, in the heat of the day, you encounter a young woman nursing a small baby in the shade of a tree. You approach her and discover that her name is Viola Armstrong. You smile and tell her, "In 39 years, your baby boy will be the first man to walk on the Moon." She would certainly look at you in wide-eyed disbelief...

Yes, the clock is indeed ticking, and the baby who will walk on Mars (boy or girl) has already been born. Giancarlo's book will help it all happen.

Giovanni Bignami
Accademia dei Lincei and International Academy of Astronautics
Milano, Italy
May 2016

Acknowledgements

The author is coordinating the study group of the International Academy of Astronautics (IAA) whose aim is to study a global human Mars Mission (SG 3.16). While acknowledging the contribution given by all the members of the group and, in particular by the coordinators of its various sections, the ideas in this book are my own and do not involve either the study group or the IAA.

I am deeply indebted to all my colleagues at the IAA and the university who made suggestions, criticism, and bibliographical indications that greatly assisted my writing. In particular, I would like to thank (in alphabetical order) Giovanni Bignami, Claudio Bruno, Alain Dupas, Richard Heidmann, Les Johnson, Nick Kanas, Julien Alexandre Lamamy, Susan McKenna-Lawlor, Maria Antonietta Perino, Giuseppe Reibaldi, Andreas Rittweger, and Jean-Marc Salotti. Obviously the responsibility for any errors or omissions is fully mine.

Many students worked for their theses and undertook PhD research on themes related to Human Mars Exploration. The work of Marco Dolci, Federica Maffione, and Cristiano Pizzamiglio was essential in writing some parts of this book.

I have made every effort to obtain permission from the copyright holders of the figures, but I apologize for cases, in particular for figures taken from the internet, where I have not been able to achieve my objective. If any reader has appropriate information, please contact the publisher and I shall happily include a credit in a future edition of this book.

And I must express my sincere thanks to Clive Horwood of Praxis Publishing in the United Kingdom, Maury Solomon of Springer in New York, and David M. Harland, their appointed editor, for constructive criticism and suggestions which greatly improved the project.

Last, but not least, this book could not have been written without the support, encouragement, criticism, and suggestions by my wife Franca – my advisor, critic, editor, companion, and best friend for over 45 years – who read the manuscript several times.

Author's preface

On 20 July 1969 two humans landed for the first time in history upon a celestial body: the Moon. This was the fulfillment of the dreams of many pioneers of astronautics and appeared to open a new era. As Konstantin Tsiolkovsky wrote, "Earth is the cradle of humanity, but one cannot live in a cradle forever." In the new era we would start our true life as citizens of the Universe. In a less poetic manner, nowadays we would say we would start a spacefaring civilization.

That sooner or later we would land on the Moon was not taken for granted by everybody. Forty-three years earlier, in 1926, a British scientist A.W. Bickerton wrote, "This foolish idea of shooting at the Moon is an example of the absurd lengths to which vicious specialization will carry scientists. To escape Earth's gravitation, a projectile needs a velocity of 7 miles per second. The thermal energy at this speed is 15,180 calories [per gram]. Hence the proposition appears to be basically impossible."[1]

What Bickerton considered impossible was leaving the Earth's gravitational well. In his view, reaching the Moon was as impossible as going to Mars or any other body in our solar system. Now that we have proven him wrong in the case of the former, we can make plans to go farther.

Space travel seemed to belong more to fiction, particularly to science fiction, than to science or technology. Perhaps more so in the opinion of scientists aware of the difficulties involved in such an enterprise, as opposed to the men-of-letters who were much interested in it. As an example of the latter, in 1952 the Italian writer Carlo Emilio Gadda wrote about the exploration of the Moon, Venus and Mars, and defined these celestial bodies as the New Indies, a definition which implies not only exploration but also colonization.

Unfortunately once the technical problems seemed to have been solved, other problems, mostly to do with politics and economics, forced a stop to the creation of a spacefaring civilization.

---

[1] This much quoted sentence, however, likely was meant to deny the possibility of literally shooting to the Moon using a gun, as in Jules Verne's famous novel. Bickerton probably did not mean that traveling to the Moon using a rocket would be impossible.

This false start has still to be fully understood. Perhaps our technology was still inadequate, or the projected costs were too high, or the motivations for the lunar adventure were too bound up with the Cold War and had to fade away once one of the two parties had demonstrated it could beat the opponent in this area. It is clear that the technological advances made since the time of Apollo, and those that are predicted for the near future, will make human space exploration much easier, safer, and less costly, thereby enabling us finally to realize the dreams of recent decades.

There is no agreement on the target for this renewal of exploration beyond low Earth orbit. Some hold that we should return to the Moon, saying this is the natural candidate for our first experience of colonization, the first place beyond the Earth where humankind can live, work, create communities, and prosper both culturally and economically. They insist that the exploitation of lunar resources is sufficient reason in itself, because these will be necessary to venture farther and settle Mars and other more distant destinations.

Others argue that humans should focus directly on the exploration of Mars because the Moon, with its lack of an atmosphere, low gravity, and a day that is almost one month long, is too inhospitable for human colonization. Mars, on the contrary, is a true planet that could be made suitable for human life.

A third group argues for creating space stations either in Earth or lunar orbit, or in the gravitationally neutral Lagrange points of the Earth-Moon system. Their short-term goal is to colonize cislunar space. They say the first missions into deep space should be to asteroids, which offer immense resources.

Actually these three ways of seeing the future of humankind in space are not as different as it might appear, at least as far as long-term goals are concerned. The difference is more about the early priorities than the ultimate goals, as it is likely that humankind will ultimately settle both the Moon and Mars, plus many other celestial bodies, and that many people will live permanently in space habitats at various locations. The short-term difficulty is in choosing the best programs on which to concentrate our scarce resources.

Although there is little real doubt that we must return to the Moon as soon as possible and that asteroid missions might be pursued in a more or less far future, the next important destination for human exploration will be Mars. In this sense therefore, Mars is the next stop in our travel toward the stars.

In the opinion of the author, it is not a matter of whether human explorers will reach the Red Planet, but when this will occur, who will do it, and how they will do it. Advances in technology will make this increasingly possible and the growth of the world economy will make it increasingly affordable.

As Donald Rapp points out in the preface of his book *Human Missions to Mars* [22], in science there are roles for both advocates and skeptics. The former play an important role in imagining what might be, and stubbornly pursue a dream which might be difficult to realize but in the end be achievable. The latter identify the difficulties, the barriers, the pitfalls, and the unknowns that impede the path and point out the technical developments needed to enable such dreams to be fulfilled. Skeptics therefore play an essential role in the study of an enterprise as complex as a human mission to Mars. Their viewpoint must be accurately weighted so that we can proceed safely.

The aim of my book is to discuss in detail the problems, the opportunities, and the alternatives for performing the first human Mars exploration mission, possibly in a not-too-distant timeframe.

To explore a whole planet like Mars is an enormous, costly, dangerous task, the more so because it will be the first planet the human species attempts to explore. Of course I am not dismissing the Moon, but the distance to Mars, its size, and its complexity make it a much tougher objective. To speak of organizing a mission to Mars is necessarily reductive, because to explore Mars we must initially mount a campaign that consists of at least three coordinated missions, the organization of which will be as complex as the technological, scientific, and human aspects.

To proceed with this military jargon, we will speak of a 'campaign' made of a number of missions. The first one might be just a 'sortie' (as often a flyby or a short-stay mission is defined) but subsequent ones must be full-blown missions in which humans will spend more than one year on Mars. Since the cost, complexity, and risks of a 'sortie' are only marginally less than those for a longer mission, it has been suggested that even the first mission should have humans on the planet for more than a full year. Right at the beginning therefore, we must establish an 'outpost' on the planet which later can be permanently inhabited and become the nucleus from which the colonization of the Red Planet will start.

This book has 14 chapters and four appendices. The first chapter summarizes briefly the history of the various early projects that were devised with the aim of starting the human exploration of Mars. The motivations that justify resuming our operations beyond low Earth orbit, and in particular mounting a campaign which aims at human exploration of Mars, are described in Chapter 2. The environments humans will face when on Mars and on its satellites are described in Chapter 3. The issues related to both backward and forward contamination are briefly discussed.

In order to reach Mars (and later to come back) it is necessary to cross a large span of interplanetary space, with all the related problems due to radiation which constitute the biggest of the difficulties humans will encounter in this enterprise. Chapter 4 discusses interplanetary space and the dangers presented by this very harsh environment. Crew issues are very important in all human space missions, and will be particular so for a long and difficult voyage to Mars. These are dealt with in some detail in Chapter 5.

Chapter 6 focuses upon the journey to Mars. This is one of the most critical aspects and is right at the frontier of modern technology. Chapter 7 reviews the design of human Mars missions and the choices we must make in preparing to mount such an enterprise. How many astronauts must travel to Mars? How long must they remain on the planet? And which kind of propulsion ought we to use? These, and several other design choices are dealt with.

The explorers will need a place to live on Mars, in particular in case of long-stay missions. They will need a power plant and a number of other infrastructure items and devices. These are dealt with in Chapter 8. Then Chapter 9 considers devices which will allow astronauts to move around on Mars, crawling on the ground, flying in the atmosphere, and hopping from one place to the other.

Although apparently a less important issue, a number of infrastructures on Earth will be instrumental in undertaking a human mission to Mars. These include the communication network, the ground control centers, the astronaut training facilities, and laboratories needed to perform simulations of the various devices. This theme is the focus of Chapter 10.

An enterprise like a human mission to Mars cannot be improvised. It requires a long preparation and a roadmap must be studied in detail and implemented in the due timeframe, as described in Chapter 11. In this chapter the need to return to the Moon on our way to Mars, and the construction of an outpost (perhaps even a Moon Village) are discussed.

Chapter 12 deals with some more futuristic possibilities that may make it easier to reach Mars, to build an outpost there, and ultimately to colonize the Red Planet.

Chapter 13 contains examples of Mars missions designed using different criteria that address various different requirements. Missions of different types that use different types of propulsion are considered and compared.

Some conclusions are drawn in Chapter 14, and a reference section lists some of the books that have been published on this and related subjects. For the technically minded, there are appendices explaining astrodynamics and the issues of mobility on Mars.

Giancarlo Genta
Torino, Italy
May 2016

Acronyms

ABS	Antilock system (in German)
AC	Alternating Current
ACR	Anomalous Cosmic Rays
ALARA	As Low As Reasonably Achievable
AM	Additive Manufacturing
aRED	advanced Resistive Exercise Device
ARM	Asteroid Redirect Mission
ATV	All-Terrain Vehicle
ATV	Automated Transfer Vehicle
AU	Astronomical Unit
BLEO	Beyond Low Earth Orbit
CAPS	Crew Altitude Protection Suit
CEV	Crew Exploration Vehicle
CEV	Constant Ejection Velocity
COSPAR	COmmittee on SPAce Research
COTS	Commercial Orbital Transportation Services
COUPUOS	Committee on the Peaceful Use of Outer Space
CSTS	Crew Space Transportation System
CVT	Continuously Variable Transmission
DC	Direct Current
DOD	Department of Defense
DOE	Department Of Energy
DRA	Design Reference Architecture
DRM	Design Reference Mission
DSN	Deep Space Network
ECLSS	Environment Control and Life Support System
EEG	ElectroEncephaloGram
EDL	Entry Descent and Landing
EGR	Exhaust Gas Recirculation

EMPIRE	Early Manned Planetary-Interplanetary Roundtrip Expeditions
EMU	Extravehicular Mobility Unit
ERTA	Electro-Raketniy Transportniy Apparat (Russian)
ERV	Earth Return Vehicle
ESA	European Space Agency
ESOC	European Space Operations Centre
ESTRACK	European Space TRACKing network
EVA	Extra Vehicular Activity ·
FAR	Federal Acquisition Regulations
FLEN	Flyby Landing Excursion Module
FY	Fiscal Year
GCR	Galactic Cosmic Rays
GDP	Gross Domestic Product
GES	Global Exploration Strategy
GER	Global Exploration Roadmap
GN&C	Guidance, Navigation & Control
HEAB	High Energy AeroBraking
HIAD	Hypersonic Inflatable Atmospheric Decelerator
HMI	Human-Machine Interface
HMMFI	Human Mars Mission Feasibility Index
HMMWV	High Mobility Multipurpose Wheeled Vehicle
HRP	Human Research Program
HSTI	Human Space Technology Initiative
HT	High Thrust
IAA	International Academy of Astronautics
IAC	International Astronautic Congress
IAF	International Astronautic Federation
ICAMSR	International Committee Against Mars Sample Return
ICE	Internal Combustion Engine
ICME	Interplanetary Coronal Mass Ejections
IMF	Interplanetary Magnetic Field
IMEF	International Mars Exploration Forum
IMLEO	Initial Mass in LEO
ISAS	Institute of Space and Astronautical Science
ISECG	International Space Exploration Coordination Group
ISRU	In Situ Resource Utilization
ISPP	In Situ Propellant Production
ISP	Specific Impulse
ISRO	Indian Space Research Organization
ISS	International Space Station
JPL	Jet Propulsion Laboratory
LED	Light Emitting Diode
LEO	Low Earth Orbit
LEVA	Lunar Extra-vehicular Visor Assembly
LMO	Low Mars Orbit

LH2	Liquid Hydrogen
LN2	Liquid Nitrogen
LOC	Loss Of Crew
LOX	Liquid OXygen
LRV	Lunar Roving Vehicle
LOM	Loss Of Mission
LOP	Loss Of Program
LT	Low Thrust
MARIE	Mars Radiation Environment Experiment
MARPOST	MARs Piloted Orbital STation
MAV	Mars Ascent Vehicle
MAVR	MArs-VeneRa
MEK	Mars Expeditionary Complex
MEM	Mars Excursion Module
MEPAG	Mars Exploration Program Analysis Group
MIT	Massachusetts Institute of Technology
MMH	MonoMethylHydrazine
MO	Mars Orbit
MOI	Mars Orbit Insertion
MORL	Manned Orbiting Research Laboratory
MPK	Mars Piloted Complex
MSR	Mars Sample Return
MSSR	Mars Surface Sample Return
MTV	Mars Transfer Vehicle
NASA	National Aeronautics and Space Administration
NCRP	National Council on Radiation Protection
NEA	Near Earth Asteroids
NEP	Nuclear Electric Propulsion
NERVA	Nuclear Engine for Rocket Vehicle Application
NIMF	Nuclear rocket using Indigenous Martian Fuel
NTO	Nitrogen TetrOxide
NTP	Nuclear Thermal Propulsion
NTR	Nuclear Thermal Rocket
OPS	Oxygen Purge System
PEL	Permissible Exposure Limit
POF	Probability Of Failure
PLSS	Portable Life Support System
PVA	PhotoVoltaic Arrays
PVT	Psychomotor Vigilance Test
PWM	Pulse Width Modulation
RCS	Reaction Control System
RCU	Remote Control Unit
REID	Risk of Exposure Induced Death
RWGS	Reverse Water Gas Shift
RTG	Radioisotope Thermoelectric Generator

SAA	South Atlantic Anomaly
SAIC	Science Applications International Corporation
SAS	Space Adaptation Syndrome
SCB	Sample Collection Bag
SEI	Space Exploration Initiative
SEP	Solar Electric Propulsion
SETV	Solar Electric Transfer Vehicle
SHAB	Surface HABitat
SI	International System of units
SLS	Space Launch System (NASA heavy launcher)
SM	Service Module
SNAP	System Nuclear Auxiliary Power
SPE	Solar Particle Event
SRC	Sample Return Capsule
STCAEM	Space Transfer Concepts and Analyses for Exploration Missions
STEM	Science, Technology, Engineering and Mathematics
STP	Standard Temperature and Pressure
TEI	Trans-Earth Injection
TMI	Trans-Mars Injection
TMK	Heavy Interplanetary Spacecraft (in Russian)
TPS	Thermal Protection System
TRL	Technology Readiness Level
UNOOSA	United Nations Office for Outer Space Affairs
UDMH	Unsymmetric DiMethylHydrazine
UMPIRE	Unfavorable Manned Planetary-Interplanetary Roundtrip Expeditions
VASIMR	Variable Specific Impulse Magnetoplasma Rocket
VDC	Vehicle Dynamics Control
VEV	Variable Ejection Velocity
ZBO	Zero-Boil-Off

1

Half a century of projects

Mars has always been a source of fascination for humankind, and dreams of traveling to the Red Planet are common in literature. In the last 65 years however, increasingly realistic plans have been proposed. The history of projects for human Mars missions is a long one. The most important cases are summarized in this chapter.

1.1 THE NINETEENTH CENTURY MARS

The Red Planet has for centuries fostered dreams and legends. Galileo, who in 1609 was the first man to clearly observe geographical features on an extraterrestrial body, the Moon, aimed his telescope also to Mars without succeeding in detecting anything except the fact that its disc was slightly flattened at the poles. The first person to claim to have seen something on Mars was Francesco Fontana who, in 1636, drew a rough map of the planet. Unfortunately it was later realized that the features he saw were optical illusions.

In the following centuries, generations of astronomers tried to map the surface of the Red Planet. Christian Huygens and Giandomenico Cassini succeeded in measuring the length of its day, which is now known as a *sol*. Cassini's value of 24 hours 40 minutes is remarkably close to the correct 24 hours 39.6 minutes. He also discovered the southern polar cap.

But distinguishing details on Mars was very difficult, and beyond the performance of the telescopes of those times. When telescopes powerful enough to see details on the planet's disc became available, new surprises were at hand. The features were changing over time. In particular, the ice caps at the poles extended in winter and contracted in summer. Variations in the colors of the surface suggested the presence of vegetation. Darker areas were interpreted as seas. As a whole, Mars seemed to be a smaller sister of Earth: a living planet inhabited by an unknown flora and, perhaps, fauna. Now we know that most of the changes we see on the surface of Mars are due to sand and fine particles being blown around by the wind, but at that time there was no way to ascertain this.

In 1867, Richard Anthony Proctor drew a detailed map on which he assigned names to the various features. In 1869, Jules Janssen, using a spectroscope, concluded that although the atmosphere was thin there was water on the surface of the planet.

© Springer International Publishing Switzerland 2017
G. Genta, *Next Stop Mars*, Springer Praxis Books, DOI 10.1007/978-3-319-44311-9_1

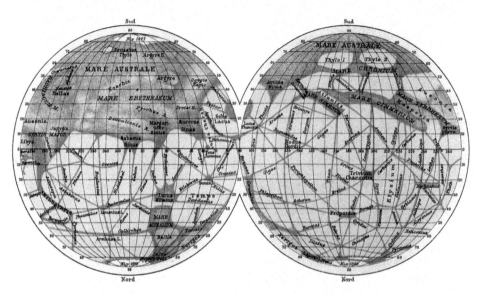

Figure 1.1 A map of the two hemispheres of the planet Mars drawn by Schiaparelli following his observations during six oppositions between 1877 and 1888.

It was presumed that not only living creatures but also intelligent beings – more or less similar to ourselves – roamed on the surface of Mars.

In the latter half of the nineteenth century, three great astronomers, the Italian Giovanni Schiaparelli, the Frenchman Camille Flammarion, and the American Percival Lowell made a series of contributions both to the scientific knowledge and mythology concerning Mars. The former plotted a number of maps (e.g. Figure 1.1) that remained an important reference until the first images to be received from a space probe completely changed our understanding of the planet.

As shown in the figure, the maps drawn by astronomers were oriented in the same manner that they saw the planet in their telescopes, namely with south toward the top and the western limb on the right. Hence to compare the map of Figure 1.1 with the modern one in Figure 3.4, the former must be rotated 180°.

Schiaparelli was the first to detect some thin dark features on the surface of the planet. He described these lines using the Italian word canali, which can be used both for artificial and natural water courses. However, the translation into English as "canals" was limited to artificial waterways and this led to many speculations about the civilization that might have undertaken such gigantic works of engineering, supposedly to survive the rapid process of desertification on their planet.

The idea that there were intelligent beings on Mars prompted many novels, ranging from *The War of the Worlds* by H.G. Wells to *Under the Moons of Mars* by E.R. Burroughs, and from *Out of the Silent Planet* by C.S. Lewis to *The Martian Chronicles* by Ray Bradbury… to name just a few.

Although some of the classical misconceptions were corrected in the first half of the twentieth century – notably, because there was neither oxygen nor water vapor in substantial quantity in the atmosphere, there could be very little water on the surface and the

Figure 1.2 Hypothetical forms of Martian life. Drawing by Douglas Chaffee for an article by Carl Sagan in *National Geographic* in 1965.

canals were an optical illusion – our impression of the Red Planet continued to bear a striking resemblance to that of Schiaparelli and Lowell, except with it being a dry world possessing a thin atmosphere which was probably inhabited at least by some primitive forms of life. Intelligent beings, if still present, must have sought refuge underground. A common feature of many descriptions were the "atmosphere machines," huge artifacts built by the intelligent Martians to maintain for as long as possible the conditions necessary for their survival. In many descriptions even some canals survived the likelihood that they were merely optical illusions.

This was the planet described by Wernher von Braun [1] when popularizing his 1950s project for a human expedition to Mars.

By the 1960s, further astronomical work showed that if the planet hosted any form of life, that could only be the most primitive of species. Nevertheless, when Carl Sagan published an article in the *National Geographic* in 1965 and suggested that Mars lacked an ozone layer, he illustrated his article with hypothetical forms of life that had developed a protective layer against radiation from the Sun (Figure 1.2).

1.2 THE DISAPPOINTMENT OF THE PROBES

In 1960, just three years after launching Sputnik 1 as the first Earth satellite, the Soviet Union attempted to send two probes to fly past Mars. Mars 1960A and Mars 1960B each weighed 650 kg and carried a variety of scientific instruments. Both were lost when their rockets failed. At the next launch opportunity in 1962, the Russians launched three more probes. Two failed to start their interplanetary voyage and the third, designated Mars 1, was lost en route to the Red Planet. A further attempt in 1964 with the Zond 2 probe also failed.

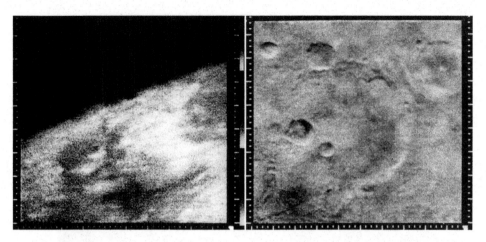

Figure 1.3 The Mariner 4 probe in 1965 gave us our first close view of the surface of Mars. (NASA images)

In 1964, America tried its hand at Mars exploration by launching Mariner 3 and Mariner 4. Built by the Jet Propulsion Laboratory (JPL), these probes were to fly close to the planet. The first mission failed, but the second, launched on November 28, reached its target on July 14, 1965. Twenty two pictures were recorded on tape on board and later transmitted to Earth over a period of 4 days. Altogether the images recorded about 1 percent of the surface of the planet.

Two of the pictures are shown in Figure 1.3. The one on the left is the first image ever received from Mars. It is an oblique view of the limb and covers an area of about 330 km by 1,200 km.

The images were shocking and really disappointing because they showed many impact craters, some of which appeared to indicate traces of ice. The overall impression was of a planet unsuitable for life. Mars was apparently a desolate place, rather similar to the Moon. Other data showed that the planet has only a very weak magnetic field (evaluated at about 0.1 percent of the strength of Earth's). The manner in which the radio signal was attenuated as the probe crossed the limb of the planet revealed the atmosphere to be made almost entirely of carbon dioxide rather than, as had been believed, mostly nitrogen.

In 1969 Mariner 6 and Mariner 7 made similar flybys of the planet and transmitted a larger number of pictures which were better than those of Mariner 4. Although these results essentially confirmed what had been learned in 1965, it was also realized that Mars is not so similar to the Moon as was initially suspected. In particular, the south polar cap appeared to be primarily solid carbon dioxide, and the mean atmospheric pressure at the surface of 6 to 7 millibars was lower than expected.

The next mission was Mariner 9. This entered orbit around the Red Planet in order to map it. Launched in 1971, the probe arrived when most of the planet was covered by a global dust storm. When the dust settled, the many images transmitted over a period of almost a year revealed that Mars is much more complex than the impression gained from the limited coverage of the flyby missions. It has huge volcanoes and canyons and, above

Figure 1.4 Chryse Planitia viewed from ground level by the Viking 1 lander. (NASA image)

all, dry river beds. For sure, if Mars today is a dry and dead world, then earlier in its history it must have been both warmer and wetter. There was the tantalizing prospect that life originated in that earlier era, then perhaps it may still survive today.

The next step was to attempt a landing on the planet. In the late summer of 1975 a pair of Viking missions were launched, carrying for the first time scientific instruments to analyze the surface of Mars in order to search for life. Both landers, each with a mass of 576 kg, touched down safely: the first on the western slope of Chryse Planitia (Figure 1.4) and the second on Utopia Planitia, located on the opposite side of the planet.

Besides taking pictures and collecting other science data, the two landers conducted three biology experiments designed to identify the presence of living organisms. These experiments revealed unexpected and enigmatic chemical activity in the Martian soil, but provided no clear evidence for the presence of life at the landing sites. Apparently, Mars' surface is sterile owing to a combination of solar ultraviolet radiation, the extremely arid conditions, and the oxidizing nature of the soil.

The results from the probes of the 1960s and 1970s led many scientists to pessimistic conclusions not only about finding life in the solar system but also of humans being able to explore Mars with a view to ultimately colonizing it.

The Viking results completed the shift of the paradigm about Mars. The Red Planet of the astronomers had yielded to the Mars of the space probes. In later years other missions were sent by NASA, Roscosmos, ESA, ISAS and ISRO, and a succession of orbiters, landers and rovers gradually refined our impression of the planet (Figure 1.5). Although the results have unveiled many of the mysteries, thousands of important details remain to be clarified before humans will be able to land there.

We are now sure that liquid water flowed on the surface of Mars in the distant past, at a time when it was much less desolate than today. The low pressure and temperature, and the composition of the atmosphere together with the results of the experiments aimed at searching for life, eliminated all hopes of finding higher forms of life and, for the majority of scientists, even the prospect of finding bacteria seems bleak.

Figure 1.5 A view of Mount Sharp at the center of Gale Crater, taken by the Curiosity rover on September 9, 2015. (NASA image)

Phobos and Deimos, the small satellites of Mars, have also attracted attention. Prior to probes providing images, some scientists, for instance, explained the low density of the former by assuming it to be an artificial satellite or, better still, a large space station built by the dying Martian civilization as a sort of library or museum to preserve its legacy. Several probes have imaged the two satellites, showing them to be irregular shapes with cratered surfaces. There have been several unsuccessful attempts to land instruments on Phobos, the larger of the pair. These bodies play an important role in some plans for human Mars exploration.

Since the first attempts in the early 1960s, not all of the probes launched to Mars were successful. Some failed to leave Earth. Some fell silent during the interplanetary voyage. A few remained operational but flew by the planet at too great a distance to make any proper observations. Some made perfect flybys. Some failed to achieve Mars orbit. Some achieved the wrong orbit. Some worked perfectly. Some intended landers missed the planet entirely. Others crashed. A few reached the surface and functioned perfectly. Landing on Mars was a formidable challenge, but the success rate has increased over the years and nowadays we are able to make precision landings with reasonable safety.

Many unknowns remain. We know that Mars hosts a large amount of water – although likely not in liquid form. It is present in the form of subsurface ice, but we are still unsure of where it is located and at which depth. We have clues to the existence of large caves. These might be useful as bases when settling the planet, but we aren't sure about how many there are. We still need to ascertain the radiation dose to which humans will be subjected while on the surface. There are many such questions. Many automatic missions are still required and, above all, there will need to be sample return missions before anyone is able to set foot on the Red Planet. The dream of sending a human mission to Mars is becoming ever more feasible, and at some time in the not too distant future the first ambassador from Earth will send back a message from its surface.

1.3   THE EARLY PROJECTS (1947–1972)

1.3.1   General considerations

This early period starts with the first detailed studies carried out in the immediate aftermath of World War II, and lasts until the setback of human space exploration which terminated the Apollo lunar program. As stated above, fictional accounts of voyages to Mars can be found in the literature much earlier, namely since the last decade of the nineteenth century.[1]

Often the voyage to Mars – in many case the very first expedition – is described in detail, but sometimes the problem of how this is achieved is completely bypassed. For instance, John Carter, the main character of Burroughs' novels, simply falls asleep (or dies) on Earth and awakens (or is re-embodied) on Mars. Even where an actual journey is described, that is a fantasy and therefore has little to do with the subject of this book.

Essays written by true space travel pioneers like Konstantin Tsiolkovsky, R.H. Goddard, Hermann Oberth and several others accurately represent the basic concepts that will make a flight to Mars possible, but they do not elaborate the details of the design of such a mission.

To qualify as a project for a mission, a study must contain at least the main details required to actually implement the mission, and must at least touch on basic issues regarding the feasibility, safety, and possibly the cost of sending the crew to Mars and home again. If possible, it should also account for the emergencies which may take place, and how these might be dealt with.

The earliest detailed studies began in 1947, and were based on the assumption that Mars was like the nineteenth century astronomers described, namely a dry but not completely dead world, with an atmosphere which humans could not breathe but neither would they require a full space suit (as would be needed on the Moon), and an air density which, although low, was sufficient for aerodynamic flight.

But this optimistic picture was ruled out by the first close pictures taken of the planet by Mariner 4 in 1965. Slowly it was realized that the atmosphere was only marginally better than the vacuum of the Moon, that astronauts would require full space suits, and that using a glider in order to land would be impracticable. Thus 1965 marks a turning point.

In most early designs, the idea was that the crew would reach Mars without much previous knowledge of the planet, and that they would spend their first days in orbit mapping in order to select their landing site. Perhaps they might send automatic probes down to the surface to get some idea of what they would find when they themselves landed. Otherwise they would make the first sortie without any knowledge of what they would find, including whether there might be hostile or possibly even friendly Martians. As an alternative, they might send down teleoperated probes that would be driven from orbit, a possibility which was rightly deemed to be within the scope of predictable technology.

The need for humans to control and maintain spacecraft meant that some people would have to remain in orbit while others explored the surface. This was the way that the Apollo lunar missions were conducted.

---

[1] Several are described in https://en.wikipedia.org/wiki/Mars_in_fiction

After 1965, the Martian surface was considered a dangerous place, perhaps more so than space. In case of an accident, the astronauts would require to take off, enter orbit around the planet and then wait for the correct time to start the return journey: space was a safe haven. Following this idea, short stay missions or even flybys without landing were considered more expedient than long stay missions (see Section 7.2). Indeed, there was also the possibility of staying on Phobos or Deimos and teleoperating robots on the planet. Phobos orbits closer in and the time delay for teleoperating robots on Mars is just 40 ms; much less than the delay from Deimos (134 ms).

Even if knowledge of conditions on Mars was clearly insufficient to correctly design a mission, astrodynamics was very well known and (even without the powerful computers we have now) it was feasible to compute trajectories in enough detail to deal with that aspect of the mission design in a satisfactory manner.

1.3.2   Von Braun's project

A list of the studies for human Mars missions carried out between 1947 and 1972 is given in Table 1.1.[2] This lists complete projects along with partial studies, but is far from complete. It also specifies the name of the author (or the company or agency), the type of propulsion used for the interplanetary transfer, the crew size, and the Initial Mass in Earth Orbit (IMLEO) in tons (t).[3]

The first detailed study of a human mission to Mars was undertaken by Wernher von Braun between 1945 and 1948, and he published the results in 1949 as *Das Marsprojekt*. This technically-sound analysis established the feasibility of reaching Mars using a technology that was likely to become available in the not too distant future. An English translation of the study was published in 1952 [1].

The expedition was to be performed by a fleet of ten 3,720 t spaceships with a total crew of 70 astronauts. All interplanetary spacecraft would be powered by chemical rockets, operating on storable propellant (hydrazine and nitric acid). The ships would be assembled in Earth orbit, and this preliminary phase would require a total of 950 launches of huge multistage rockets of a size that would now be defined as heavy lift launchers.

An interesting solution was devised for landing on Mars. After all ships had entered Mars orbit, a winged craft equipped with skis would glide down to land on the north polar ice cap. This method was selected because the prevailing opinion at that time was that the atmosphere was much denser than it later proved to be. Consequently, von Braun designed entry vehicles as gliders with very large wings (Figure 1.6) which had to land horizontally like airplanes in a similar manner to the large military gliders of World War II.

Then those people who had landed would engage in a 4,000 km trip using a tractor to reach a suitable place near the equator at which to build a runway to permit two other winged ships to land with other landing parties and the materials to build an outpost which would allow the expedition to live on Mars for more than a year. After their wings had been removed, the landers would lift off vertically to reach Mars orbit. The entire crew would then occupy those ships designated for the return journey and head home.

---

[2] More detailed information can be found on *Encyclopedia Astronautica* (www.astronautix.com).

[3] The SI symbol for ton (t) is used throughout this book.

Table 1.1 Studies for human Mars missions performed between 1947 and 1972.

Year	Expedition	Country	Author	Prop.	Stay	Crew	IMLEO
1947	von Braun Mars Ex.	USA	W. von Braun	C	L	70	37,200
1956	MPK	USSR	Tikhonravov	C	L	6	1,630
1956	von Braun Mars Ex.	USA	W. von Braun	C	L	12	3,400
1957	Stuhlinger Mars	USA	Stuhlinger	NEP	L	200	6,600
1959	TMK-1	USSR	Maksimov	C	F	3	75
1960	TMK-E	USSR	Feoktistov	NEP	L	6	150
1960	Bono Mars Vehicle	USA	P. Bono	C	L	8	800
1960	Mars Ex. NASA	USA	NASA Lewis	NTP	S	7	614
1962	EMPIRE A	USA	Aeronutronic	NTP	F	6	170
1962	EMPIRE L	USA	Lockheed	NTP	F	3	100
1962	EMPIRE G.D	USA	Gen. Dyn.	NTP	S	8	900
1962	Stuhlinger Mars	USA	Stuhlinger	NEP	S	15	1,800
1963	Mavr Mars flyby	USSR	Maksimov	C	F	3	75
1963	Faget mars Ex.	USA	M. Faget	NTP	S	6	270
1963	Faget mars Ex.	USA	M. Faget	C	S	6	1,140
1963	TRW Mars Ex.	USA	TRW	C	S	6	650
1964	Project Deimos	USA	P. Bono	C	S	6	3,996
1964	UMPIRE C	USA	G. D. & Convair	C	L	—	—
1964	UMPIRE D	USA	G. D. & Douglas	NTP	S	6	450
1965	MORL Mars Flyby	USA	Douglas	C	F	3	360
1966	KK Mars Ex.	USSR	K. Feoktistov	NEP	S	3	150
1966	FLEM Mars Ex.	USA	R.R. Titus	C	S	3	118
1968	IMIS Mars Ex.	USA	Boeing	NTP	L	4	1,226
1969	von Braun Mars Ex.	USA	W. von Braun	NTP	S2	12	1,452
1969	MEK Mars Ex.	USSR	Chelomei	NEP	S	6	150
1971	NASA Mars Ex.	USA	NASA	C	S	5	1,700
1972	MK-700	USSR	Chelomei	NTP	S	2	1,400
1972	MK-700	USSR	Chelomei	C	S	2	2,500

(C: Chemical; F: flyby; S: Short stay; L: Long stay; S2: Short stay, 2 months; IMLEO is in t).

The total travel time of such a mission would be about 3 years, so this mission can be defined as a long stay one.

Although technologically consistent, this project did not take into account the likely costs and, as perhaps was inevitable for a first attempt, it was not economically sustainable. An artist's impression of the orbital assembly of the huge winged vehicles which were to land on Mars is shown in Figure 1.6.

Von Braun also wrote a science fiction novel (probably in the early 1950s) based on the mission to Mars described above, but this was not published until 2006 [20]. Apart from the very large crews involved, what is most striking for the modern reader is the description of the planet. It was very close to the late-nineteenth century impressions of Schiaparelli and Lowell. The intelligent beings living on Mars are somewhat more civilized than those of Burroughs' novels, but not all that different from them. Other points that might surprise modern readers, but were typical of the time in which von Braun was writing, are that the 70 person crew was all male and that psychological problems are not addressed. A mission to Mars was felt to be not much different from any wartime mission of the navy, particularly in submarines.

Figure 1.6 An artist's impression of the assembly in Earth orbit of the winged spacecraft designed by von Braun for landing on Mars. (NASA image)

Von Braun's project was updated and simplified by its author in 1956 and then again in 1962. In 1969 he proposed a smaller project in which the number of spacecraft was reduced to just two, one being a winged landing module, and the crew was cut to twelve people. By then, the Saturn V, a veritable superheavy lift launcher, was operational, sending Apollo crews to the Moon. This could certainly be used to assemble a Mars expedition in Earth orbit, but during the effort to reach the Moon even larger launchers had been studied. One of these, called Nova, was intended to directly launch a large spacecraft to land on the Moon and lift off for the return to Earth without involving a rendezvous in lunar orbit. Far larger than the Saturn V, this rocket would be capable of placing 250 t in LEO. Although abandoned for the Moon, the Nova was a natural choice for going to Mars.

1.3.3   Early Russian projects

Russians too were interested in reaching Mars. In 1956 a study designated MPK (Martian Piloted Complex) was proposed by Mikhail Tikhonravov. A 1,630 t spacecraft would be assembled in LEO by launching 25 N1 rockets (as developed for a human lunar venture). The spacecraft would have carried a crew of six to Mars on a 900 day mission. So this would also have been a long stay mission.

Other projects were developed in both the US and the USSR, some of which envisaged using either Nuclear Electric Propulsion (NEP) or Nuclear Thermal Propulsion (NTP).

The Russian acronym TMK (Heavy Interplanetary Spacecraft) was the designation of a number of interplanetary crewed spacecraft intended for Mars and Venus (Figure 1.7). The TMK-1 was a comparatively small mission on which a crew of three would make a Mars flyby in the 1971 launch opportunity. The 75 t spacecraft would have taken slightly more than 3 years to achieve the voyage. During the flyby, a number of probes would have been dropped onto the planet to be remotely operated by the crew during their return journey. The propellants for the propulsion system were to be kerosene and LOX. The leader of this project was G.U. Maksimov.

The TMK-E spacecraft was much larger, with a crew of six, and was to be propelled by NEP employing a xenon thruster. The crew were to land on Mars and the entire mission was to have lasted about 1,000 days. The project leader was Konstantin Feoktistov.

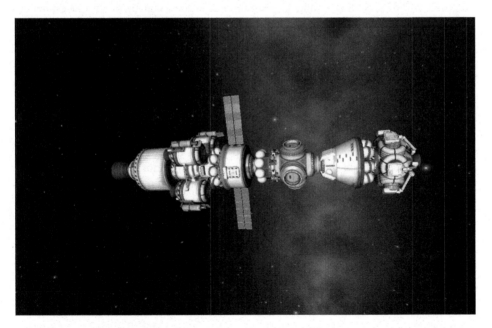

Figure 1.7   One of the Russian TMK Mars spacecraft. (http://i.imgur.com/9oLNM2J.png)

Figure 1.8 A nuclear electric spacecraft designed in 1962 for a human Mars mission. (Based on a NASA image)

1.3.4 Nuclear electric spacecraft

In America, Ernst Stuhlinger suggested using NEP for Mars in 1954. The mission that he proposed in 1957 (Stuhlinger Mars Expedition) envisaged a convoy of ten spaceships for a total crew of 200 astronauts.

Perhaps the best known of these early proposals was that which Stuhlinger developed in 1962 [3] (Figure 1.8), a few months after Kennedy's address advocating both a mission to the Moon and the development of nuclear propulsion. The mission was planned for the 1980s and was much more realistic than the previous one, involving only five spacecraft, powered by a 115 thermal MW nuclear reactor, with a total crew of 15. The project was quite detailed, and included both rotating the spacecraft at 1.3 rpm to produce 0.1 g of artificial gravity, and radiation shelters to which the crew could retreat during solar storms.

A total mission duration of 572 days was predicted, but if the astronauts joined their spacecraft after the initial acceleration phase and left them prior to the final deceleration this could be reduced to about 490 days. Since the time spent on the planet was assumed to be a mere 29 days, the mission could be classified as a short stay one. Some of the design parameters were quite optimistic; for instance, 2 kg/kW was assumed for the specific mass of the generator. The spacecraft included details that are still discussed today, including artificial gravity, a radiation shelter, and a huge thermal radiator.

1.3.5 Flybys and other projects

In a paper presented to the International Astronautical Congress (IAC) held in Rome in 1956 Gaetano Arturo Crocco suggested a fast trajectory for a Mars flyby with a total travel time of about 1 year.[4] The drawback of passing quite far from the planet could be avoided

---

[4] Gaetano A. Crocco, "One-Year Exploration-Trip Earth-Mars-Venus-Earth," Seventh Congress of the International Astronautical Federation, Rome, Rendiconti, pp. 227–252, Sept. 1956.

by using a gravity assist trajectory which, after Mars, visits Venus in a sort of Mars-Venus grand tour. A second gravity assist at Venus would enable the spacecraft to return to Earth in just about one year. Whilst not a minimum energy trajectory, the use of gravity assists would reduce the propellant requirements and hence the cost. This was not a mission design study, merely an astrodynamics analysis which showed that a Mars flyby can be performed with a cheap and small spacecraft.

A detailed design of a low cost mission to Mars using a single vehicle directly launched from Earth was performed in 1960 by Philip Bono, then at Boeing, and is widely known as the Bono Mars Expedition. The 3,800 t launch vehicle was based on a parallel stages architecture and would use cryogenic propellants (LOX/LH2). The launch in 1971 was to be direct to Mars, with no waiting in LEO. The spacecraft comprised an orbiter and a winged lander that would directly enter Mars' atmosphere carrying the crew of eight to the surface. The time spent on Mars was to be 479 days. Then a section of the lander, including the glider, would lift off and rendezvous with the orbiter. The united spacecraft would then head to Earth, where the glider would make a direct atmospheric entry.

In 1960 the NASA Lewis Center presented an alternative project for a Mars mission (NASA Mars Expedition) using the 1971 launch opportunity. The chief designer was Max Faget, the designer of the Mercury spacecraft. This project is interesting because it is the first to envisage using NTP. The 614 t spacecraft was to be assembled in LEO from several Saturn V payloads. Upon reaching Mars, all seven astronauts would descend to the surface for 40 days (making it a short stay mission). The Earth return would have ended with a winged spaceplane making a direct atmospheric entry. In the whole design, much thought was given to the radiation that the crew would have to endure, with a mass trade-off between a slower journey with a heavier radiation shelter and a faster one that would carry lighter shielding.

In his famous speech to a joint session of Congress on May 25, 1961, the same speech in which he challenged his nation to land a man on the Moon, President Kennedy also called for allocating $30 million to "…accelerate development of the Rover nuclear rocket. This gives promise of someday providing a means for even more exciting and ambitious exploration of space, perhaps beyond the Moon, perhaps to the very end of the solar system itself."

The opinion that nuclear propulsion would have to be developed to venture beyond the Moon was widespread, and serious action was taken to develop NTP. At that time, the NEP option was less promising because the development of large electric thrusters (ion or plasma) was considered to require much more time. In those days, nuclear generators were under development not to provide propulsion but for power generation in space. For example, the American SNAP (System Nuclear Auxiliary Power) and similar Russian generators were tested several time in space – the first SNAP 10 unit was launched in 1965. When electric thrusters did eventually become available, these were able to be powered by scaling up the auxiliary systems already tested.

In 1962, perhaps influenced by Crocco's proposal, a new project for a Mars flyby was developed: the EMPIRE (Early Manned Planetary-Interplanetary Round-trip Expeditions). Actually, the NASA Marshall Center funded three study contracts. These were assigned to Aeronutronic, Lockheed, and General Dynamics and all were based on NTP.

The Aeronutronic study was based on a spacecraft with an initial mass of 187.5 t that would be carried into LEO by a Nova rocket and then make a flyby of Mars. The Lockheed study was based on a smaller spacecraft that would be able to be launched using a Saturn V. The General Dynamics project was much bigger. A crew of eight would be carried aboard a spacecraft which would be accompanied by two cargo ships. Once in orbit around Mars, three automatic vehicles would land on the surface of Mars, Phobos and Deimos in order to collect specimens. After this first phase, a lander would deliver two astronauts to Mars to conduct a very short stay mission.

In 1963, NASA initiated another study by Max Faget. Two missions were proposed, both of which envisaged a six person crew spending 40 days on Mars. The first was based on two spacecraft with cryogenic (LOX/LH2) propulsion: one was a Mars Excursion Module (MEM, Figure 1.9) to be launched directly from Earth to the surface of Mars. After 40 days, the crew would lift off and proceed to rendezvous in solar orbit (instead of the more common solution of rendezvous in Mars orbit) with a second spacecraft in order to return to Earth. In this way the second spacecraft would not have to slow down at Mars. The mass to be carried in LEO would have been 1,140 t. The second mission concept envisaged a much smaller spacecraft of 270 t using NTP. Both missions would have needed the Nova launcher to achieve LEO prior to setting off for Mars.

Figure 1.9 An artist's conception of the Mars Excursion Module (MEM) proposed in a NASA study in 1964. F.P. Dixon, "Summary Presentation: Study of a Manned Mars Excursion Module," Proceeding of the Symposium on Manned Planetary Missions: 1963–1964 Status. Huntsville, Alabama: NASA George C. Marshall Space Flight Center.

In 1963 TRW designed a Mars expedition using LOX/LH2 propulsion, aerobraking at both Mars and Earth, and a gravity assist from Venus on the way home. Only 10 days would have been spent on the surface of Mars. Seven Saturn V launches would have been needed to assemble the spacecraft in LEO, and it would provide artificial gravity by rotating the crew compartment using a tether.

Also in 1963, the Russians worked on a variation of the TMK mission for a combined Mars-Venus flyby. This was to be propelled by LOX/kerosene combination, with the Venus flyby occurring during the trip home. It was named Mavr (MArs-VeneRa) and the project was chaired by G.U. Maksimov.

One year later, Philip Bono proposed Project Deimos. This envisaged a huge spacecraft based on the proposed Rombus single stage to orbit launch vehicle. After being launched from Earth it would be refueled with LOX/LH2 in LEO. The Mars-bound spacecraft would have an initial mass of 3,966 t. The spacecraft which would enter orbit around Mars at an altitude of 555 km would have a mass of 984.75 t. Three astronauts would remain aboard the orbiting spacecraft while their three colleagues descended in the MEM for a 20 day stay on the planet. The total round-trip time would be 830 days and for a consumption rate of 1.27 kg/day/person the total provision for the crew would have been 6,300 kg of food, oxygen and water. A novel aspect of this scenario was that most of the hardware, including the launcher, was intended to be reusable.

NASA launched other studies in 1963, this time to investigate traveling to Mars during an unfavorable launch opportunity in which the opposition of the planet occurs when it is not at its minimum possible distance from Earth. As companions to the EMPIRE studies, these were the UMPIRE (Unfavorable Manned Planetary-Interplanetary Roundtrip Expeditions) studies. A study by General Dynamics and Convair showed that only long stay missions with a duration of 800–1,000 days could be performed in these instances. However, another study by General Dynamics and Douglas found that if NTP was used then it would be possible to undertake a short stay mission of 30 days even at an unfavorable opposition. This study proposed to use the Rombus launch vehicle to place into LEO a nuclear spacecraft for a crew of six.

1.3.6 The new ideas on Mars

A study by Douglas in 1965 proposed using the Douglas Manned Orbiting Research Laboratory (MORL) as a habitation module for a Mars flyby. Four Saturn V rockets would deliver to LEO three fully fueled S-IVB stages plus the MORL, an Apollo command module with a crew of three, and a retro module that would slow the command module for re-entry at the end of the mission. The trajectory was essentially of the type proposed by Crocco, and the plan was to release robotic probes and landers while close to Mars. The advantage of this scenario was that practically all of the hardware required was being developed for the Apollo program and the IMLEO was only 360 t.

A new version of the TMK-E was proposed in 1966. This KK Mars expedition would be smaller, involving a crew of just three and 30 days on Mars (transforming it from a long stay mission to a short stay one). Two N1 rockets would be sufficient to launch the total IMLEO of 150 t, and owing to the high specific impulse the total propellant required was just 24 t.

A new approach meant to reduce the mass IMLEO was the Flyby-Landing Excursion Module (FLEM) of a proposal developed in 1966 by R.R. Titus at United Aircraft. The main spacecraft would perform a Mars flyby and drop a lander 60 days prior to reaching Mars. The lander would adopt a faster trajectory in order to land before the parent craft made its approach. After 19 days, the lander would lift off and dock with the main spacecraft in solar orbit. As in the Max Faget proposal described above, this strategy eliminated the need for the interplanetary ship to slow down at Mars; it performs a simple flyby. In this manner it would be possible to land three people on Mars for a stay of just 19 days with an IMLEO of just 118 t – which could be achieved by a single Saturn V launch.

An unmanned version of FLEM dubbed the Mars Surface Sample Return (MSSR) was proposed as a sample return vehicle for the NASA Planetary Joint Action group flyby plan developed in 1966–1967.

1.3.7   Using the NERVA NTR

The IMIS (Integrated Manned Interplanetary Spacecraft) designed by Boeing in 1968 was to use the NERVA nuclear rocket. This was the result of a 14 month study which was, in a way, the high point of more than two decades of design of Mars expeditions, and it served as the starting point when studies of human Mars missions resumed in the 1980s.

Five modular NERVA nuclear thermal stages were to launch on an interplanetary trajectory several unmanned probes, a manned MEM lander, a Mission Module crew compartment, and an Earth Entry Module for use at the end of the mission. The approach was toward a modular and flexible design that could accomplish a number of long stay and short stay missions to Mars and Venus at different launch opportunities, starting in the mid-1980s. The design of the lander was the first to incorporate the Mariner 4 data on the very low density of the Martian atmosphere. The total IMLEO for sending a crew of six to Mars was 1,226 t, which was to be carried aboard Saturn V rockets. The development cost through the first two missions to Mars was estimated at $30 billion, which was about 50 percent greater than the overall cost of the Apollo program.

After the first Moon landing in 1969, NASA began to make plans for something that would follow the Apollo Application Program, for which the Skylab space station was being created. Von Braun proposed a human Mars mission to be launched in 1981 with the goal of carrying a crew of 12 to Mars orbit, six of whom would land on the planet. On August 4, 1969 he presented what would be his final Mars proposal to the Space Task Group (STG), chaired by Vice President Spiro Agnew.

The design was based on an NTP spacecraft referred to as the Nuclear Shuttle, which was based on the NERVA rocket. This reusable spacecraft would have been used for lunar exploration, for setting up a lunar base, and then for a Mars mission. For redundancy, there would be two ships, each consisting of three Nuclear Shuttles (two of which were to resume LEO after helping with the initial acceleration of the departing vehicle), and a Mars vehicle that comprised a Planetary Mission Module (PMM) habitation section and a Mars Excursion Module (MEM). Since the recently discovered low density of the planet's atmosphere made the winged glider of von Braun's earlier concepts impracticable, the landing would involve a parachute, a ballute, and finally retro-rockets. A 2 month stay on Mars would require a Venus flyby on the way home, during which four probes would be released to study that planet. Once safely back in Earth orbit, the spacecraft would dock at a space station.

A cost of $7 billion was estimated for the first year, with an $8–10 billion annual spend thereafter. After approval by the Space Task Group and NASA, the proposal was passed on to President Nixon on September 15, 1969. However, the plan was rejected, in part owing to its high cost but also because the results from the flyby probes of that year confirmed the view of Mars as a dead world, not much different from the Moon. And of course the general political situation was not very conducive to initiating another major human space program.

In that same year, Russia proposed its own expedition to Mars. The Mars Expeditionary Complex (MEK) would use NEP to take a crew of three to six to Mars and back in 630 days. The design was named Aelita, after the Queen of Mars in a famous Russian science fiction novel and movie. The chief designers of the three leading design bureaus – Vasily Mishin, Mikhail Yangel and Vladimir Chelomei – began competitive designs for such expeditions.

In the final design the spacecraft would carry a crew of six, half of whom would spend 30 days in orbit around Mars whilst the others descended to spend one week on the surface. The total IMLEO was 150 t, which would have been launched by two N1 rockets, and the electric power required for propulsion was 15 MW.

By the end of 1969, Mishin and Yangel had withdrawn from the competition. Because the N1 rocket had failed on both of its test flights that year, it was evident that the advanced programs that required it would have to wait. However, the first step would clearly be to test the TMK in Earth orbit and it seemed wise to begin with a simpler Mars mission. In the end, the program was delayed to the next century.

In 1971, after Nixon's decision to terminate all designs of Mars missions, NASA set out to wrap up the work that it had achieved. So a final design was put together, now based, as far as possible, on hardware designed for the Space Shuttle. The design was based on cryogenic (LOX/LH2) propulsion and it was to place a crew of five into a highly elliptical orbit of Mars. They would then land and during a one month stay on the surface two people would explore 20,000 square km using a pair of single seat rovers. It would have required 71 Shuttle launches to achieve the IMLEO of 1,900 t. While in space, a slow rotation would have provided 1/6 g of artificial gravity. The total mission duration would be 570 days.

Meanwhile in Russia the failures of the N1 launcher imposed a pause. Chelomei was the only one to conclude an Aelita project. His design was a 1,400 t NTP spacecraft called MK-700 that would be assembled in LEO. A chemical version was expected to have an IMLEO of 2,500 t. During a mission lasting a total of 730 days, the crew of two would spend 30 days on Mars.

Different variants were suggested, the first based on a monster launcher capable of placing 750 to 800 t into LEO, sufficient to assemble the interplanetary craft with just two launches. A smaller variant was based on a 480 to 520 t launcher. The Mars spacecraft was to be tested using a Moon base, which was still in planning. About 12–15 years' of studies were deemed necessary to ensure that the crew would survive such a long voyage. The NTP technology would require 15–20 years to achieve the desired maturity. One potential problem would have been the diplomatic negotiations required in order to launch such a large reactor into orbit.

In the end, the likely protracted design time and the high cost (possibly 30 to 40 billion rubles) prompted a state commission to recommend that further work on Mars expeditions be deferred indefinitely. Nevertheless, work on NTP was permitted to continue in a test facility located 50 km southwest of Semipalatinsk, and also in another more elaborate facility. Two nuclear rockets were tested, one with a 35 kN thrust and one with a 700 kN thrust, and thirty simulated flights were conducted between 1970 and 1988 without failures.

Some sources indicate that Chelomei proposed another Mars flyby expedition in 1974 but the details are sketchy.

Of the 28 missions listed for the period 1947–1972, five (18 percent) were for Mars flybys, fifteen (53 percent) where for short stay missions, and eight (29 percent) were for long stay missions. They envisaged a wide range of IMLEO, varying from the very large masses of the early missions (up to 37,000 t) to the very low (and probably overly optimistic) masses for some flyby missions (as low as 75 t). In many cases the IMLEO was severely underestimated in these early studies.

1.4   THE POST-APOLLO ERA (1982–1990)

1.4.1   General consideration

As production of the Saturn V ceased and the Apollo program was drawn to a conclusion, the American development effort was focused on a single program: the Space Shuttle. All studies of nuclear rockets were halted and a mission to Mars receded far into the background. A poll by Gallup found that 53 percent of Americans were against such a mission and only 39 percent were in favor. This mood served only to further convince decision makers against starting new large human space exploration projects. Although the situation was not much different in the Soviet Union, studies of NTP and NEP were able to continue.

In the 1970s, when apathy to Mars exploration was at its peak, planetary scientist Chris McKay and graduate students at the University of Colorado at Boulder hosted a conference to discuss the exploration of Mars. Afterwards, an informal group called the Mars Underground set out to rekindle interest in human exploration of Mars, this time using private organizations such as The Planetary Society and later (in the 1990s) The Mars Society that was founded by Robert Zubrin, one of the prime proponents.

The projects conceived in the post-Apollo era often featured characteristics different to those of earlier years. These changes reflected a general rethinking of the problems which caused, or at least contributed to, the failure of the latter. The first reason for this failure was often attributed to the unsustainability of the huge projects that had been proposed. So the first improvement was to reduce the cost and the time required to implement the project. In particular, it was decided to 'live off the land' by exploiting resources on Mars to make some of the consumables for the mission. In many cases, the quantity of consumables for the crew (air, water, food) and for the rockets (propellant) had been underestimated. Nevertheless, the amount to be carried to Mars proved to be very large. Most of the projects devised during this second period included at least one of the various forms of what came to be known as either In Situ Resources Utilization (ISRU) or In Situ Propellant Production (ISPP).

Many missions studied in this period were 'split' missions, meaning that the cargo was intended to be carried to Mars by a different spacecraft (sometimes a completely different type) from that carrying the crew. In several cases, the cargo would be sent to Mars on an earlier launch opportunity.

To reduce costs, in particular when chemical propulsion was planned, aerodynamic forces in the atmosphere would be used to decelerate while entering orbit, and then again later when landing.

Another significant realization was that, even though Mars is an inhospitable place, it is better than being in space. In particular, due to the lack of Van Allen Belts on Mars, Low Mars Orbit (LMO) is as rich in radiation as interplanetary space. So if something goes wrong, it is better to seek refuge on the surface than in space. As a result, most of the projects that were developed during this period adopted the concept of a 'safe haven' on the planet.

A related issue was the emphasis placed on long stay missions versus short stay ones or, even worse, flybys. That is, because at least one year is required in order to travel to Mars and return to Earth and because short stay missions usually require much longer times to be spent in space, it would be better to stay a longer time on Mars. Thus the emphasis of these studies swung away from short stay missions, because during the long time spend in transit the crew would soak up radiation and be at increased risk of psychological problems owing to having nothing useful to do. The balance of emphasis therefore shifted toward long stays.

In addition, human exploration of Mars started to be viewed as the high point of a long period of exploration involving robotic vehicles, including orbiters, landers, rovers, and other specialized devices. On landing on the planet, humans should be aware of what they will find and so not be completely exposed to the unknown. In particular, a Mars Sample Return (MSR) mission was seen as an essential prerequisite to a human mission to the surface. Furthermore, recent advances in the field of robotic and autonomous agents have made it possible to have robots cooperate with astronauts in the task of planetary exploration.

1.4.2   Starting again

Table 1.2 lists the studies of human missions to Mars that were carried out between 1982 and 1990.

The first one, published in 1982, was a British study known as Mars via Solar Sail, but this was a misnomer because the mission was to be performed by a cargo spacecraft and a ship for the crew. Only the cargo vehicle would be slowly propelled by a solar sail. The crew would use chemical propulsion for a faster voyage. Of the total crew of eight, four were to land on the planet for several weeks while their colleagues remained in orbit. There were to be two crew ships and two landers for redundancy. It would take 53 launches of the Space Shuttle to place the IMLEO of 1,300 t into LEO, with that vehicle requiring some modifications.

The Planetary Society commissioned the Science Applications International Corporation (SAIC) in 1983 to undertake a study for a human Mars mission. The result was a split mission powered by chemical propellants (LOX/LH2) that would have an IMLEO of 460 t. Artificial gravity of about 1/4 g would be provided. The total duration of 721 days would include the crew of four spending a short stay (30 days) on the surface of the planet.

Table 1.2 Studies for human Mars missions performed between 1982 and 1990.

Year	Expedition	Country	Author	Prop.	Stay	Crew	IMLEO
1982	Mars via Solar Sail	GB	—	S.Sail	S	8	1,300
1983	Planetary Soc. Mars Ex.	USA	SAIC	C	S	4	460
1984	Case for Mars II	USA	R. Zubrin	C	—	6	—
1985	Lagr. Int. Shuttle Vehicle	USA	—	C	—	—	—
1986	Mars 1986 Mars Ex.	USSR	NPO Energia	NEP	S	4	365
1986	Pioneering the Sp. Front.	USA	T. Paine	C	—	—	—
1987	Ride Report	USA	S. Ride	C	—	—	—
1988	Mars Evolution 1988	USA	NASA	C	L	8	330
1988	Mars Ex. 88	USA	NASA	C	S	8	1,628
1988	Phobos Ex. 88	USA	NASA	C	O	4	765
1989	90 Day Study	USA	NASA	NTP	L	4	1,300
1989	Mars Ex. 89	USA	NASA	C	S	3	780
1989	Mars Evolution 1989	USA	NASA	C	—	4	—
1989	Mars 1989	USSR	NPO Energia	SEP	S	4	355

(C: Chemical; O: Mars orbit or satellites; S: Short stay; L: Long stay; the IMLEO is in t).

Case for Mars II was a workshop held in July 1984 which considered the creation of a permanent space infrastructure to enable human Mars exploration. A cycler spacecraft would be placed into a solar orbit that would allow a crew of six to commute back and forth between Earth and Mars without the need for expendable hardware. The parameters of the flybys of the planets would be chosen to enable the cycler to maintain a useful trajectory.

Another infrastructure proposal was the first to envisage establishing an in situ facility to produce propellants and other consumables on the Martian surface.

The cycler would depart the vicinity of Earth for a six month journey to Mars every 20 to 30 months. People bound for Mars would be transported from Earth orbit to the cycler on a small Crew Shuttle. At Mars they would separate from the cycler in that vehicle, aerobrake into orbit around Mars and then land. The outpost on Mars would be a permanent base that would be occupied by a succession of crews transported between planets using one or more cyclers. All vehicles would use chemical propulsion (LOX/LH2). The total mission time of 1,825 days would include 730 days on the planet, which was more than most proposed long stay times.

Another solution was proposed in 1985, in which a large Interplanetary Shuttle Vehicle could be assembled and fueled in the L1 Lagrange point of the Earth-Sun system. From this point, the Lagrangian Interplanetary Shuttle Vehicle could be flown to the corresponding L1 point of the Mars-Sun system, using a number of Earth and Moon flybys. The concept is not much different from a cycler, but is considered to be more flexible. It was demonstrated in 1998 using the ISEE-3 satellite.

In the Soviet Union, NPO Energia started to develop the superheavy lift rocket which is now known simply as the Energia launcher. Because this could be used to carry the large IMLEO required to mount a Mars mission, between 1978 and 1986 the Soviets conducted another series of studies for a human Mars mission. Called Mars 1986, this was partly derived from the 1969 study, in that the NEP system was to be powered by a 15 MW nuclear plant and would provide a short stay mission. An IMLEO of 365 t for a crew of four was planned. The stay on the planet was expected to be 30 days, with a total mission time of 716 days.

1.4.3   ISRU and the role of the Moon in Mars exploration

Meanwhile in America, the National Commission on Space had been formed. It was chaired by former NASA Administrator Thomas Paine and its membership included various prominent scientists and aerospace experts such as pilot Chuck Yeager, who was the first to exceed the 'sound barrier,' and several astronauts, most notably Neil Armstrong. In 1985 it produced a report called *Pioneering the Space Frontier* that contained a study for a human Mars mission which was not dissimilar to the 1969 concept. On 28 January 1986, just one month before the date on which the report was scheduled for publication, the Shuttle Challenger disintegrated during its ascent to orbit. The Commission then realized that all hopes of using the Shuttle as a low-cost and safe launcher for a Mars mission had vanished. It recommended development of new spacecraft based on those previously designed by SAIC and Eagle Engineering. The plan suggested the development of a space station by 1992, an Orbital Transfer Vehicle by 1998, and an orbital Spaceport for assembly of Mars and lunar spacecraft by 1998. It was expected that large new cargo launchers would become available by 2000. A permanent base on Mars would follow over the next 25 years.

The possibility of manufacturing propellant and other consumables on Phobos was also envisioned, as was the use of the Moon on which habitats, laboratories and a spaceport could be built. Overall, this plan was very ambitious and its cost was estimated at $700 billion. However, owing to its very high cost it was not taken seriously.

The Commission chaired by astronaut Sally Ride after the Challenger disaster produced a document in 1987 called *NASA Leadership and America's Future in Space: A Report to the Administrator*. Commonly referred to as the Ride Report, it proposed four main initiatives:

- *Mission to Planet Earth* proposed a number of investigations of our planet from orbit, and included a space station.
- *Exploration of the Solar System* proposed a number of robotic deep space missions, envisaging more missions than were currently planned.
- *An Outpost on the Moon* called for establishing a lunar base that would grow to host a crew of 30 by 2010.
- *Humans to Mars* envisaged following a series of robotic missions to the planet with a human mission, possibly as soon as 2010, with the long term objective of starting the construction of an outpost in the 2020s.

For all these activities, a Shuttle-derived cargo launcher had to be built to diversify the fleet of launchers.

Aware that the Soviet Union was studying a human Mars mission, in 1988 NASA initiated four studies aimed at the planet.

One, called the Mars Evolution 1988 study, was a long term study of the development of a self-sufficient, sustained human presence beyond LEO. The sequence of steps was to build a lunar base, then an infrastructure on Phobos and Deimos to produce propellant, and finally the outpost on Mars. Propellant and other materials would be ferried between these locations by NEP craft that used argon as their propellant. Other spacecraft for crews of eight would be propelled by LOX/LH2 rockets. Their oxygen would be liberated from the metal oxides that are common in the lunar regolith. If it turned out that water could be extracted, then it would also be possible to produce the hydrogen on the Moon.

A second project, Mars Expedition 88, was developed by NASA in 1988 as a short stay mission. It was also based on LOX/LH2 propulsion and had a split philosophy in that a cargo mission would deliver to Mars the habitat, the ascent vehicle and all of the equipment required. The propellant for the ascent would be carried rather than being manufactured in situ. Fifteen months later, a vehicle carrying eight astronauts would be launched on a fast trajectory. The two vehicles would rendezvous in Mars orbit, four people would transfer to the Mars Lander Vehicle and descend to the surface. After a stay of 20 days, the explorers would lift off and rendezvous with the mothership, which would then head home to Earth. The total mission would last 440 days. Quite fast interplanetary trajectories were planned in both directions.

A third study performed by NASA in 1988 envisaged a human mission to Phobos. Its basic rationale was that such a mission could be implemented as early as 2003 and would serve as a precursor to a Mars landing mission. Furthermore, humans in orbit or on Phobos could teleoperate robots on the surface of the planet to undertake much scientific work in preparation for a manned expedition, including retrieving samples for analysis. This was a split mission in which a cargo ship on a minimum energy trajectory would transport all the necessary equipment into Mars orbit and to the surfaces of Mars, Phobos and Deimos. The cargo flight was scheduled for February 2001. About 18 months later a crew of four would depart Earth on a 9 month trajectory. After aerobraking into Mars orbit, two members of the crew would remain in orbit to teleoperate robots while their colleagues went to Phobos for exploration activities that would include a total of 24 hours of Extravehicular Activity (EVA). After 30 days they would all return to Earth using a very fast 4 month trajectory, with a total mission time of 440 days. The Phobos mission was to be the first of four missions. In a way, this would be a precursor for missions to the surface of the planet. Its IMLEO of 765 t would require 20 to 30 launches from Earth.

In 1989, to mark the twentieth anniversary of the Apollo 11 landing, President George H.W. Bush gave a speech in which he proposed three goals for his nation's space program:

"First, for the coming decade, for the 1990s, Space Station Freedom, our critical next step in all our space endeavors. And next, for the next century, back to the Moon, back to the future, and this time, back to stay. And then a journey into tomorrow, a journey to another planet, a manned mission to Mars…"

In response, NASA Administrator Richard H. Truly created a task force to perform a 90 day analysis of this Human Exploration Initiative (also known as the Space Exploration Initiative; SEI). Published in November 1989,[5] this study estimated that to undertake all three directives in the presidential address would cost about $500 billion over a period of

---

[5] *Report of the 90 Day Study on Human Exploration of the Moon and Mars*, NASA-TM-102999, November 1999; http://history.nasa.gov/90_day_study.pdf

two or three decades. The part related to human exploration of Mars was evaluated at about $258 billion. It envisaged using NTP for both of the cargo and crew vehicles and aerobraking into orbit at Mars. The total IMLEO of 1,300 t would require 140 launches of a Shuttle derivative dubbed Shuttle Z that was to be capable of placing 87.5 t into LEO. The crew of four was scheduled to remain on Mars for 600 days and the total duration of the mission would be 1,000 days. The mission architecture was based on developing an ISRU capability. The cost was so great, and the political opposition so strong, that the 90 day study attracted almost no supporters.

So a new and much less costly plan was drafted as Mars Expedition 89. In this case LOX/LH2 propulsion was to be used, the reduced crew of three people would travel together with the cargo, and in addition to 10 days spent in orbit around Mars the surface expedition would be just 20 days. No ISRU was planned. After a total mission duration of 500 days there would be a direct entry into the Earth's atmosphere. A total IMLEO of 780 t would need six launches of a superheavy lift rocket.

The Mars Evolution plan of 1988 was revived the following year. This time the basic infrastructure was a free-flying orbital crewed fixture that would be positioned close to the space station where the various spacecraft intended for Mars would be assembled. Similarly, an infrastructure had to be built at the other end, on Phobos, to produce propellant. Three operational phases of emplacement, consolidation, and utilization, were expected to progress from the first mission to the construction of an inhabited outpost on Mars.

The first phase would perform a 30 day short stay mission with a crew of four. Next there would be a long stay mission that would spend 16 months on the surface. Another three flights would complete the consolidation phase. This plan proved to be almost as costly as the Martian part of the 90 day study. Also some of its points were questionable; for instance, the proposal to produce propellant on Phobos. This satellite is in a circular orbit around Mars, but incoming spacecraft would be on highly elliptical orbits and circularizing the orbit would itself burn a large quantity of propellant, which strongly undermines the benefit of producing propellant on Phobos. As with the 90 day study, this new proposal was widely dismissed.

The final study performed in the years under consideration in this section is the Mars 1989 Russian project by NPO Energia. It was essentially a follow-up to the 1986 study, with the difference that instead of using a nuclear reactor to power the electric thrusters, now they would be powered by solar arrays based on those developed for the Salyut 7 and Mir space stations. With a 355 t IMLEO delivered to LEO by five Energia launchers, the Solar Electric Propulsion (SEP) spacecraft would carry a crew of four, two of whom would spend a week on Mars. The total mission time was 716 days. The solar arrays, measuring 200×200 m, would supply 15 MW at Earth. This would diminish with the square of the heliocentric distance as the vehicle receded from the Sun. The specific mass of the arrays of $\alpha = 1$ kW/kg was rather optimistic. Ion thrusters using xenon as the propellant were intended.

Thus four of the fourteen studies considered during this second period were not aimed at a specific mission. Of the remaining ten, one (10 percent) was to reach Mars orbit and satellites, six (60 percent) were for short stay missions, and three (30 percent) were for long stay missions. The IMLEO estimates ranged between 330 and 1,630 t.

1.5   THE LAST 25 YEARS (1990–2015)

1.5.1   General considerations

The advocates of Mars exploration learned many lessons from the debacle of the SEI, above all that proposing pharaonic programs that incorporate everything, perhaps with the objective of gaining the widest possible support from the many lobbies, interest groups, and scientific communities can be counterproductive. In proposing a program, choices must be made and it must be realistic, affordable, and properly assess the sources of funding.

Stating that a laboratory, an infrastructure, and a spaceport on the Moon must be built as a prerequisite to going to Mars, and also that in order to undertake operations on the Moon it will first be necessary to create a shipyard with the related infrastructure and perhaps a station in LEO will inevitably result in that Mars program being scrapped.

In a sense, 1989 marked a turning point, particularly in the USA, because after that most human space projects aimed at Mars were split architectures involving distinct cargo and crew missions, and most at least considered the production of consumables (mainly propellant) on Mars. In the case of chemical propulsion, and sometimes even nuclear propulsion, arrival at Mars would involve braking into orbit by aerodynamic means, and therefore much greater attention was paid to aerocapture and aerobraking techniques.

Table 1.3 lists the studies for human Mars missions that were performed between 1990 and 2015.

1.5.2   Mars Direct

The Mars Direct [10] project was proposed in 1991 by Robert Zubrin, who would later found The Mars Society. This was able to demonstrate that a human Mars mission could be affordable. The selection of aerocapture, In Situ Propellant Production (ISPP), direct launch from Earth and direct return, together with the choices for some options (long stay, split mission, and a safe haven on Mars) changed permanently the overall background. Zubrin chose chemical propulsion because waiting for the technology developments required for nuclear propulsion would severely delay any attempt to reach Mars.

On a given launch opportunity a purposely developed heavy lift launcher would send the material and the equipment required on Mars (a total of 40 t) directly from the surface of Earth to the surface of Mars (hence the name Mars Direct). Soon after arrival, the ISPP plant would start to produce methane and oxygen from the atmosphere, using 6 t of hydrogen brought from Earth. Powered by a 100 kWe nuclear reactor, this operation would fill the tanks of the Mars Ascent Vehicle (MAV) that would also function as the Earth Return Vehicle (ERV).

At the next launch opportunity (after 2.2 years), presuming that the MAV was fully fueled, a similar launcher would send four people to Mars with the same direct operating modes. During the interplanetary cruise, the spacecraft would be rotated for artificial gravity using a tether and a spent rocket stage as the counterweight. After a stay of 550 days on the surface of Mars, the crew would lift off in the MAV straight for Earth, where it would conclude the mission with a direct entry into the atmosphere. The total mission duration would be 880 days.

Table 1.3 Studies for human Mars missions performed after 1990.

Year	Expedition	Country	Author	Prop.	Stay	Crew	IMLEO
1991	Mars Direct	USA	R. Zubrin	C	L	4	220
1991	STCAEM chem.	USA	NASA	C	S	4	800
1991	STCAEM SEP	USA	NASA	SEP	S	4	410
1991	STCAEM NEP	USA	NASA	NEP	S	4	500
1991	STCAEM NTR	USA	NASA	NTP	S	4	800
1991	Synthesis Study	USA	NASA	NTP	L	6	1,080
1991	Mars Semi-Direct	USA	NASA	C	L	4	220
1992	ERTA	Russia	RKK Energia	NEP	—	—	—
1993	Des. Ref. Mission 1	USA	NASA	NTP	L	6	900
1994	Mars 1994	Russia	Kurchatov I.	NTP	S	5	800
1994	Mars Together	USA-Rus.	Energia-JPL	SEP-NEP	—	—	—
1996	Athena	USA	R. Zubrin	C	F	2	100
1996	Des. Ref. Mis. 3	USA	NASA	NTP	L	6	550
1998	Combo Lander Mis.	USA	NASA	SEP+C	L	4	280
1998	Des. Ref. Mis. 4 SEP	USA	NASA	SEP	L	6	400
1998	Des. Ref. Mis. 4 NTR	USA	NASA	NTP	L	6	400
1999	Dual Lander Mis.	USA	NASA	SEP+C	L	6	600
1999	Mars Society Mis.	USA	CalTech	C	L	5	900
2000	Marpost	Russia	L. Gorshkov	NEP	S	6	400
2005	European Mars Mis.	Ger.	Mars S. Ger.	C	L	5	120
2006	Mars Oz	Austr.	Mars S Aust.	C	L	4	694
2008	Des. Ref. Arch. 5	USA	NASA	NTP	L	6	850
2015	Humans orbiting Mars	USA	Plan. Soc.	SEP+C	—	4	—

(C: Chemical; F: Flyby; S: Short stay; L: Long stay; the IMLEO is in t).

The launch from Earth using the purposely built Ares heavy lift launcher [10] and the Trans-Mars Injection (TMI) maneuver were to be performed with LOX/LH2 propellants. No propulsive maneuver was planned on arrival at Mars and the ascent from Mars would be made using methane and oxygen extracted from the atmosphere. An NTP stage was considered as an option for a later phase of the program. By that time the explorers might be traveling around on the surface using a 'hopper' powered by a nuclear rocket which used atmospheric carbon dioxide as its propellant, a concept known as Nuclear rocket using Indigenous Martian Fuel (NIMF).

Mars Direct was more than the plan for a single mission. By repeating the launches at successive launch opportunities it would have allowed a semi-permanent human presence on the planet (Figure 1.10).

Although this proposal was generally considered as optimistic, Mars Direct triggered a number of new studies and allowed NASA to produce a number of versions of what is known as a Design Reference Mission (DRM) in the years 1993–2009. This effort culminated in 2009 with the Design Reference Architecture (DRA5) [24].

Also in 1991, NASA proposed another short stay mission study. This STCAEM (Space Transfer Concepts and Analyses for Exploration Missions) was developed by Boeing and had variants for cryogenic/aerobraking (CAB), SEP, NEP, and a nuclear thermal rocket (NTR).

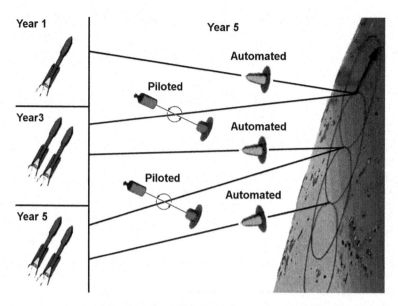

Figure 1.10 Mars Direct, involving five launches in three successive launch opportunities. (Based on a presentation by The Mars Society)

The first concept served as the NASA reference vehicle and was basically a development of the projects studied in the previous years. Its major innovation was the development of High Energy AeroBraking (HEAB) for planetary capture.

Since LOX/LH2 were to be used in the whole mission, these would have to be stored for a long time. Because these cryogenic liquids are very volatile, a low-boil-off technology would have to be developed.

The IMLEO of 800 t would be carried to LEO using eight launches of a rocket with a payload capability of 140 t. The total mission duration would be 580 days and a crew of four would spend 30 days on Mars.

The second concept in this family would use SEP. Of the four alternatives, it offered the lowest IMLEO at 410 t. The spacecraft would be assembled in LEO and then slowly spiral away from Earth unoccupied. The crew would board when it was close to achieving TMI. Upon reaching Mars, the braking-capture would be performed by SEP. The total power of 10 MWe was to be provided by 35,000 m^2 of solar arrays with a specific mass α = 3.9 kg/kW. The same crew as for the chemical propulsion spacecraft was planned, but the stay on Mars would be only 20 days and the total mission duration would be 550 days.

The third STCAEM concept was based on NEP. It envisaged a spacecraft quite similar to the SEP plan, with the difference being the power plant and the higher specific impulse of the electric thrusters (10,000 seconds). The total mass of the power system (with shielding) was about 188 t. Since the power produced was 40 MW, the specific mass was α = 4.7 kg/kW. This figure took into account the fact that the reactor was duplicated for safety, but this was not a large mass penalty since the reactor in itself was just 7.4 t. Also, the turboalternators were redundant (three pairs, plus two backup pairs). As usual with this type of machinery, most of the problems involved the thermal radiator. The engine assembly was made of 40 individual ion thrusters (including 10 spares) configured in a 5 × 8 rectangular array. The crew would spend 30 days on Mars and the total mission duration was 430 days. In this case the IMLEO was 500 t.

The last concept involved NTP, which after the ground tests performed in the NERVA project was considered a fully proven technology. A specific impulse of 1,050 seconds was assumed, because it seemed likely that nuclear rockets would be able to achieve that by the time of the mission. In overall terms the mission was similar to the others, with fully propulsive maneuvers and reusable hardware. The timing was similar to the NEP plan, being just 10 days faster. The predicted IMLEO was 800 t.

In May 1991, President Bush declared that he would support a human Mars mission, and NASA drew the results of the STCAEM studies into a single Synthesis Study for NTP that was considered by some as a concession to the Los Alamos National Laboratory. It was a split mission of the long stay type and included the use of ISRU. The total mission duration of 900 days would include 460 days on the planet. The IMLEO was 1,080 t for a crew of six. When Congress canceled further funding for Mars studies in FY 1991, NASA's Exploration Office was dissolved.

Also in 1991, a project that appeared to be a low cost version of the STCAEM was drafted. Because this fell between Zubrin's Mars Direct and NASA's concepts it became known as Mars Semi-Direct. It can be thought of as a bridge between Mars Direct and the Design Reference Mission 1.0 issued by NASA in 1993. Two Mars Direct Ares launchers would lift the IMLEO of just 220 t. The total mission duration of 900 days was to allow a crew of four to spend 550 days on Mars.

In 1992 in Russia, NPO Energia designed a small-to-medium-sized NEP space tug called the Elecktro-Raketniy Transportniy Apparat (ERTA) that was to enter service by about 2005. The nuclear reactor would provide 150 kW for propulsion and about 10–40 kW for powering the systems of unmanned planetary spacecraft. The 7,500 kg reactor would provide electrical power for up to 10 years and thrust for 1.5 years. By 1994 it was expected that such tugs would be used by other space agencies, possibly launched by Titan or Ariane 5 rockets. Larger versions for human missions to Mars were also possible. A reactor in the 5–10 MW class would be suitable for a 150 t spacecraft, powering it for 1.5 years, or even longer at a reduced power level. Because the design was modular, it would be possible to use multiple units to satisfy specific mission requirements.

1.5.3   NASA Reference Mission 1

In 1993 the International Academy of Astronautics (IAA) issued a Cosmic Study on Human Mars Exploration [7]. However, it is not listed here because it was a general study and not a specific mission design.

In the 1990s both NASA and Roscosmos launched new robotic missions to Mars. While the US Mars Observer (1993) and the Russian Mars 96 (1996) failed, the Pathfinder-Sojourner mission (1997) was a huge success, not only from the technological and scientific viewpoint but also as a media event. As a result of this latter mission, there was a renewal of interest in Mars exploration, both by space enthusiasts and the public opinion in general. In July 1997, for the first time, public support for a human Mars mission exceeded 60 percent; although it fell back below 50 percent a few months later.

In the 1970s robotic space exploration was seen as an alternative to human exploration, but many of the more recent robotic missions to Mars were explicitly meant to pave the way for human exploration of the planet.

In 1993 NASA issued a new study that differed significantly from its predecessors. This NASA Reference Mission 1.0 could be seen as a NASA version of Mars Direct which involved larger spacecraft and NTP (Figure 1.11) rather than chemical propulsion. Aerobraking at Mars was retained. The crew of six would head for Mars only after the cargo part of the mission was known to be successful, with the MAV fueled by methane and LOX propellants produced in situ. However, instead of using just one spacecraft for the whole return journey – and thus using for the whole trip the propellant produced on Mars – the MAV would rendezvous with an ERV in orbit in order to provide more space for the crew during the long flight home. The overall mission duration of 880 days would include 610 days on the planet. The IMLEO was 900 t, and would require several flights by a 240 t superheavy lift launch vehicle of the same class as the old Nova.

In 1994 Russia had an NTR in the final design stage, ready to be applied to a human Mars mission. This RD-0410 was a bimodal engine, in that it supplied a thrust of 200 kN and could simultaneously supply a thermal power of 1,200 MW. While not providing thrust during the cruise, it could deliver about 50–200 kWe. When tested over an operating time of 5 hours it provided a specific impulse of between 815 and 927 seconds, which was slightly higher than that of the NERVA engine. It required a radiator of 600 m². A cluster of three to four of these engines would have a mass in the range 50–70 t. The Kurchatov Institute, which had designed the engine, started to design a mission to apply it. In addition, the Keldysh Institute drew up a Mars mission called Mars 1994. In both designs, the reactor would be

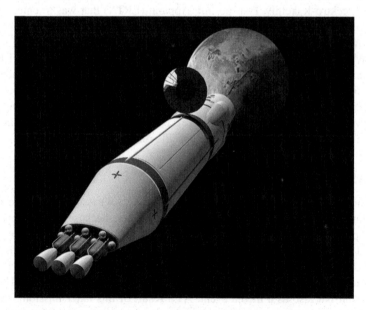

Figure 1.11 The spacecraft of the NASA Reference Mission approaching Mars. It is provided with three NERVA-type nuclear thermal rockets. (NASA image)

located as far as possible from the crew compartment, and the latter would be shielded by locating the liquid hydrogen tank around the habitat. A crew of five would stay on Mars for 30 days, and the total mission duration would be 460 days. The proposal did not include the split mission concept, braking at Mars was done using the thruster, and ISRU was not included. Nine Energia rockets would be required to lift the IMLEO of 800 t.

1.5.4   The first idea of a joint Russian-American mission

As the political situation improved during the 1990s, it created the opportunity to consider a joint Russian-American Mars mission. This was appropriately dubbed Mars Together. RKK Energia and NASA's Jet Propulsion Laboratory studied the problem of how to mount a joint human Mars mission in 1994. They studied both NEP and SEP alternatives for providing the electric thrusters with 30–40 kW of power. But they concentrated on preliminary studies and suggested launching a reduced scale model of about 150 kg equipped with about 30 m^2 of solar arrays capable of producing 3 kW in order to undertake basic studies of a SEP spacecraft for Mars and attain some science from orbit. Although a low cost approach, this did not gain any support. The name Mars Together would be used repeatedly for proposed robotic missions to Mars in the following years.

In 1996 Robert Zubrin proposed a low cost Mars flyby mission, dubbed Athena. The trajectory was innovative in that the spacecraft would remain in the vicinity of Mars without entering orbit around the planet. It would remain in heliocentric orbit, but spend about a year located at one of the Mars-Sun Lagrangian points. The trajectory would provide two normal flybys: the first upon arrival at Mars when the path would be deflected toward the Lagrange point and the other upon departure when the trajectory would be deflected back toward Earth. A total mission duration of about 2.5 years was planned. Notwithstanding the small size of the spacecraft, which would carry a crew of two, it would rotate using a tether to provide them with artificial gravity. When in the vicinity of Mars they would teleoperate four rovers on the planet's surface without the substantial time delay that is imposed by performing this function from Earth. It would use cryogenic LOX/LH2 propulsion and the plan called for using one of the launch opportunities in 2001, 2003 or 2005.

The rationale of the Athena mission was to demonstrate the feasibility of low cost human Mars missions and pave the way to more elaborate missions. With an IMLEO of a mere 100 t, the two Shuttle launches and four Proton launches needed to get it underway (plus four Delta 7925 or Molniya launches for the rovers to Mars) gave an all-in price of just over $2 billion, making this one of the cheapest Mars missions.

The discovery in 1996 of what appeared to be fossils in a meteorite believed to have come from Mars rekindled interest in the Red Planet, so NASA reviewed its earlier proposals and in 1997 published Design Reference Mission 3 [12]. This retained the same basic layout but greatly simplified the mission. The original IMLEO of 900 t was cut to 550 t in order to require only six launches of a smaller rocket – a Shuttle derivative called Magnum – rather than a newly designed superheavy launcher.

In the spring of 1988, NASA conducted a study of a Mars mission which could be launched by three heavy-lift rockets. The idea was to use a SEP tug to reach a high energy Earth departure orbit, from which the spacecraft could set off for Mars using chemical propulsion (LOX/LH2). This was called the Combo Lander mission – the term Combo referred to the combination of two landers and not to the combination of different propulsion systems. It didn't use nuclear power, but it involved storing LH2 for up to four years and thus would require zero-boil-off technology to be developed. One strange aspect of the plan was the use of electric propulsion deep in the Earth's gravity field and chemical propulsion for the subsequent maneuvers. The crew was reduced from six to four, and the total mission duration of 970 days would provide 580 days on Mars. The IMLEO of 280 t would require three launches of the Magnum rocket, hence its name as the Three-Magnum Mars Mission. The crew would launch on a Shuttle after the SEP tug had performed the orbit raising maneuver.

Two further studies were made in 1998 starting from DRM 3.0. These were the Design Reference Mission 4 SEP and Design Reference Mission 4 NTR. The first would require the Solar Electric Transfer Vehicle (SETV) developed by NASA Lewis. This 123 t vehicle used two Russian-built Hall thrusters. As in the case of the Combo Lander mission, the SETV was to raise the orbit so that a chemical stage could perform the TMI burn. The total IMLEO was 400 t. The second design utilized a bimodal NTR which cut the IMLEO by about 2.5 percent, reducing it to just short of 400 t.

The Combo Lander mission was reformulated in 1999 to make it more realistic, since it had proven to be too lean. The result was the Dual Lander Mission. The crew was restored to six and the split arrangement was reintroduced, along with ISRU. The IMLEO was now of 600 t, which was too heavy to be lifted by three Magnums. No final report was issued for this study.

In 1999, The Mars Society asked the California Institute of Technology to produce a mission design as an alternative to the NASA reference mission. The resulting Mars Society Mission was based on LOX/LH2 propulsion, a split architecture, ISRU, and a crew of five. However, it assumed the development of a new heavy launcher family based on the Delta IV known as the Qahira Interplanetary Transportation System (QITS) and the IMLEO of 900 t would need seven such launches. The proposal envisaged spending 570 days on Mars and a total mission duration of 850 days.

1.5.5   The New millennium

The first Mars mission proposed after the turn of the millennium was the MARs Piloted Orbital STation (MARPOST) proposed by Leonid Gorshkov of RKK Energia. It was first offered in 2000 as an alternative to Russian participation in the ISS, then revised in 2005.

With an IMLEO of 400 t, the spacecraft would be assembled in orbit using four Energia launches and then be propelled in space by arrays of hundreds of nuclear powered ion thrusters using xenon as propellant. The crew of six would be comfortably hosted in a huge vehicle with a diameter of 6 m and a length of 28 m. Three of the six astronauts would spend about 30 days on the surface of Mars, and the mission would last a total of 730 days. The spacecraft was to be reusable. A new lander was designed, and a model was successfully tested in the wind tunnel. The estimated cost of $10 billion, spread across 10 years, was a really modest price tag. In 2001, an International Science and Technology Committee was established to turn MARPOST into an international project.

Meanwhile ESA developed the Aurora project, which included many robotic missions to Mars with the long term objective of preparing a human mission.

In 2005 The Mars Society Germany proposed a European Mars Mission (EMM) capable of being launched using an improved Ariane 5 booster. Like the missions proposed by The Mars Society in the USA it was a split mission with ISRU and chemical propulsion, but was based on the semi-direct launch philosophy that would involve a rendezvous on Mars orbit on the way home. The total duration was about 800 days for a stay on the surface of 600 days, and with a crew of five the total IMLEO was just 120 t.

Also The Mars Society Australia produced a study called MARS-Oz (actually MARS-Oz is the name of the Mars Analogue Research Station in the outback of Australia). The overall mission architecture was similar to the Mars semi-direct mission proposed by Zubrin, except for the decision not to use nuclear power on Mars: instead, solar cells producing 40 kW in sunlight would have the same mass as a nuclear reactor producing 100 kW continuously. Like in the semi-direct approach, propellant would be produced on Mars from the atmosphere using hydrogen carried from Earth. The power limitations constrained the crew size and the amount of propellant that could be produced. For a crew of four the IMLEO would be 694 t.

NASA concluded its series of reports on the Reference Mission in 2008 (although it wasn't published until 2009) with Design Reference Architecture 4 [24], in which several variants of the already studied design were proposed. The baseline design was still quite similar to its predecessors (NTP, a crew of six, use of ISRU, a long stay, and so on) but with an IMLEO of slightly less than 850 t. Other alternatives were also studied.

Out of the 20 detailed mission designs reported in Table 1.3, one (5 percent) was a flyby mission, six (30 percent) were short stay missions, and thirteen (65 percent) were long stay missions. The IMLEO figures ranged from a very optimistic 100 t for Athena to 400 t or more for most projects. The highest IMLEO was about 1,100 t. For comparison the mass of the ISS is about 400 t, so the mass to be carried into LEO is of the same order of magnitude as that, or even larger in the more conservative estimates.

In 2012 the International Academy of Astronautics (IAA) launched a new study on the theme of Global Human Mars System Missions Exploration. The resulting Cosmic Study was published in 2016.

Several events that occurred during the first decade of the millennium may have a major bearing on human Mars missions. In particular, the loss of Shuttle Columbia in 2003 raised a question in America about whether human spaceflight was actually worth the cost and risk. The question was settled with the decision later that year to continue and the announcement by President George W. Bush in January 2004 of a Vision for Space Exploration having the ultimate objective of landing humans on Mars. This plan would use the Moon as a stepping stone to allow new technologies to be mastered whilst remaining within financial constraints. However, this program was launched with insufficient budgetary resources to be able to meet its objectives.

When the Obama administration entered office it was clear that the new president did not consider human space exploration as a priority, and the Constellation program initiated by his predecessor was canceled. And declaring "We've already been there," Barack Obama canceled a human return to the Moon. As a result, the nation was left without a clear roadmap for Mars exploration. It was reasoned that various technologies would require to be developed before it would be possible to commit to a human Mars program. But this technology innovation was to be pursued in general terms, rather than specifically for Mars; it might be tested by other missions to other destinations, such as asteroids. Nevertheless, Congress insisted that NASA continue the development of two essential assets of the canceled program, namely the Orion spacecraft and the heavy launch vehicle that became the Space Launch System (SLS). At the time of writing (2016) their development is progressing satisfactorily. Another aspect of the new space policy was that routine space transportation to LEO should be entrusted to private ventures through Commercial Orbital Transportation Services (COTS) contracts.

A new approach emerged in the Humans Orbiting Mars workshop which was held in March 2015 by The Planetary Society [31]. In line with the earlier concepts by NASA, it was still based on an essentially American mission, although there were vague statements about international cooperation. It took into account affordability as never before. A new strategic approach was defined, centered not on a single mission but on a campaign which would lead to a set of three missions: the first in Mars orbit with the possibly of visiting Phobos (Figure 1.12), then a short stay mission to the Red Planet, and finally a long stay mission that would serve as the first step to creating a permanent settlement. A timeline (Figure 11.1) called for launching the first human mission in 2033 and the third mission ten years later. Because the basic hardware was the Block II SLS and the Orion spacecraft, no new technology developments were needed. The cargo spacecraft was based on the SEP tug designed for the Asteroid Redirect Mission (ARM). No exploitation of in situ resources was planned. Detailed analysis of the costs indicated that the entire campaign would be feasible with a NASA budget that increased only to compensate for inflation.

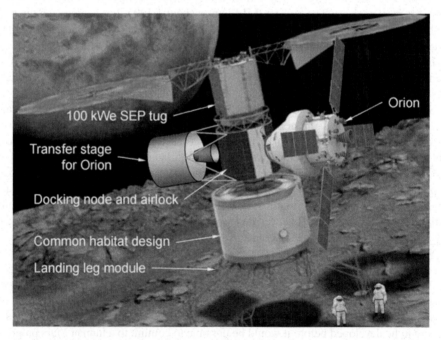

Figure 1.12 Two astronauts on Mars' satellite Phobos. The first mission of the Humans Orbiting Mars campaign. (NASA image)

As a final comment, little consideration was given to the human factors, particularly for the first mission where there would be no landing on Mars. The astronauts will be subjected to a long period in space with little protection against radiation and no artificial gravity. There is the possibility of severe psychological problems due to the very small habitable volume (see Chapter 5), even with such a small crew size.

1.6 INTERNATIONALIZATION OF SPACE EXPLORATION

After the turn of the millennium, new spacefaring nations began to give thought to human missions to Mars. The Aurora project of the European Space Agency spawned a number of robotic precursors, some of which were in partnership with Russia as well as with the USA.

Emerging from the financial woes that followed the collapse of the Soviet Union, Russia resumed its proactive role. Its MARPOST was considerably more than a simple declaration of interest. Russia showed a more general interest in human space exploration, starting with lunar exploration. After all, they aren't held back by the American attitude of "We've already been there."

China is showing a general interest in space exploration, and it is likely they will begin with human lunar exploration and later turn their interest toward Mars. And India, which succeeded in inserting a probe into orbit around Mars at the first attempt, is also showing interest in Mars exploration.

The construction and operation of the ISS is proving that it is possible for a number of space agencies to cooperate on difficult missions in space. This cooperation has been quite resilient, even surviving the tensions which developed between some of the participating nations, and this is an important signal. At present, funding for the ISS has been extended to 2024, but by that time its systems will be approaching their operational limits. Ultimately it will become necessary to dispose of the facility – either by deorbiting it in a controlled way or by raising its orbit so that it can be abandoned in situ. Afterwards there will be the issue of whether it should be replaced by another such facility or whether the opportunity should be taken to initiate an entirely new international large space project, and a human Mars mission is a candidate. However, it is possible that by the time we have to dispose of the ISS we won't be ready to start an International Mars Exploration Initiative (or whatever it will be called). In that case, it is possible that an intermediate initiative will act as a bridge between the ISS and a Mars mission. Perhaps an International Lunar Base would satisfy this role. It would be useful for developing technologies and operating procedures of value to future missions to Mars.

At any rate, several space agencies participated in the International Space Exploration Coordination Group (ISECG),[6] and this produced a Global Exploration Roadmap (GER) (Figure 1.13) [26]. This roadmap considered three destinations for exploration missions: the

---

[6] The 14 partners of ISECG are, in alphabetical order: ASI (Italy), CNES (France), CNSA (China), CSA (Canada), CSIRO (Australia), DLR (Germany), ESA (European Space Agency), ISRO (India), JAXA (Japan), KARI (Republic of Korea), NASA (United States of America), NSAU (Ukraine), Roscosmos (Russia), UKSA (United Kingdom).

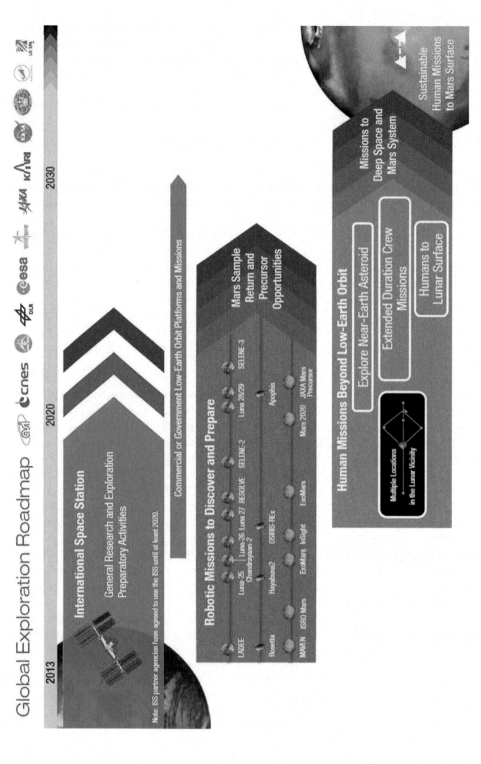

Figure 1.13 The Global Exploration Roadmap published by ISECG [26].

Moon, cislunar space and nearby asteroids, and Mars. Although human Mars exploration was only a long term goal, several 'stepping stones' were identified, one being a strong program of robotic exploration. The final goal of the roadmap was 'Sustainable Human Missions to Mars Surface' and the attached timeline envisaged this at some time in the mid-2030s (the opposition of 2035 will be very favorable). One key stepping stone toward this goal would be the inclusion of 'Mars sample return and precursor opportunities' among the robotic missions to be performed in the late 2020s. Other stated goals for about this same time were 'Explore near-Earth asteroids,' 'Extended duration crew missions,' and 'Humans to lunar surface.'

But if a credible roadmap to the Martian surface is to be devised, more detailed stepping stones will need to be defined. For instance, if NEP or NTP is chosen, a thorough testing in space of those technologies must be undertaken. On the other hand, if it is decided to proceed with chemical propulsion, there must be thorough testing of aerobraking and aerocapture for large spacecraft. To omit these activities from the roadmap will simply mean that no way of reaching Mars will be ready when needed. Similarly, there are statements about using ISRU, but until detailed testing of the relevant technologies is incorporated into programs that are actually carried out, ISRU will remain a paper concept. The same applies to artificial gravity, radiation shielding, and so on.

1.7   EMERGING PRIVATE INITIATIVES AND NEW PLAYERS

Until recently, the common understanding was that human space exploration was the domain of space agencies, possibly cooperating with each other to mount international endeavors like the ISS. Private organizations such as The Planetary Society and The Mars Society were only able to act as advocacy groups, performing support tasks and supplying fresh ideas.

The introduction of COTS and the idea that space agencies could accomplish their goals more cheaply by entrusting the transportation of materials and people to LEO (and then perhaps beyond) to private companies which operate under typical market rules, has produced a semi-private approach to space exploration.[7]

For sure, the savings that can be achieved in this way are large. Indeed, a recent study [29] has calculated this saving at about 90 percent. NASA estimated that the development of the Falcon 9 launch vehicle and the Dragon reusable spacecraft under the cost-plus methods of the Federal Acquisition Regulations (FAR) would have cost $3.977 billion. However, the figure declared by SpaceX for achieving this development was $433 million. In the same study, it is stressed that there is little new in this result: the private development of the two SpaceHab flight units for carriage aboard the Space Shuttle cost $150 million, whereas that the same job would have cost NASA $1.2 billion if performed under traditional government practices and methods.

---

[7]G. Genta, "Private space exploration: A new way for starting a spacefaring society?" *Acta Astronautica* vol. 104, pp. 480–486, 2014.

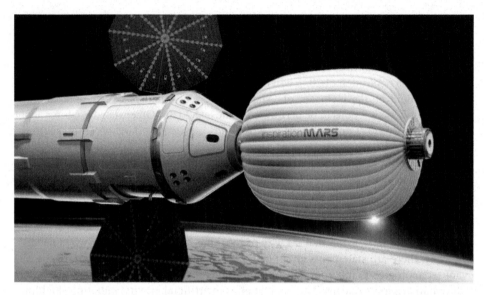

Figure 1.14 The Inspiration Mars spacecraft ready to leave Earth orbit for its Mars flyby.

Similar reasoning by Robert Zubrin [11] has shown that a human Mars mission could be performed at a fraction of the cost stated by NASA.

Recently, however, private companies and non-profit organizations have declared their intentions of mounting space exploration expeditions and/or starting space exploitation tasks.

The simplest mission has been proposed by the non-profit organization Inspiration Mars Foundation.[8] This is a 'space cruise' for two people on a 501 day Mars flyby. The main goals are to generate excitement about space travel and test technologies that will be needed to land on Mars in the future. It is thus an extreme 'space tourism' enterprise that places considerable emphasis on outreach and advertising.

The mission would combine a commercial space capsule and an inflatable module to form a single spacecraft (Figure 1.14) that will carry a man and a woman, perhaps a married couple, to Mars and back. The proposed timeframe was extremely tight, because the 2018 (or 2020) launch opportunity was suggested. It would be challenging for two people to live harmoniously in such confined quarters for so long. More seriously, some of the technical aspects are questionable.

The Dutch non-profit organization Mars One has announced its intention to land four colonists on Mars by 2025.[9] They designed their campaign with multiple one-way voyages, because colonists go out to set up home at their destination. These would be the ultimate in long stay missions! The first trip was expected to cost $6 billion. This would be paid for with a reality-TV event built around the mission, from astronaut selection to the settlers' first years on the Red Planet. However, the plan is quite questionable. While a one-way trip is much simpler than a two-way one, designing a life support and an ISRU system able to ensure the indefinite survival of the crew is very difficult and either at or

---

[8] http://marsfoundation.org/

[9] http://mars-one.com/

Figure 1.15 An artist's impression of the Red Dragon spacecraft landing on Mars.

perhaps beyond the limits of present technology. In a way, aiming for an indefinite stay displaces the difficulties from the transportation system to the survival system. Upon first impression, it appears to introduce complexities in the latter that are larger than those which are removed from the former.

A study was conducted at MIT which focused on ISRU and the Environmental Control and Life Support System (ECLSS).[10] This established that the mission is, in the opinion of its authors, not feasible with present technology. There is little doubt that eventually humans will colonize Mars, but at the moment an indefinite stay mission does not seem possible.

Nevertheless, Elon Musk of SpaceX looks forward to establishing a colony on Mars in the not too distant future.[11] On April 26, 2016 the company announced its intention to launch a Red Dragon unmanned spacecraft to Mars as early as 2018 (Figure 1.15) using a Falcon Heavy launcher. The long term goal would be to establish a settlement of up to 80,000 people. In a lecture at the International Astronautical Congress in September 2016, Elon Musk presented a huge spacecraft, the Mars Colonial Transporter, able to carry 100 colonist to Mars in a single launch. The spacecraft would take off from the Earth surface, carrying about 500 t to LEO. After being refueled in LEO, it would fly to Mars and land directly on the planet. In the lecture no detailed reference to a launch date was made, except that it wouldn't occur before 2024. This vast enterprise is expected to be made feasible by the reusable rockets that SpaceX is in the process of developing. Musk's declared goal is to contribute to the birth of a multiple-planet spacefaring civilization.

Other private enterprises have plans to exploit the resources of Mars' satellites. This strategy could create synergies with the exploration of the planet. There is little doubt that space resources utilization of the Moon, of the asteroids, and of Mars, will eventually be a strong facilitator for human space exploration in general and for human Mars exploration in particular.

---

[10] S. Do, A. Owens, K. Ho, S. Schreiner, and O. de Weck, "An independent assessment of the technical feasibility of the Mars One mission plan – Updated analysis," *Acta Astronautica* vol. 120, pp. 192–228, 2016.

[11] http://www.spacex.com/

2

Reasons for human Mars exploration

A human mission to Mars is a costly, dangerous, and difficult enterprise which cannot be undertaken unless we really understand the motivations for doing so. The natural urge which drives human beings to explore the world that surrounds us, is insufficient to perform such an expedition, and above all to undertake it within a short period of time. For sure, a mission to Mars is a great scientific enterprise, but not even the pursuit of science provides adequate motivation. The discussion concerning why we should go there is still open. The different opinions expressed on the subject are reflected in the various approaches and priorities of the many actors and stakeholders of this enterprise.

2.1 A RATIONALE FOR HUMAN EXPLORATION BEYOND LEO

The innate drive to explore, which is so common in the human species, was synthesized by George Mallory when responding to a question about why he desired to climb Mt. Everest: "Because it's there." Unfortunately, there is no way that we can justify a difficult, costly and risky human mission to Mars with such a simple statement.

There have been many attempts to justify undertaking a human mission to Mars in the near future, as yet without clear success. The question is part of the wider issue of human space exploration in general. Following the termination of the Apollo program in 1972, many arguments against the continuation of human space exploration were raised, but both of the main spacefaring nations of the time – America and the Soviet Union – decided to push on, and now, working together and with new players, they are able to maintain a more or less permanent human presence in space.

But this human presence is confined to LEO. Future exploration in deep space appears only in vague and sporadic declarations without anything being realized. As related in the previous chapter, there have been many proposals to return to the Moon and to head to Mars. While none of these moved past the report stage, this long period of time at least allowed ideas for mission architectures to mature. This cleared out some early misconceptions (e.g. the preference for a short stay, or opposition, scheme; and the view of the Mars' surface as being more dangerous than space itself). It also identified several technical

© Springer International Publishing Switzerland 2017
G. Genta, *Next Stop Mars*, Springer Praxis Books, DOI 10.1007/978-3-319-44311-9_2

innovations as important enabling technologies (e.g. using aerocapture at Mars for large payloads; In Situ Propellant Production, and so on).

A recent study by the National Research Council of the National Academies in the USA [28] identified what were called two enduring questions motivating human spaceflight:

- How far from Earth can humans go?
- What can humans discover and achieve when we get there?

Most of us are sure that humans can go to Mars (and well beyond). However, the second question has still fuzzy answers.

The study subdivided the rationales for human spaceflight – and hence also for human missions to Mars – into two wide categories:

- Pragmatic rationales:

  1. Economic benefits.
  2. Contributions to national security.
  3. Contributions to national stature and international relations.
  4. Inspiration for students and citizens to further their science and engineering education.
  5. Contributions to science.

- Aspirational rationales:

  1. Eventual survival of human species.
  2. Shared human destiny.
  3. Aspiration to explore.

In retrospect, we can see that human exploration of the Moon fulfilled most of these rationales. Apart from points 2 and 3, linked with the then-prevailing Cold War climate, the economic benefits (point 1) were huge, primarily due to the returns from the technological innovations that were achieved – indeed, the returns were several times the cost. Point 4 was achieved; although the wave of inspiration has long since subsided. Point 5 has always been controversial, since many have argued that scientifically the Apollo missions were a failure. What Apollo taught us about the Moon is certainly not negligible. And of course, in terms of its contributions to technology the program was highly successful.

Little can be said about aspirational rationales, primarily because they are difficult to measure. And because times have changed, it is not an easy matter to compare the 1960s with the situation today. Moreover, the modern reluctance to develop new technologies for space exploration may cause us to miss one of the greatest results of earlier endeavors in this realm. What is done in the name of saving money, may ultimately merely reduce the returns which might accrue.

Generally speaking, it is not easy to find one single reason among those listed above that is strong enough to justify human exploration beyond LEO, but taken as a whole they make a strong case.

The motivations are very different depending on whether the case is argued by a space agency, a private company, or a non-profit organization. The latter, for instance, would be likely to stress the aspirational rationales or perhaps the scientific returns, while a company must concentrate on economic returns. A space agency can afford a more balanced view but is very exposed to changes of governmental or parliamentary climate.

2.2   WHY GO TO MARS?

Mars is not Earth's nearest destination in space, because the Moon is incomparably closer and in fact Venus is also somewhat closer than Mars. In a way, the Moon is not an alternative destination to Mars. Being so different, they constitute two complementary destinations.

The distance to the Moon is almost constant at about 384,000 km because it is in orbit around Earth. Mars and Earth travel individual orbits around the Sun, so their separation is highly variable, ranging from less than 60 million km at opposition to over 380 million km when they are located on opposite sides of the Sun. However, a spacecraft can't travel in a straight line, at least at our technology level, so such figures mean very little except for the time taken by a radio transmission to reach its destination.

In spite of the greater distance, the energy required to reach Mars is not much more than that needed to reach the Moon. Moreover, since the Moon is an airless world whilst Mars has an atmosphere, albeit thin, if the spacecraft is designed to perform an aerobraking maneuver on arrival at Mars, or at least is able to use aerodynamic forces to decelerate, then the energy required to reach Mars might actually be less than that required to travel to the Moon and land on its surface. This does not mean that it is easier to reach Mars than the Moon, because that is not true, particularly in the case of human exploration. In fact the above statements hold only for a minimum energy transfer that can reach the Moon in just a few days but takes several months to reach Mars. To avoid remaining in space for almost 9 months on such a transfer and suffering the problems associated with lengthy exposure to radiation and microgravity, much higher energies are required.

Venus, which is closer than Mars, has little interest for human exploration owing to the unbearable conditions on its surface. However, as we shall see, flybys of Venus can greatly assist with missions to Mars.

The large Earth-Mars distance, the long travel time, and the long communication delay cause medical (radiation and microgravity exposure) and psychological problems. An added problem is related to the launch opportunities, which for Mars occur at intervals of about 26 months and are narrow. In an emergency (technical or medical) a return from the Moon can be achieved at almost any time but returning from Mars outside the relevant launch window is practically impossible.

Planets travel around the Sun in elliptical orbits, with the ellipticity of Mars' orbit being much greater than that of the Earth's orbit, and this causes successive launch opportunities to differ. That said, there are some regularities which recur every 7–8 synodic periods (15–17 years) and others that recur every 37 synodic periods (79 years).

Another difference between the Moon and Mars, is that the latter is considerably larger and therefore its gravity is stronger – more than double. Furthermore, the period of Mars' axial rotation is only about 40 minutes longer than the terrestrial day, whereas the Moon holds one hemisphere facing Earth and takes a month to rotate on its axis, with the result that the lunar days and nights each last 14 terrestrial days. These characteristics will make it easier to adapt to living on Mars.

In the long term, Mars could even be terraformed [9] by renewing its atmosphere and warming its surface to make living on Mars very like living on Earth, apart from the fact that the gravity is 1/3 as strong. Given adequate resources the Moon could be terraformed too. Its atmosphere would require constant replenishment though, because the gas would leak away in the 1/6 gravity. However, the very long day would make living in the open on the Moon a distinctly alien environment.

Mars has much more resources than the Moon, or to be more explicit it offers different resources. For instance, because its surface is exposed to the solar wind, the Moon has large (at least in comparison to our potential needs) amounts of helium-3. This may be important when controlled nuclear fusion becomes feasible as a power source on Earth. The Moon has a great deal of oxygen, mostly in the form of metal oxides in the regolith. Recently, water was discovered in the polar regions as ice crystals in the regolith at shallow depth. Mars also has large quantities of water in the form of ice, some on the surface in the polar caps and the rest elsewhere at a shallow depth. And its atmosphere is almost entirely made of carbon dioxide, from which methane, oxygen, and hydrogen (from water) can be readily created. Some of the resources found on the Moon will be shipped to Earth, but those found on Mars, because of its greater distance, will be used solely to develop that planet and make it habitable. In time, other valuable resources are sure to be found.

Although the Moon will play its part in turning the human species into a multiple-planet civilization, it is Mars which holds out the greatest promise. This is because, in addition to its offering a more familiar environment in terms of gravity and diurnal cycle, it is the nearest solar system body capable of becoming a Second Earth.

2.2.1 Scientific and technological motivations

While it is evident that the scientific goals of Martian exploration do not, by themselves, justify the sharp increase of resources required in order to switch from obtaining science through robotic missions to obtaining science through both robotic and human exploration, science nevertheless remains a key driver for Mars exploration. This is particularly so for missions managed by space agencies. A 'private' mission can be less centered on scientific motivations, except for selling experiment opportunities to scientists.

Since the scientific community is the principal customer to satisfy, it is of the utmost importance for a project to be created in partnership with scientists and that it be centered on scientific objectives. The participating scientists ought to appreciate that such a collaboration will be a strong growth factor for their activity. Moreover, it must be clearly stated that the human presence won't replace robots but, to the contrary, will enhance the productivity of the robots. Having the science community endorse the objectives of a mission and then provide their support is an essential prerequisite for success.

Amongst the various aspects of scientific research which can be performed on Mars, is astrobiology. This is difficult to perform using robotic devices. From statements made in the past, it might be understood that the search for extraterrestrial life is the only, or perhaps the principal, scientific goal of human Mars missions. But this is not only incorrect, it is also a dangerous misunderstanding. It is wrong because astrobiology is only one of many sciences involved, and it is dangerous because a failure to find life on Mars, which to be honest is the most likely outcome, could then be interpreted as a failure of a specific mission and hence a good reason to cancel further missions.

Moreover, life on Mars, or indeed extraterrestrial life in general, is a sensitive subject with both the scientists and the general public. It stirs up both hopes and fears, and the very fact that an organization such as the International Committee Against Mars Sample Return (ICAMSR) exists is significant. Actually it is not calling for a generic ban on Mars Sample Return (MSR), rather it insists that specimens from Mars be thoroughly examined either on Mars or in space prior to being transported to the Earth's surface. On the other hand, if Mars proves to be sterile, there will inevitably be some people who interpret this as a failure of the entire exploration program, and claim there is no justification for sending humans to Mars. These aspects will be discussed in more detail in Section 3.7.

Realistic scientific objectives for human Mars missions must also be consistent with the technical means available. The demands of the science community are many and often well in advance of what is affordable in terms of resources (mass, energy, cost, etc.) or development risks. Consequently, frustrations in the science community, and misunderstandings between scientists and engineers must be avoided if a project is to be a success.

Another key issue is the pacing of the program. The route to a first human landing on Mars should be paved by sequential robotic missions and should be coordinated with other human space activities such as in cislunar space, on the Moon, and on nearby asteroids. The technological and economic consistency of such missions with the Mars objective should be evaluated in each case.

2.2.2 Economic motivations

It is recognized that a big techno-scientific program that ventures into the unexplored and to the limits of our know-how constitutes one of the most powerful tools available to a government to advance the innovative capabilities of its nation and improve its economy.

By involving industrial strategies, high technology and very expensive facilities, activities related to Mars exploration may generate highly paid jobs which cannot be easily outsourced. Even technologies which have reached a reasonably high Technology Readiness Level (TRL), may need to be pushed to their limits and then combined to create the most complex space system ever conceived in order to conduct such a mission.

The ingenuities of research organizations and the capabilities of the aerospace industry must be driven to increased quality and efficiency, and new and better products that give a pay-off in terms of increased market share. The forceful impact of the program on economic development should be documented in terms of the assessment of a number of indicators, the most important of which are:

- Identification of domains where progress will occur.
- Forecasts of industrial innovation and new application areas.
- Workforce enrolled.

If a mission is to be promoted and operated by a private company, or perhaps a group of companies, then strong economic motivations must be offered. For example, awarding rights for the exploitation of space resources will act as a strong motivator. However, this example is more appropriate for a mission to the Moon or to an asteroid than it is to a mission to Mars.

Finally, an important contribution to economic development can be attained by motivating youngsters to obtain higher education and pursue STEM (Science, Technology, Engineering and Mathematics) careers. This is particularly so in nations which lack young specialists in these areas.

2.2.3 Political motivations

Owing to its outstanding performance in monitoring the natural resources and the environment, security, defense, science and commercial applications, space has become a strategic arena for nations at all stages of development.

Those nations willing to maintain or reinforce their international influence cannot neglect any of the main sectors of this new domain of activity. So much so, in fact, that when one of the leading space powers moves in a new direction, the others have little option but to follow. When this happens in an area of space exploration that is perceived as an endeavor in the name of all humankind and a tool for peace and global development, this is even more likely.

It is widely admitted that a program to undertake a Mars mission that is initiated by a single nation will inevitably turn into an international collaboration. Programs that involve international cooperation should satisfy a number of criteria that relate to efficiency and to long term robustness, namely:

- The program structure and conceptual organization should be designed with special care paid to facilitate sharing work and responsibilities, while preserving overall efficiency.
- International partners should be able to satisfy their national interests, which may be many, varied and very specific. Some will desire to participate in order to strengthen their leadership, others to upgrade their rank, others to acquire technologies, others to reinforce their diplomatic links, and so on.
- Presenting the program as an effort to enhance peace and global development ought to be made explicit when defining project objectives and public communication policy.

The strategic character of innovation and of entrepreneurship is becoming more evident as globalization leads to accelerated diffusion of know-how and activities. Even if not all the concerned nations are yet up to the challenge, each is aware that a sufficient level of research and development effort is necessary in order to defend its economy and, in the end, its wealth.

It is necessary for the public to understand and accept that a fraction of their taxes must support such long term objectives. This is not easy and requires significant confidence in the future, as well as a clear demonstration of the resulting benefits. From these two points of view, starting a comprehensive Mars exploration program that offers to the public, and in particular the younger generation, the prospect of exciting discoveries, of new activities, and a reason to dream, would be a smart move. Thus, when defining the program, much attention must be given to factors that will enhance:

- The level of public acceptance and support in major spacefaring nations and other participating countries.
- The interest of youths, and more specifically the involvement of college and university students in the project.

Political feasibility requires that each participating nation takes the decision to enter the enterprise, and that its governing bodies ratify and sustain this commitment for the required time. Taking the decision could depend upon the political climate at that time. The level of commitment might be jeopardized by a decline in the international situation. Interestingly, it was the political climate of the Cold War that drove progress in the early years of the Space Age. In that climate, international cooperation in space projects was impossible. It was only after the collapse of the Soviet Union that tensions relaxed sufficiently to pursue the creation of the ISS. If this high level of cooperation persists, then it might be possible to undertake an international enterprise such as a human Mars mission.

It is encouraging that the decline in the relationship between the USA and Russia during recent years did not jeopardize their close collaboration on the ISS. This suggests that a large international venture in space that is already running can survive fluctuations in the political climate. However, during a crisis it would be difficult to achieve the relevant agreements to initiate such a project.

The likelihood of maintaining the commitment of all players for the duration of a large international project in space depends on the political structure of the various states and on the stability of their governments.

Economic feasibility is an important point in convincing decision makers. And avoiding cost increases as the program progresses is crucial in preventing partners from dropping out. As technological and industrial development can cause a decrease of the costs – even in fields outside aerospace – it is predictable that the introduction of new technologies and the increase of the role of private companies in space will make things easier.

Another important point is popular support. This requires a serious outreach activity by the organizations that advocate space exploration and, specifically human Mars exploration. It also requires an increase of scientific and technological literacy in the general population – a difficult task for nations afflicted with a subculture that disdains science and technology. The general development of commercial activities in space and the development of private space ventures will provide a strong incentive for nations to take the appropriate political decisions.

There are a number of aspects of future human space exploration which might generate opposition in some sectors of the public opinion. One of them is the use of nuclear energy. At present, two types of nuclear power systems have been employed on spacecraft: Radioisotope Thermoelectric Generators (RTG) and fission reactors. In addition, many spacecraft contain Radioisotope Heat Generators (RHG), but these are small and usually do not raise concern. From time to time, there have been protests against the use of nuclear devices in space, most notably in 1997 with attempts to prevent the launch of the Cassini spacecraft because it was powered by RTGs which, it was claimed, would pollute the environment with plutonium if the launch rocket were to fail or the spacecraft were to crash on Earth during a flyby.

The United Nations Office for Outer Space Affairs (UNOOSA) recognizes "that for some missions in outer space nuclear power sources are particularly suited or even essential owing to their compactness, long life and other attributes" and asserts "that the use of nuclear power sources in outer space should focus on those applications which take advantage of the particular properties of nuclear power sources." It has adopted a set of principles applicable "to nuclear power sources in outer space devoted to the generation of electric power on board space objects for non-propulsive purposes." The allowed units include both radioisotope systems and fission reactors.[1]

About 40 satellites powered by nuclear reactors were placed into orbit by the USSR. The better known Russian reactor is the TOPAZ-II, having a power of 10 kW. In 1965 the USA tested the SNAP 10 reactor in space. There is a general consensus that nuclear power will be necessary for human missions to Mars. If nuclear power is used for propulsion or for power generation while in space, the reactor must be started in Earth orbit prior to departing for Mars, but a reactor that powers an outpost on the planet will not be started until it is on the surface.

There is no agreement about the minimum orbital altitude at which a nuclear reactor can operate. The SNAP 10 worked in an orbit at about 900 km altitude and most satellites powered by nuclear reactors had orbits between 800 and 900 km. However, the orbit of Cosmos 1932 had a periapsis of 256 km and an apoapsis of 279 km. Often it is asserted that the altitude at which Nuclear Thermal Rockets (NTR) can be started is higher than that at which nuclear generators and hence Nuclear Electric Propulsion (NEP) can be used, but there is no real basis for this. The tests of NTR in the early 1970s were performed on the ground, and open to the environment. As there is no international agreement prohibiting this from being done, it has even been proposed that rockets based on NTR ought to be available as launchers from the Earth's surface [23].

Even if the first human mission to Mars could be performed using only chemical rockets and solar power, it is important that nuclear power and nuclear propulsion be developed as soon as possible for both Mars and Moon exploration, and it is essential that this be explained to the public in the correct manner.

As already stated, another point that could attract the opposition of public opinion is the possible presence of life on Mars and the dangers of human contact with it. This issue must be defused so that a mission will have the widest possible public support. Until it is definitively ascertained that there is no life on Mars, all people and objects returning from the planet will probably undergo a period of quarantine, in much the same manner as was done for the early Apollo missions even though the likelihood of there being any life on the Moon was extremely low.

2.2.4 Long terms goals

Although it is a controversial view, it has been claimed that the true motivation for a human mission to Mars is to start out on the road which will eventually transform humankind into a multiple-planet spacefaring civilization. We will have an outpost or possibly a colony on the Moon, but it will be Mars that becomes our Second Earth. It is this long term vision that will ensure the continuity of funding required for the exploration effort.

---

[1] Resolution 47/68 of the General Assembly, 14 December 1992; http://www.unoosa.org/oosa/en/ ourwork/spacelaw/treaties.html

A continuity of support will prevent the fate that befell the Apollo program. The main source of funding continuity is linked with the exploitation of Mars for economic purposes, therefore that is of primary importance in achieving the sustainability of the entire effort. The long term goals include providing humankind with a safe haven in case Earth suffers a major disaster such as a catastrophic asteroid strike.

By themselves these goals cannot justify a human mission to Mars, but they can act to reinforce other motivations. The consensus view is that no single motivation is sufficient to justify such a complex, costly, and risky enterprise; it requires the convergence of several powerful motivations.

2.3   THE TIMEFRAME

Perhaps the challenge isn't so much to find a rationale for human spaceflight or for human exploration of Mars, but to find a rationale which places them within a given timeframe. With the advancement of technology, reaching Mars will become ever simpler. It will become ever less critical to justify human exploration. The point is not to understand whether humans will explore Mars, but when this will occur.

The expected timeframe of such a mission is influenced by, and in turn influences, the expected Technology Readiness Level (TRL) of the means by which it will be undertaken. This is particularly true because in the case of Mars it is impossible to imagine a concentrated effort to develop new technologies as rapidly as was done for the Apollo program.

In this regard there are two opposing points of view. Some people argue that there is a window for human expansion in space. If humankind fails to start a spacefaring civilization and to exploit extraterrestrial resources within a given timeframe, the lack of resources will make it impossible to do so and civilization will collapse. An even more pessimistic form of this reasoning says the rate of technological advances is slowing down, and that it may halt entirely in the not too distant future.[2] In that scenario, we cannot rely on future technological advances to make it easier to expand into space.

The second point of view states the situation will not deteriorate at all; actually, it will either continue to improve or at least not deteriorate so quickly as argued by the pessimists, and Earth's resources will support civilization long enough for technological developments to make space exploration and exploitation much easier. In this case, while expansion into space is a necessity for humankind, to try to do it with an insufficient technology exposes us to the risk of abandoning the enterprise even though the early missions succeed, just as happened with the Apollo program. However, in this scenario there is an incentive not to waste time; not due to the pessimistic consideration that a long delay will cause the window of opportunity to close, but to the optimistic one that expanding civilization into space will be such a bonanza that it makes sense to start it as soon as possible.

---

[2] P. Musso, "How Advanced is ET," 7th Trieste Conference on Chemical Evolution and the Origin of Life: Life in the Universe, Trieste, September 2003.

For human Mars exploration, those who adopt the first approach tend to favor the use of chemical propulsion, whereas the others hold that even the first missions should be performed using nuclear propulsion. Perhaps, as it will be seen later, there is an intermediate way, based on solar electric propulsion.

As already stated, a human Mars mission is the occasion to mount a large international enterprise as a follow-up to the ISS. However, the ISS will be decommissioned sometime in the 2020s (at present it is funded to 2024, and further continuation is possible) but this is likely to be too early for starting a collaboration aimed at a human Mars program. It is thus possible that the ISS will be followed by an international collaboration which constructs an outpost on the Moon.

Johann-Dietrich Wörner, the director general of ESA, suggested in 2015 that the follow-up to the ISS could be a Moon Village. Although a goal in itself, this could become a stepping stone toward Mars because many of the technologies that will have to be developed to live on Mars can be developed at a lower cost and with less risks in the context of a lunar outpost. An international cooperation aimed at Mars exploration may therefore be seen as a follow-up to one aimed at establishing a lunar outpost.

Several roadmaps for human Mars exploration have been proposed which incorporate the realistic assumption that a number of steps must be performed before an enterprise of such scale can be attempted (see Chapter 11).

2.4   RISKS

As Herodotus said, "Great enterprises are carried forward amid great dangers."[3]

Exploration is never free of risk. In the Age of Discoveries, all expeditions suffered casualties and often entire expeditions were completely lost at sea – in the modern argot we would say they suffered Loss of Crew (LOC). Mutiny was a fairly frequent occurrence and the sailors responsible for such acts were usually either put to death or abandoned at sea. Even the explorations of the nineteenth and early twentieth century such as polar explorations claimed many lives.

The acceptability of risk is a cultural and political issue and our culture, particularly in the case of the western society, is increasingly risk averse. The precautionary principle is at the same time a symptom and a consequence of this trend. Although this exists in a strong and a weak form – and has several formulations – in essence it states that any action or policy that has a potential to cause harm to the public or to the environment should not be carried out in the absence of scientific proof that no harm will be done. In practice, instead of prohibiting actions that are proven to be dangerous, it prohibits actions which have not been proven to be harmless. All too often the meaning of the potential risk is not clearly defined, and it may be very difficult, if not actually impossible, to prove that an action is completely harmless. In its strongest formulation the precautionary principle is arguably capable of hampering all progress and innovations even when, by the weaker formulation, the associated risks are deemed to be acceptable. Indeed, we can wonder whether fire would ever have been accepted if prehistoric humans had adopted the precautionary principle!

---

[3] Herodotus, *The Histories*, VII, 50.

But there can be significant contradictions. We accept that some people indulge in very dangerous activities that are useless as a whole, such as some really dangerous extreme sports, but similar risks in useful activities are considered to be unacceptable.

In terms of human spaceflight, the risk of Loss Of Crew on an Apollo lunar mission was calculated at about 5 percent, whereas for a Shuttle mission a value of 2 percent was accepted. Although it is impossible to make a statistical analysis on such a small number of cases, these values were respected. Of the seven Apollo lunar flights, Apollo 13 was the only instance of Loss Of Mission (LOM) and there were no of LOC. For 135 Shuttle flights there were two cases of LOC (1.5 percent). Significantly, even though the losses were within what had been deemed acceptable, each accident prompted calls for program cancellation.

What about human missions to Mars? It has been suggested that a LOC risk of about 2 percent would be reasonable, taking into account both the difficulties associated with such a complex mission into the unknown and the risk that a LOC will prompt the termination of the whole program – Loss Of Program (LOP). This problem is particularly severe in the case of a mission performed by space agencies or other public organizations, and less in case of private missions that are likely to be better able to tolerate a failure so long as it is not so costly as to cause that entity to cease its activities. On the other hand, a program motivated by the slogan 'Mars or Bust' would be dangerous and should be avoided.

Of course there are other potential risks associated with such a major program, mostly of economic or political nature. These include cost overruns, delays, scientific returns that are over sold, activities that turn out to be dead ends, partners that withdraw their support, and international cooperation which proves unmanageable. Any of these events may lead to LOP. For instance, although Apollo had no LOC and only one LOM, other forces caused LOP in which missions were canceled and the surplus space vehicles became lawn ornaments at the nation's leading space centers.

The risks of initiating a program and of undertaking missions must be balanced against the likely returns, and measures must be formulated in advance to keep the risks at bay. A sound proposal must optimize the technical and operational design of the project. Ideally, a project should address forceful political requirements, and the risks should be convincingly identified and reduced to an acceptable level.

2.5   AFFORDABILITY

Of course affordability must be carefully assessed. Space exploration is clearly expensive, and human missions to Mars will be particularly so, but activities in space are not as expensive as many other activities. In 2012 the total expenditures of all space agencies amounted to US$ 42.0 billion,[4] of which $17.9 billion was spent by NASA. This amounted to just 2.6 percent of the total military expenditures in the same year of $1,617.7 billion,[5] of which the American share was $739.3 billion. Hence space activities are comparable with the annual cost of other major infrastructure projects. Table 2.1 lists the governmental expenses in 2013 for civilian space activities of the fifteen countries that have the highest space budgets.

---

[4] Organisation for Economic Co-operation and Development, 2012.

[5] Stockholm International Peace Research Institute, 2012.

Table 2.1 Governmental expenses (in $million) for civilian space programs in 2013 in the fifteen countries which have the highest space budget.

Country	Expenses	Country	Expenses	Country	Expenses
USA	19,770	Germany	1,886	Brazil	505
Russia	6,414	India	1,173	South Korea	313
France	2,418	Italy	915	Spain	279
Cina	2,384	UK	559	Belgium	277
Japan	1,957	Canada	507	Kazakhstan	244

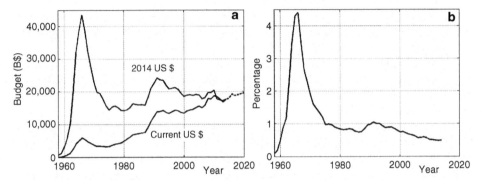

Figure 2.1 NASA expenses in the years, from its beginning to present. (a) In current and 2014 US$. (b) As a percentage of the federal budget.

The NASA expenses, as obtained from the US Office of Management and Budget (OMB), are reported as a function of time in Figure 2.1a. These data come from different sources and often are not directly comparable; for example, NASA's budget includes funding for activity linked to aeronautics and not space. Figure 2.1b reports the NASA budget as a fraction of the federal budget. It is evident that (in 2014 dollars) space expenditure in the USA had a peak of 4.5 percent in the mid-1960s, during the build-up of Apollo. It has been decreasing steadily since the 1990s. In fact, since 1995 it has remained consistently between 0.5 and 1 percent of the federal budget.[6]

Since year 2000, the NASA budget has grown only to match inflation and therefore is decreasing in terms of percentage of the general federal budget. One recent document about human exploration of Mars [31] says it would be unreasonable to expect a larger growth in the likely timetable for attempting such a mission. This document also warns that it cannot be presumed that sufficient public support will develop to justify the level of increased funding that would be needed. It is interesting that whilst there has been considerable fluctuation in public support over the last 50 years, there has always been a hard core of supporters at around 35–38 percent. Only in July 1997 was support above 60 percent and that was attributable to the Pathfinder-Sojourner mission.

---

[6] US Office of Management and Budget.

In these conditions, the only way that a single nation could undertake a human mission to Mars (as advocated by [31]) would be by limiting its scope to the bare minimum, not using all the launch opportunities, and minimizing the necessary technological develop-ments. It is really questionable whether such an 'affordable' mission would be viable; for instance, the production of the launchers and the other hardware would not benefit from the economies of scale resulting from operating a production line, and eschewing new technologies might introduce risks which could lead to the program being prematurely terminated. Such a mission would be little more than a 'flag and footprint' affair – with the flag and the footprint being on Phobos rather than on the planet.

Given that the increase in the budget that would be required for a human Mars program is impossible for a single nation, the inescapable conclusion is that it will require interna-tional cooperation. In this regard, the highly positive experience of the ISS is encouraging. It is also clear that there will be a role for private organizations.

Figure 2.2a gives the governmental civilian space funding (in 2012) as a percentage of the Gross Domestic Product (GDP) of the ten countries which spend most (in percentage) on space. While in absolute terms the country with the highest space budget is the USA, in relative terms it is Russia.

The expenditures on space exploration are given as a percentage of the government budget in those same countries in Figure 2.2b.[7] The two countries which traditionally spend most are the USA (0.18 percent) and Russia (0.12 percent), with China (0.07 per-cent) coming third. The other spacefaring countries are 0.02–0.03 percent. Space

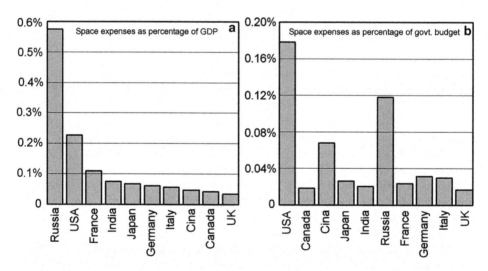

Figure 2.2 (a) Governmental civilian space expenses as a percentage of the Gross Domestic Product (GDP) of the ten countries which spend most (in percentage) in space. (b) Expenses for space explo-ration as a percentage of the government budget in the same countries.

---

[7] G. Reibaldi, M. Grimard, "Non-Governmental Organizations Importance and Future Role In Space Exploration," *Acta Astronautica* vol. 114, pp. 137–137, 2015.

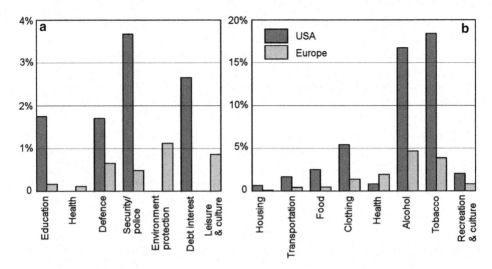

Figure 2.3 Expenses for space exploration as a percentage of the main public and private expenditures in the USA and Europe.

Table 2.2 List of the richest private space investors (from Forbes).

Forbes rank	Name	Worth (B$)	Company	Space investment
1	Bill Gates	79.1	Microsoft	Kymeta
15	Jeff Bezos	34.8	Amazon	Blue Origin
19	Larry Page	29.7	Google	Planetary Resources
43	Charles Ergen	20.1	Satellite TV	DISN Network
51	Paul G. Allen	17.5	Microsoft	Scaled Composites, Stratolaunch Systems, Vulcan Aerospace
56	Ma Huateng	16.1	Tencent Holding	Satellogic, Moon Express
81	Sheldon Adelson	31.4	Casinos	Space IL
100	Elon Musk	12.2	payPal, Tesla motors, Solar City	SpaceX
137	Eric Schmidt	9.1	Google	Planetary Resources
330	Richard Branson	4.8	Virgin	OneWeb, Virgin Galactic
557	Yuri Milner	3.2	Facebook	Planet Labs, Breakthrough

exploration is compared with public and private expenditures in the USA and Europe in Figures 2.3a and 2.3b. The only cases in which space exploration is not at least an order of magnitude smaller are those for alcohol and tobacco (and even those only in the USA).

In the last decade, many private investors have entered the aerospace business in general and in particular for space exploration and exploitation. They include some of the richest people in the world as surveyed by the Forbes list. Table 2.2 lists the eleven richest investors in space. They include people who are actively involved in space exploration. Elon Musk, the PayPal billionaire who went on to found SpaceX, has declared an interest in the exploration of Mars. Some who work in the space tourism sector, like Jeff Bezos and Richard Branson, are creating opportunities for space tourism. Larry Page and Eric Schmidt have started companies aimed at mining extraterrestrial resources.

As earlier pointed out, the efforts of private investors will strongly hasten and make the exploration of space beyond LEO both simpler and more affordable.

Although space budgets will have to be increased significantly in order to undertake a human mission to Mars, the expenditures will be spread out over many years. Human Mars exploration, particularly if it is a truly international program, will impose less strain on the budgets of the participants than was borne by the USA in the 1960s when it mounted the Apollo program.

To reach economical affordability it is essential that the various players, particularly the industrial partners, be able to make a profit from their investment in mounting the mission. A true space economy is developing for space tourism. It may soon involve the exploitation of space mineral resources [30]. Human Mars exploration must therefore find its place in this context.

In particular, the exploitation of the Moon is very important because it will promote the development of technologies for building and operating surface habitats in conditions that are even more harsh than those on Mars. Certainly, ISRU and ISPP systems created for the Moon will differ from those for Mars. To obtain oxygen on the Moon it is necessary to process the regolith, whereas on Mars it is simpler to process the atmosphere. What little water there is on the Moon is at the poles in the form of crystals within the regolith, whereas there is plenty of water on Mars, both as ice on the surface at the poles and elsewhere as a permafrost at a shallow depth. Although not identical, the processes are similar in both cases. Much that is learned on the Moon will be able to be applied to Mars. Another issue in undertaking such operations will be generating the necessary electrical power. Although solar power will be available on the Moon, at places away from the poles it will be interrupted by the lunar night which lasts a fortnight, therefore nuclear reactors will probably be necessary. This is another technology that will transfer to Mars.

What is perhaps most important, is the development of an industrial infrastructure for space exploitation. Irrespective of whether Mars exploration is pursued on a public or private basis, it will be necessary to involve companies whose core businesses are in space. As recent experience has shown, private ventures can supply services to space agencies at a fraction of the cost it would take the agencies to do so themselves. And of course, in an entirely private program the companies would directly undertake the exploration activities.

Only if the various missions (except perhaps the very first one) are successful in creating economic returns, will it be possible for human Mars missions to avoid suffering a LOP such as befell Apollo. A successful program on the scale of a lunar colony and human missions to Mars stands a good chance of transforming our society into a spacefaring civilization.

3

Mars and its satellites

Mars is a small, arid planet of major geological and topographical complexity. Conditions on the surface in the remote past appear to have been rather more favorable to life than they are today. Whilst it is unlikely Mars still hosts life, particularly on its surface, things may have been different earlier on. The search for life, in particular for fossil life, is one of the reasons for exploring Mars, but it is not the main reason, and therefore if it turns out that life is absent, this must not be seen as a failure of Mars exploration.

The actual destination of a human mission to Mars is usually thought of as a selected place on the surface of the planet, but several studies have been conducted in which the targets for the early missions were either low Mars orbit (LMO) or one of the two natural satellites, Phobos and Deimos. Although missions that do not reach the surface of the planet might have some advantages, particularly in terms of easier implementation, recently the satellites are being less emphasized and so they will be described only briefly here.

3.1 ASTRONOMICAL CHARACTERISTICS

Mars is the fourth planet of the solar system, located just beyond the orbit of Earth. Its radius is slightly more than half of Earth's radius but, by lacking oceans, its land surface is roughly similar to Earth. As its orbit is considerably more elliptical than that of Earth, its distance from the Sun varies much more dramatically.

The main astronomical features of Mars are reported in Table 3.1 and compared with those of Earth. The characteristics of Venus are also included because there are trajectories for Mars missions that include making a flyby of Venus.

The two-way communication delay with Mars is very variable, determined by the positions of the planets in their solar orbits, ranging from about 6.2 minutes when Mars is in opposition to about 44.6 minutes when it is in conjunction. However, taking into account the relative motions of the planets and the lack of telecommunication satellites in Mars orbit at that time, the 1996 Pathfinder lander had a communications window of only 5 minutes per day.

© Springer International Publishing Switzerland 2017
G. Genta, *Next Stop Mars*, Springer Praxis Books, DOI 10.1007/978-3-319-44311-9_3

Table 3.1 The main features of Mars compared to those of Earth and Venus. The surface acceleration at the equator, taking into account the planet's rotation, is also included.

	Mars	Earth	Venus
Mass (10^{24} kg)	0.64185	5.9736	4.8675
Volume (10^{10} km^3)	16.318	108.321	92.843
Equatorial radius (km)	3,396.2	6,378.1	6,051.8
Polar radius (km)	3,376.2	6,356.8	6,051.8
Ellipticity (flattening)	0.00648	0.00335	0
Topographic range (km)	30	20	15
Mean density (kg/m^3)	3,933	5,515	5,243
Surface gravity (m/s^2)	3.71	9.81	8.87
Surface acceleration (m/s^2)	3.69	9.78	8.87
Escape velocity (km/s)	5.03	11.19	10.36
Solar irradiance (W/m^2)	589.2	1367.6	2613.9
Orbit semimajor axis (10^6 km)	227.92	149.60	108.208
Sidereal orbital period (days)	686.980	365.256	224.701
Perihelion (10^6 km)	206.62	147.09	107.477
Aphelion (10^6 km)	249.23	152.10	108.939
Synodic period (days)	779.94	–	583.92
Mean orbital velocity (km/s)	24.13	29.78	35.02
Max. orbital velocity (km/s)	26.50	30.29	35.26
Min. orbital velocity (km/s)	21.97	29.29	34.79
Orbit inclination (°)	1.850	0.000	3.39
Long. of ascending node (°)	49.579	–	76.678
Orbit eccentricity	0.0935	0.0167	0.0067
Longitude of perihelion (°)	336.04	114.27	55.186
Sidereal rotation period (hrs)	24.6229	23.9345	−5,832.60
Length of day (hrs)	24.6597	24.0000	2,802
Obliquity (°)	25.19	23.45	177.36
Min. dist. from Earth (10^6 km/light minutes)	55.73/3.1	–	38.2/2.12
Max. dist. from Earth (10^6 km/light minutes)	401.322/22.3	–	261.0/14.5

Due to the different eccentricities of the orbits of Earth and Mars, and also to the angle between the planes of their orbits, Mars' path in the sky does not repeat every synodic period (779.94 terrestrial days, about 26 months) as would be so with circular, coplanar orbits (see Section A.1). As the planets travel their independent orbits, there is a time when Mars comes closest to Earth. At this time it is said to be in opposition because it lies opposite the Sun in the sky. At its greatest distance it is said to be in conjunction because it is located on the far side of the Sun.

Table 3.2 list the Earth-Mars oppositions for the period 2016 to 2061, plus the relative angular position $\Delta\nu$ of Venus with respect to Earth and Mars. It is useful to plan trajectories which include a Venus flyby. The distances between the two planets at the oppositions from 2014 to 2185 are reported in Figure 3.1.

The closest distance between the two planets thus varies from a minimum of slightly less than 60 million km for an opposition that occurs when Mars is close to perihelion (making this a perihelic opposition) to about 100 million km when an opposition occurs with Mars close to aphelion. As the orbit of Earth is much more circular than that of Mars, the position of Earth in its orbit is of little importance in this regard.

Table 3.2 Earth-Mars oppositions from 2016 to 2061. Also the relative angular position of Venus with respect to Earth and Mars Δv is reported. The latter is useful in planning trajectories that include a Venus flyby.

Date	Hour (UMT)	Min. Dist. (AU)	Min. Dist. (Gm)	$\Delta v(°)$
May 22, 2016	11:11	0.50321	75.28	170.0
Jul 27, 2018	05:07	0.38496	57.59 (perihelic)	304.3
Oct 13, 2020	23:20	0.41492	62.07	82.9
Dec 8, 2022	05:36	0.54447	81.45	207.5
Jan 16, 2025	02:32	0.64228	96.08	320.0
Feb 19, 2027	15:45	0.67792	101.42	71.2
Mar 25, 2029	07:43	0.64722	96.82	181.1
May 4, 2031	11:57	0.55336	82.78	296.9
Jun 27, 2033	01:24	0.42302	63.28	61.4
Sep 15, 2035	19:33	0.38041	56.91 (perihelic)	204.0
Nov 19, 2037	09:04	0.49358	73.84	333.1
Jan 2, 2040	15:21	0.61092	91.39	90.2
Feb 6, 2042	11:59	0.67174	100.49	200.4
Mar 11, 2044	12:44	0.66708	99.79	310.8
Apr 17, 2046	18:01	0.59704	89.32	64.2
Jun 3, 2048	14:45	0.47366	70.86	183.7
Aug 14, 2050	07:46	0.37405	55.96 (perihelic)	321.4
Oct 28, 2052	06:28	0.44091	65.96	98.5
Dec 17, 2054	22:09	0.57015	85.29	218.0
Jan 24, 2057	01:26	0.65552	98.06	330.8
Feb 27, 2059	05:25	0.67681	101.25	82.1
Apr 2, 2061	12:47	0.63199	94.54	191.7

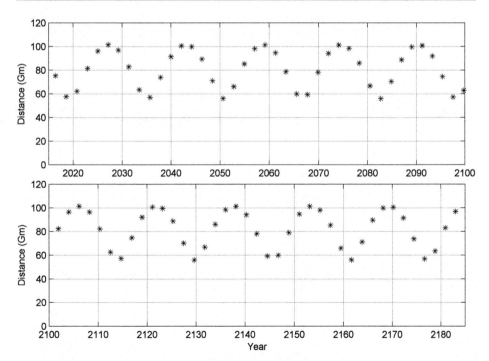

Figure 3.1 The distance between Earth and Mars for oppositions from 2015 to 2185.

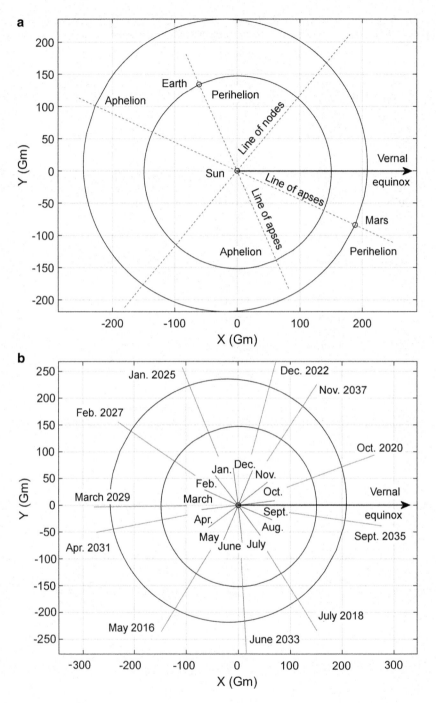

Figure 3.2 (a) Orbits of Earth and Mars, with the planets drawn at perihelion. (b) Oppositions of Mars between 2016 and 2037.

The orbits of Earth and Mars are shown to scale in Figure 3.2a. The reference frame is on the ecliptic plane (by definition it is the plane of Earth's orbit about the Sun) with the x axis in the direction of the vernal equinox (about September 23). The perihelion and the aphelion of both Earth and Mars, and the line of nodes of the orbit of Mars are also plotted. The line of apses is that which connects perihelion and aphelion; i.e. the line along which the major axis of the ellipse lies. The line of nodes is that which connects the intersections of the orbit with the xy coordinate plane. Note that because the selected coordinate plane is the plane of the ecliptic, in which Earth travels, there is no line of nodes for Earth.

In Figure 3.2 it is obvious to the eye that the orbit of Mars is much more elliptical than that of Earth. This has a significant effect on missions because it is easier to reach the planet when it is at opposition. The energy of an interplanetary transfer is minimum at a perihelic opposition. Figure 3.2b shows the oppositions from 2016 to 2037.

3.2 MARS' SURFACE

Mars has high mountains and deep depressions with a topographic range of about 30,000 m; considerably more than Earth. It has the largest volcano in the solar system, Olympus Mons, which rises more than 20 km tall. Although some of the Martian volcanoes may have issued lava flows as recently as several million years ago, no activity is currently underway. There is a canyon, Valles Marineris, that is likely the deepest and widest in the solar system; certainly it dwarfs the Grand Canyon on Earth. And Mars bears the traces of dramatic events in the past which have modeled its surface. The northern lowlands, Vastitas Borealis, are probably the result of a major impact that created the largest known basin in the solar system; fully four times the size of the South Pole Aitken basin on the Moon. The Tharsis Bulge, which is probably of volcanic origin, boasts three huge volcanoes: Pavonis, Arsia and Ascraeus Mons. And there are impact craters of all sizes, chasms, and mountains.

The highest spot is the summit of Olympus Mons, which stands at 21,282 m (above an arbitrary zero). The lowest point is in the Hellas plain, at −8,180 m. Figure 3.3. shows Valles Marineris, the northern polar cap, and the volcanoes atop the Tharsis Bulge.

The poles possess seasonal ice caps. The northern ice cap is primarily water ice, but the southern one has a layer of frozen carbon dioxide over a deposit of water ice. About 25–30 percent of the atmosphere condenses during a polar winter to create thick slabs of CO_2 ice. These sublime when the pole is again exposed to sunlight. This drives huge wind storms from the poles with velocities up to 400 km/h, but because the density of the atmosphere is low the dynamic pressure of such winds is low. A figure of two million cubic kilometers of water ice has been estimated to be contained in the northern ice cap.

The differences between the two ice caps can be attributed to a flattening of the globe in the polar zones, orbital eccentricity and axial inclination, all of which are higher than Earth's. This combination makes the seasons far more extreme in the southern hemisphere than in the northern. That is, Mars nears perihelion when it is summer in the southern hemisphere, and nears aphelion when it is winter. This is believed to account for the occurrence of violent dust storms that last for several months and, in extreme cases, can spread across the entire globe. The Martian day is slightly longer than the terrestrial day at 24 hours, 39 minutes, 35 seconds and is usually referred to as a sol. The tilt of the axis of

Figure 3.3 A picture of Mars taken by a robotic probe. (NASA image)

rotation is similar to that of Earth (25° and 23° respectively), producing seasons which are similar to ours but last longer because the Martian year is 687 terrestrial days.

Mars is smaller than Earth, its surface area being similar to the sum of all continental regions of our planet. A simplified map of the planet is shown in Figure 3.4. On the map the approximate landing sites of the probes are shown, with successful landings being indicated with a '+' sign.

Figure 3.4 A simplified map of the Martian surface (from a NASA image). The approximate sites of successful landings are indicated by '+' signs. These are 1: Viking 1 (July, 20, 1976). 2: Viking 2 (September 3, 1976). 3: Mars Pathfinder-Sojourner (July 4, 1997). 4: Spirit (January 3, 2004). 5: Opportunity (January 24, 2004). 6: Phoenix (May 25, 2008). 7: Mars Science Laboratory-Curiosity (August 6, 2012). Crash landings are indicated by '*' signs. These are 1: Mars 2 (November 27, 1971). 2: Mars 3 (December 2, 1971). 3: Mars 6 (March 12, 1974). 4: Mars Polar Lander (December 3, 1999).

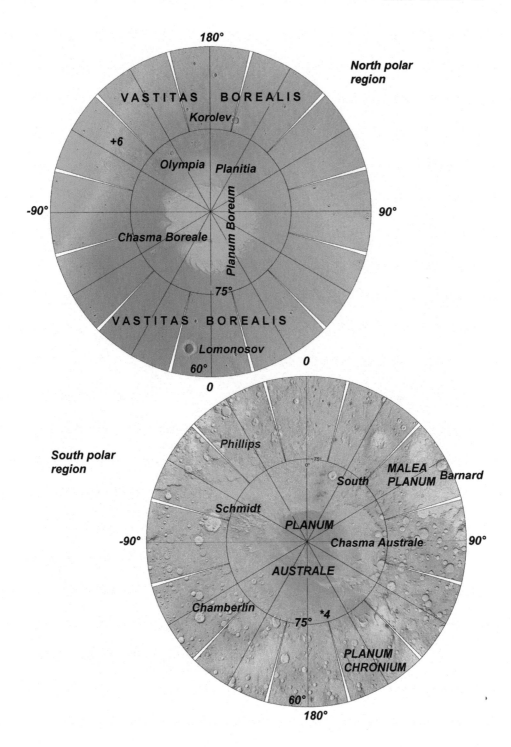

North polar region

South polar region

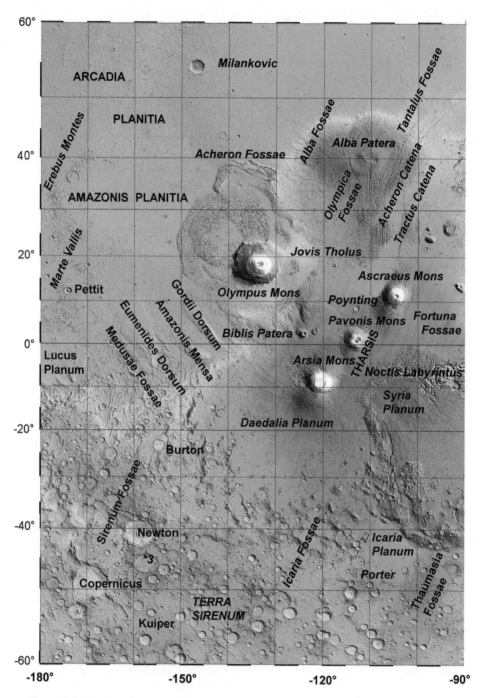

Figure 3.4 (continued)

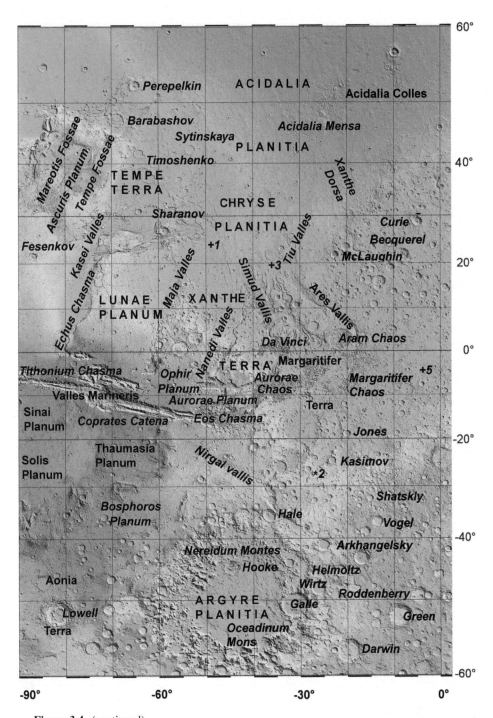

Figure 3.4 (continued)

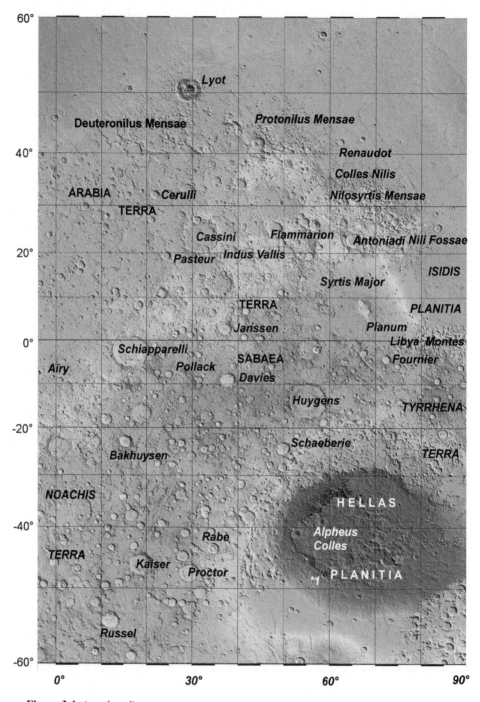

Figure 3.4 (continued)

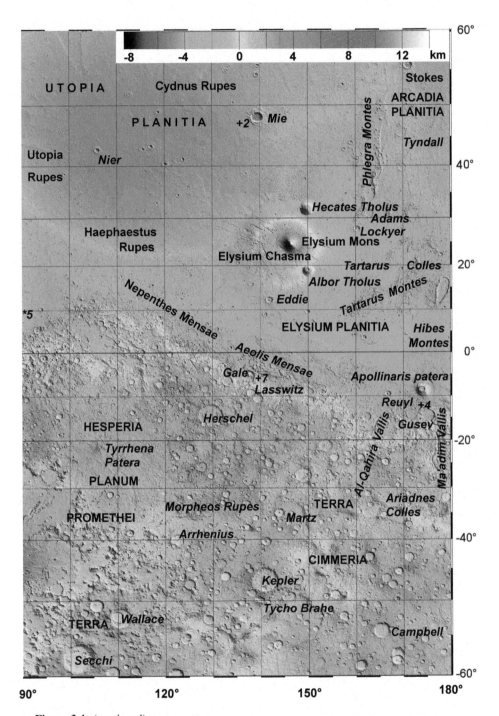

Figure 3.4 (continued)

3.3   ATMOSPHERE

The Martian atmosphere is very thin. The surface pressure is less than 1 percent of that at sea level on Earth (it roughly equals atmospheric pressure on Earth at an altitude of 35 km) and varies significantly with altitude and latitude, ranging from a minimum of about 0.3 millibar on Olympus Mons to over 11.6 millibars in the depths of Hellas Planitia. The mean surface pressure is 6.36 millibars and varies with seasons from 4.0 to 8.7 millibars. The density of the atmosphere is also strongly affected by the temperature. Its average value on the ground is about 0.020 kg/m^3 (see Section D.2). This variability poses a challenge in planning the entry of a probe into the atmosphere.

Figure 3.5 gives an approximation of temperature and pressure as functions of altitude, relative to the arbitrary zero referred to above. These values, which are just averages in time and space, have been obtained from the mathematical model supplied by NASA (see Section D.2). The model is a rough approximation, and in particular the bi-linear dependence for the temperature is questionable since at a certain altitude it would give a temperature lower than absolute zero.

The composition by volume of the atmosphere is reported in Table 3.3. Although it is present only in trace amounts, methane is concentrated in several places during the northern summer. Since methane is broken down by ultraviolet radiation, for it to be present at all there must be a mechanism which produces it. Possible explanations include volcanism, cometary impacts, and the presence of methanogenic microbial life. The mean molecular weight of the atmosphere is 43.34 g/mole.

Clouds of water ice were photographed by the Opportunity rover in 2004. The fine dust in the atmosphere gives the sky a tawny color when viewed from the ground. The two Viking landers of 1976 reported wind speeds of 2–7 m/s in the summer, 5–10 m/s in the autumn, and occasionally gusts of 17–30 m/s during dust storms. Although the wind speeds are high, the aerodynamic forces are weak due to the low atmospheric density, therefore vehicles and structures on the surface will not be mechanically stressed by even the strongest winds.

The average temperature on the surface is −63°C but there are pronounced diurnal and annual variations. At the Viking 1 site variations occurred in the range −89°C and −31°C

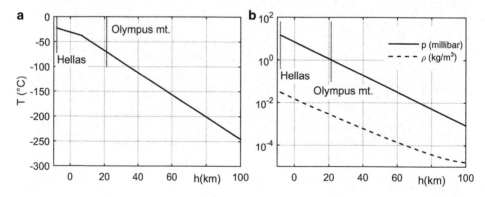

Figure 3.5   A model of Mars' atmosphere showing temperature T, pressure p, and density ρ as functions of the altitude.

Table 3.3 The average composition of Mars' atmosphere.

Gas	%
Carbon dioxide	95.97
Argon	1.93
Nitrogen	1.89
Oxygen	0.146
Carbon monoxide	0.0557
Water	210 ppm
Nitrogen oxide	100 ppm
Neon	2.5 ppm
Heavy water (in form of deuterium protium oxide HDO)	0.85 ppm
Krypton	0.3 ppm
Methane	traces

during a sol. Larger variations were recorded over a Martian year, ranging from −120°C to −14°C. Summer temperatures as high as 20–30°C above freezing have been recorded in the southern hemisphere.

Mars lost its planetary magnetic field about 4 billion years ago, so there is almost no magnetosphere. As a result, the atmosphere offers very little protection from the ultraviolet portion of sunlight, and only limited protection from cosmic rays. Heavy Galactic Cosmic Ray (GCR) particles impinging on the iron oxide in the Martian rocks release high energy alpha particles that are not stopped by the thin atmosphere. Hence from the point of view of radiation, even though the atmosphere scatters light and the environment looks more like Earth than the Moon, Mars is only a slightly better place to be than the Moon. Radiation levels at the surface would be somewhat lower than in space, and might vary significantly at different locations depending on altitude and local magnetic fields.

3.4   DUST

Winds carry large amounts of dust that is rich in iron oxide, with particles even smaller than about 1.5 µm diameter; even finer than the lunar dust. Table 3.4 indicates the distribution of particle size.[1] Table 3.5 reports its chemical composition at the Viking 1 and Pathfinder sites. Mars' regolith is rich in oxides, and Table 3.6 reports the percentage in weight recorded by both of the Viking landers and Pathfinder.[2]

---

[1] C.C. Allen, R.V. Morris, K. M.Jager, D.C. Golden, D.J. Lindstrom, M.M. Lindstrom, J. P. Lockwood, "Martian Regolith Simulant JSC MARS-1," Lunar and Planetary Science Conference XXIX, http://www.lpi.usra.edu/meetings/LPSC98/pdf/1690.pdf

[2] R.N. McGrath et al., "Red Mars - Green Mars? Mars Regolith as a growing medium," http://www. lpi.usra.edu/publications/reports/CB-1063/RedMars2.pdf, Mars Pathfinder Science Results http:// mars.jpl.nasa.gov/MPF/science/science-index.html; D.W. Ming, D.C. Golden and D.L. Henninger, "Utilization of On-Site Resources for Regenerative Life Support Systems at Lunar and Martian Outposts", SAE Technical Paper 932091, 1993; B.C. Clark, A.K. Baird, R.J. Weldon, D.M. Tsusaki, L. Schnable and M.P. Candelaria, "Chemical composition of Martian fines", *Journal of Geophys. Research*, vol. 87, pp. 10059–10067, 1982.

Table 3.4 Mars dust: grain size distribution.

Size (μm)	%
1,000–450	21
449–250	30
249–150	24
149–53	19
52–5	5
<5	1

Table 3.5 Composition of Martian regolith in terms of percentage in weight.

Element	Pathfinder			Viking 1
	A-2	A-4	A-5	
Oxygen [O]	42.5	43.9	43.2	—
Sodium [Na]	3.2	3.8	2.6	—
Magnesium [Mg]	5.3	5.5	5.2	5.0 ± 2.5
Aluminum [Al]	4.2	5.5	5.4	3.0 ± 0.9
Silicon [Si]	21.6	20.2	20.5	20.9 ± 2.5
Phosphorus [P]	—	1.5	1.0	—
Sulfur [S]	1.7	2.5	2.2	3.1 ± 0.5
Chlorine [CI]	—	0.6	0.6	0.7 ± 0.3
Potassium [K]	0.5	0.6	0.6	≤0.25
Calcium [Ca]	4.5	3.4	3.8	4.0 ± 0.8
Titanium [Ti]	0.6	0.7	0.4	0.5 ± 0.2
Chromium [Cr]	0.2	0.3	0.3	—
Manganese [Mn]	0.4	0.4	0.5	—
Iron [Fe]	15.2	11.2	13.6	12.7 ± 2.0
Nickel [Ni]	—	—	0.1	—
Not directly detected*	—	—	—	50.1 ± 4.3
Sum	100	100	100	49.9

(*Includes H_2O, NaO, CO_2, NO_x, and trace amounts of Rb, Sr, Y and Zr)

Table 3.6 Oxides in the Martian regolith in terms of percentage in weight.

Oxide	Pathfinder			Viking	
	A-2	A-4	A-5	Chryse Plan.	Utopia Plan.
Na_2OO	4.3	5.1	3.6	—	—
MgO	8.7	9.0	8.6	6	6
Al_2O_3	8.0	10.4	10.1	7.3	7*
SiO_2	46.1	43.3	43.8	44	43
SO_3	4.3	6.2	5.4	6.7	7.9
K_2O	0.6	0.7	0.7	< 0.5	< 0.5
CaO	6.3	4.8	5.3	5.7	5.7
TiO_2	1.1	1.1	0.7	0.62	0.54
MnO	0.5	0.5	0.6	—	—
Fe_2O_3	19.5	14.5	17.5	17.5	17.3
Cl	—	—	—	0.8	0.4
Other	—	—	—	2	2
Totals	≈99	≈96	≈96	≈91	≈90

(* Inferred from available data)

The very fine dust will pose a danger to machinery and human beings, so provisions to prevent it from entering any part of a habitat must be taken.

In particular, Mars' regolith is packed with several toxic substances, including:

- Perchlorates can greatly impair the thyroid gland. The first indications of perchlorates were reported by NASA's Phoenix Mars lander in 2008. There is some evidence that the odd results obtained by the Viking landers may have been caused by perchlorates. The same substances were detected in samples scooped from a sandy drift of material on the floor of Gale Crater by NASA's Curiosity Mars Science Laboratory.
- Silicates, if breathed in, can combine with the water in the lungs to create chemicals that can cause all manner of respiratory diseases.
- Gypsum, if breathed in, can have effects similar to inhaling coal dust, which is known to damage the lungs. Although it is technically not toxic, according to *New Scientist* the US National Institute for Occupational Safety and Health classifies gypsum dust as a nuisance particulate that can irritate the eyes, skin, and respiratory system, and it has set recommended exposure limits.

The experience of the Apollo missions showed that measures must be taken to prevent lunar dust from being breathed in, but Mars dust is potentially even more dangerous. Like lunar dust, it was produced by aeons of rock-shattering meteorite impacts, but Mars has an atmosphere and the resulting weathering processes have smoothed the dust grains, making them less sharp than those of lunar dust. However, one thing the grains have in common is that they will likely carry sufficient static electrical charge to stick to space suits, habitats, and vehicles. It will therefore be essential to remove dust from space suits in order to deny it entry to habitable spaces where it could block air filters and contaminate food. This discipline will be particularly strict during long stays on the planet.

During the Apollo missions, lunar dust was found to be dangerous because its abrasive characteristics can affect space suit articulations. Particular care must be given to this aspect in preparing for Mars missions, even though the Martian dust is less abrasive, because suits must be designed to function for much longer times than was the case for Apollo astronauts.

The plains of Mars are frequently crossed by dust devils, some of which are quite a large size (Figure 3.6). Particularly owing to the low atmospheric density, these dust devils are not dangerous. In some cases, they proved useful by sweeping clean the solar panels of the Spirit and Opportunity rovers, helping to maintain them in an operational state for much longer than expected.

3.5 WATER AND ICE

As can be seen from the phase diagram of water in Figure 3.7, it cannot exist in the liquid state on the surface of Mars at the temperatures and pressures mentioned above because the triple point of water occurs at 0°C and 6.12 millibars.

Figure 3.6 A dust devil about 20 km tall sweeps Amazonis Planitia during the late spring on March 14, 2007. It was imaged by the High Resolution Imaging Science Experiment (HiRISE) of the Mars Reconnaissance Orbiter. (NASA image)

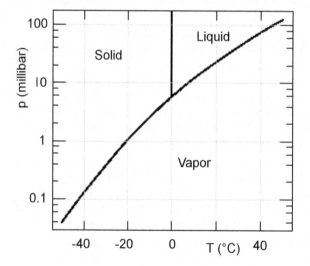

Figure 3.7 The phase diagram for water.

Surface water ice can be seen on the north polar ice cap. It also lies beneath the cap of carbon dioxide ice at the south pole (about 85 percent carbon dioxide and 15 percent water).[3] The presence of subsurface ice has been established by the data collected by

---

[3] J.P. Bibring, Y. Langevin, et al., "Perennial Water Ice Identified in the South Polar Cap of Mars," *Nature* vol. 428 (6983), pp. 627–630, 2004.

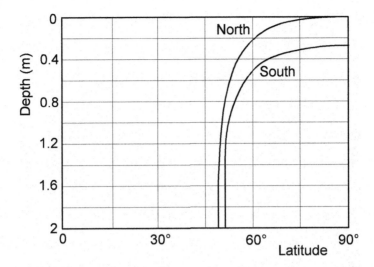

Figure 3.8 The depth at which the permafrost can be found at various latitudes in the northern and southern hemispheres.

several robotic probes indicating large quantities of hydrogen.[4] The total quantity of subsurface ice likely exceeds five million cubic kilometers, sufficient to cover the entire planet to a depth of 35 m. More is likely to exist at greater depths. Even larger quantities of water should be stored in hydrated minerals; although the amount is unknown, some results from the data obtained by the Opportunity rover at Meridiani Planum suggest the sulfate deposits there could contain as much as 22 percent water by weight.

The question is not whether there is water on Mars, it is a matter of where the water is located and how difficult it will be to extract. Water is arguably the currency of space [30], and prospecting for water will be one of the most important tasks assigned to early missions. Whilst it is possible to produce propellants on Mars just from atmospheric carbon dioxide, it will be much easier to do this if local water is available; therefore the presence of ice at a depth shallow enough for it to be readily extracted will be important in selecting a landing site.

Based on observations and mathematical models, Figure 3.8 shows the depth at which the permafrost is present at latitudes in the northern and southern hemispheres. It would seem that in order for water to be readily to hand, the outpost must be located poleward of about 50° of latitude. But things might actually be better than that because the plot refers to steady state conditions corresponding to the current parameters of Mars. In fact, conditions on the planet have changed over the last million years. Two parameters which have a large effect on the subsurface ice of Mars are the eccentricity of its orbit and the obliquity of its rotational axis – and particularly the latter. It has been shown that if the obliquity was about 30° instead of the present value of 25.2° then ice would be stable across the entire planet at a depth of several centimeters, and in the last 400,000 years this value of

---

[4] W.C. Feldman et al, "Global Distribution of Near-Surface Hydrogen on Mars," *J. Geophysical Research* vol. 109, 2004.

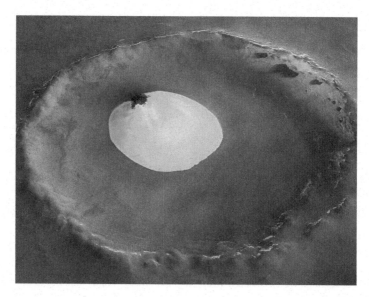

Figure 3.9 Picture of a crater located in Vastitas Borealis at 70.5° North showing clear signs of surface ice. (ESA Mars Express image)

obliquity has been attained. The issue is the length of time it takes for subsurface ice to sublimate into the atmosphere and solidify again in the polar caps when the obliquity decreases. Might latitudes substantially less than 50° still have a permafrost layer close to the surface as a relic of these old ice deposits? And might there be liquid water at a greater depth in some places? The answer is quite important, because drilling a well in order to extract liquid water will be considerably easier than mining ice. These questions are discussed in detail in [22].

At present, the search for water is performed from orbit using neutron spectrometers and gamma ray spectrometers that can detect the presence of hydrogen in the uppermost layers of the regolith. This hydrogen is believed to be incorporated into the molecular structure of ice, and from measurements of the amount of hydrogen it is possible to obtain the concentrations of water ice in the top meter of the Martian surface. In accordance with what is said above, measurements by Mars Odyssey showed ice to be abundant in the subsurface above 60° of latitude. Above 70° the concentration exceeds 25 percent and at the poles it approaches 100 percent. Below 6° there are regional patches in concentrations up to 18 percent, particularly near the Elysium volcano, Terra Sabaea, and northwest of Terra Sirenum.[5]

On July 28, 2005, the High Resolution Stereo Camera of ESA's Mars Express orbiter imaged a 35 km diameter crater in Vastitas Borealis at 103° East, 70.5° North – a position north of Utopia Planitia where Viking 2 landed – showing clear evidence of surface ice and even what was said to be an ice lake (see Figure 3.9).[6]

---

[5] J.B. Murray, B. John, et al., "Evidence from the Mars Express High Resolution Stereo Camera for a frozen sea close to Mars' equator," *Nature* vol. 434 (7031), pp. 352–356, 2005.

[6] http://www.esa.int/var/esa/storage/images/esa_multimedia/images/2005/07/perspective_view_of_ crater_with_water_ice_-_looking_east/10192616-2-eng-   GB/Perspective_view_of_crater_with_ water_ice_-_looking_east.jpg

In addition, NASA's Phoenix mission confirmed in 2008 the presence of water ice at its landing site at a latitude of 68.2° North. The ice detected by the Mars Reconnaissance Orbiter in the north polar ice cap has a volume of about 821,000 cubic kilometers; some 30 percent of the Earth's Greenland ice sheet. The same probe provided strong evidence that the lobate debris aprons in Hellas Planitia and in mid-northern latitudes are glaciers that are covered with a thin layer of rocks. These are at lower latitudes than those mentioned above.

Liquid water may occur transiently on the Martian surface today, but only under certain conditions. In 2011, NASA announced seasonal changes which occur on steep slopes below rocky outcrops near crater rims in the southern hemisphere, suggesting they were caused by salty water (brines) flowing downslope and then evaporating. Actually, although liquid water cannot exist on the surface, a brine can survive briefly because it has a lower freezing point. For example, perchlorate salts would reduce the freezing point of water from 0°C to −70°C. Changes on the surface were observed in several places with ice subliming and forming water that flowed and evaporated. And snow was seen to fall from cirrus clouds, hence at least some clouds are composed by water-ice rather than carbon dioxide ice.

Further research to determine whether there is a permafrost layer and what its depth is in a place chosen for a landing is essential, as is the development of technologies capable of efficiently extracting water from these locations.

3.6 GEOLOGICAL HISTORY

There is some evidence that, early on, Mars had plate tectonics and a planetary dynamo to produce a global magnetic field. There is evidence for this in the form of a pattern of local magnetization that is uncorrelated with the current topography. Nowadays there is no global magnetic field.

The geography of Mars is complex. Its main features are shield volcanoes (all extinct by now), lava plains (mostly in the northern hemisphere) and highlands with a large number of impact craters and deep canyons.

A total of 43,000 craters with diameters of 5 km or greater have been found, together with a large number of smaller ones. The largest canyon, Valles Marineris, has a length of 4,000 km and a depth of up to 7 km. It was formed when the regional swelling that created the Tharsis Bulge also caused a section of crust to crack and collapse to form this network of canyons. Ma'adim Vallis is one of the largest channels to have been created by water which flowed on the surface. Some 700 km long, 20 km wide and 2 km deep in places, it is much larger than the Grand Canyon in Colorado.

Entrances to large caves 100–250 m wide have been discovered by orbiting probes.[7] A cave skylight which spans 180 m on the southeastern flank of Pavonis Mons, on the Tharsis Bulge, is shown in Figure 3.10a.

Given the low gravity on Mars, it is possible there are lava tubes far larger than those on Earth. Figure 3.10b shows a structure in a valley in the Tartarus Colles region which seems to have been formed as a lava tube.[8]

---

[7] http://www.space.com/18519-mars-caves-lava-tubes-photos.html#sthash.fOwetRKn.dpuf

[8] http://www.space.com/18519-mars-caves-lava-tubes-photos.html#sthash.9l85FaNA.dpuf

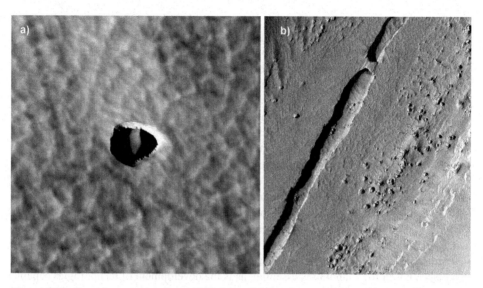

Figure 3.10 Two pictures taken by NASA's Mars Reconnaissance Orbiter. (a) A cave skylight on the southeastern flank of Pavonis Mons on the Tharsis Bulge. (b) A structure that is likely to have been formed as a lava tube located in the center of a valley in the Tartarus Colles region. (NASA images)

Caverns and lava tubes are of particular interest, not only because they may supply natural shielding from micrometeoroids, ultraviolet radiation, solar flares and the high energy particles that bombard the planet's surface, but also because they are good candidates for seeking liquid water and signs of life. Although they might be attractive as locations for human settlements, the possibility that they might host indigenous life suggests that, at least in the beginning, we should leave underground spaces untouched.

Martian rocks appear to be mostly basaltic, but a portion of the Martian surface seems to be richer in silica than is a typical basalt. The plains are rugged, similar to rocky deserts on Earth, covered by red sand, with rocks and boulders scattered all around. The most interesting places are, however, the steep slopes of mountains and canyons. These will be very difficult to access using wheeled or tracked vehicles. This suggests using rocket-powered hoppers or, more simply, helicopters for fast transportation. The soil is essentially a regolith that is rich in finely grained iron oxide dust. Its granularity (see Table 3.4) and composition are highly variable from place to place owing to wind and water erosion in ancient times when water was flowing on the surface. Values of pH $\approx$ 8.3 were measured by the Phoenix lander, making the soil basic. Together with the results of the earlier Viking landers, this seems to rule out the possibility of there being any life on the surface of the planet. Even if life is discovered underground or in locations that are shielded from radiation and direct sunlight, such as in caves or at the bottom of canyons, we can safely say there are no biological products in the regolith. This may make surface mobility easier, because on Earth the most difficult terrains are those rich in products of biological origin, where tracks instead of wheels are often needed.

The geological history of Mars is subdivided into three main periods, namely:

- The Noachian epoch (named after Noachis Terra) which ranged from 4.5 billion to 3.5 billion years ago.
- The Hesperian epoch (named after Hesperia Planum) spanning the interval 3.5 billion years ago to 1.8 billion years ago.
- The Amazonian epoch (named after Amazonis Planitia) which covers from 1.8 billion years ago to the present.

It seems that the most dramatic changes in the climate occurred very early in the geological history of the planet, mainly in the Noachian epoch (see Figure 3.11).

The formation of the Tharsis Bulge and extensive flooding by liquid water are dated in the Early Noachian epoch. Extensive lava plains were formed in the Hesperian epoch. Olympus Mons was formed during the Amazonian epoch, along with lava flows elsewhere on Mars. Thanks to the in situ observations performed by robotic spacecraft, it is now certain that in ancient times Mars had a large amount of liquid water running on its surface and geyser-like water flows. At that time, the atmosphere was much thicker. However, it is not known whether Mars was globally warmer as a result of the greenhouse effect or colder due to the Sun being fainter than it is today.

The internal heat associated with the uplift that made the Tharsis Bulge is thought to have prompted a large release of subterranean water, which ruptured the surface to form a 'chaotic' landscape of cavities from which the water emerged, then carved massive outflow channels. A much more recent (5 million years ago) outflow of water is supposed to have occurred when the Cerberus Fossae chasm developed. As noted above, there is evidence of small amounts of water flowing down slopes for brief periods even now.

3.7 POSSIBILE PRESENCE OF LIFE

In early times Mars was more suitable for living organisms than it is today. That does not imply that life actually existed, but if it did, then there might be evidence in the form of fossils. The finding in 1996 of possible fossil micro-organisms in the ALH84001 meteorite (which almost surely 15 million years ago was blasted off Mars by an impact, then traveled through space and landed on Earth) is still controversial and an exclusively inorganic origin for the alleged microfossils has also been proposed.

Water is essential for life (at least for life as we know it) and so NASA adopted the strategy of 'follow the water' to investigate whether there was once, or may still be, life on Mars. But water is not in itself a sufficient condition for life to originate. It also requires an energy source, the presence of the raw materials for the growth and reproduction of living matter, and a benign environment.

The results of in situ studies by robotic probes show that during the Noachian epoch (at least in its early phase) Mars was rich in water. For instance, Figure 3.12 is a photomicrograph taken by Opportunity of a gray hematite concretion in the rock outcrop known as El Capitan at Meridiani Planum. The process by which such concretions form requires the presence of flowing water.

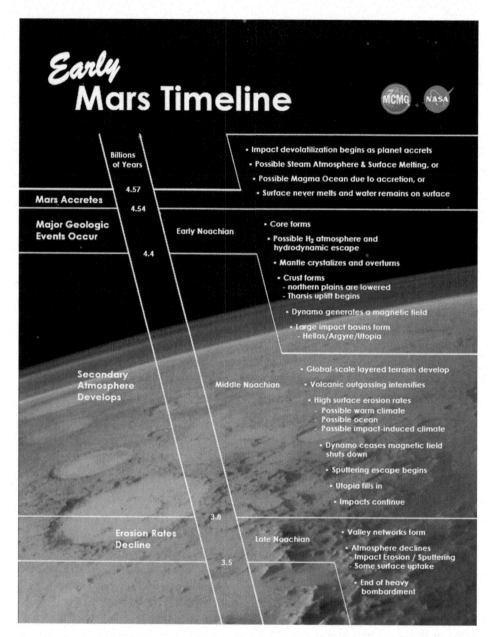

Figure 3.11 The most dramatic events in the geologic history of Mars occurred in the Noachian epoch. This shows the early timeline, as outlined by the NASA's Mars Climate Modeling Group (MCMG). (NASA image)

Figure 3.12 A photomicrograph taken by Opportunity showing a gray hematite concretion that is indicative of the past presence of liquid water on Mars. The area in this image is 1.3 cm across. (NASA image)

It seems that there were lakes, rivers and perhaps even oceans on the surface of Mars in those ancient times. This water may well have been rich in nutrients. The early magnetic field may have been sufficiently strong to protect the surface of the planet from cosmic and solar radiation. A global magnetic field would have had the additional advantage of preventing the solar wind from eroding away the atmosphere, and a dense atmosphere would have enabled liquid water to flow on its surface. But these points are debatable. There is also the possibility that the water on the Martian surface may have been too salty and acidic to support life; at least life of terrestrial type. Even if Mars possessed a magnetosphere several billion years ago, the subsequent reduction in atmospheric density might have allowed radiation to destroy all life. And of course, if the atmosphere has always been thin, life may never have developed at all.

But as stated, although the presence of water, nutrients, and a shield against radiation are prerequisites for the development of life, they are in no way sufficient conditions. The reasons for life developing on Earth are not known. It has been argued that there was no reason at all, and life originated simply by chance. Thus Mars could have remained a lifeless world even if all prerequisite for life were present.

Because we can be sure that life on Mars is not widespread on the surface of the planet now, we must consider four alternatives that are consistent with what we know:

1. Life never developed on the planet, and it is completely lifeless.
2. Life started, but has since disappeared from all the planet. In this case we might find fossils.

3. Life started, and then disappeared from the surface but remains in a few underground places. There may be fossils. It might be difficult to discover the few surviving forms of life and therefore distinguish case 3 from case 2.
4. Life started, then disappeared from the surface to remain widespread underground. In this case, it may be relatively easy to find.

In the last two cases, some people have argued that, even 2 m beneath the surface, any microbes would likely be dormant, cryopreserved by the current freezing conditions, and so metabolically inactive that they are unable to repair their inevitable cellular degradation. If this is true, then in seeking active organisms we should study fairly deep subterranean places where past volcanism may have created subsurface cracks and caves within different strata, and liquid water could have been stored. Life could have survived in such large aquifers with deposits of saline liquid water, minerals, organic molecules, and geothermal heat. The small amounts of methane and formaldehyde in the atmosphere in some areas and at certain times are cited as possible evidence for this subterranean life. Certainly some process is producing these chemical compounds because they will quickly break down in the atmosphere. But this isn't proof of life, because these compounds could instead have been produced by volcanic or other geological processes.

Establishing whether there is life on Mars is one of the most important, and most difficult, goals of Mars exploration. After the results of the experiments carried by the Viking landers, it was considered extremely unlikely that there is any life to be found on Mars; at least on the surface, and even less in places where landing is easy. Deep canyons like the Valles Marineris, where the scarce atmospheric moisture could concentrate, could be better places. However, the most suitable sites for life will be able to be reached only by ground vehicles, piloted or automatic, with the capability to climb steep slopes (see [13] and [15]).

The 'follow the water' strategy is currently judged as the best option, but this search may prove to be very difficult. It is possible that it will take many missions exploring difficult areas to discover fossil life. This possibility must be clearly stated and communicated to the public, so that lack of initial success does not cause disappointment that jeopardizes continuation of the exploration program.

Furthermore, it must be stated explicitly that finding life is not the main goal of human Mars exploration, and in its absence the exploration can succeed by addressing much wider objectives.

3.8   CHOICE OF THE LANDING SITE

The choice of landing site does not significantly influence mission design, and can be left to a later design stage. In the context of a multiple mission program, the point worth discussing is whether all missions should use the same landing site (see [24]). The NASA study deemed a single site to be best, if the ultimate goal of exploration is to prepare for colonization. If this is the goal, each mission will contribute to the build-up of a single well equipped and redundant outpost that will last a long time. Having a different landing site for each mission is the best strategy if the goal is purely scientific, because it will provide a deeper understanding of the geology of the planet and will maximize the likelihood of making discoveries, including any life that may be present.

A compromise would be to land at a number of sites which are at moderate distances from each other, but are within the range of a pressurized rover delivered by the first mission. This choice presumes a relatively flat landscape, since follow-on crews would need to move freely between sites while carrying new equipment in order to enlarge their exploration range. The downside of this strategy is the relatively small size of each outpost.

In order to make landing and living on Mars easier, a landing site should be:

- A flat place with few large boulders, so that it is not difficult to find a landing place and to travel using rovers.
- Located at low altitude in order to allow a lander additional time to lose speed in the descent, and the greater atmospheric density at low level offers also protection from cosmic radiation (see Section 4.4.6).
- At low latitude in order to simplify landing and ascending. In addition an equatorial site will not suffer such cold winters.
- At high latitude, where water ice is readily available if ISRU and ISPP involving the use of water is predicted.
- A place where it is unlikely there is life, in order to minimize contamination issues. However, in seeking life the opposite could be argued.
- A place from which interesting places can be reached, either using crewed rovers or automatic ones. However, it should be noted that automatic rovers might be landed in interesting places without the need of traveling there overland.

With these criteria, the floor of the Hellas basin is an interesting place. It is the lowest place on the planet (typically −7,152 m), hence the atmosphere is denser and the radiation is weaker. There is probably water ice, although this remains to be proved, and often there is frost. Geologically, it belongs to the Noachian epoch and is quite interesting. As drawbacks, the weather can be cold and it is a long way from interesting places such as Valles Marineris and Tharsis (see Figure 3.13).[9] Another interesting site is Argyre Planitia. This basin is less deep than Hellas, but it is slightly further south and thus it is more likely that there is water ice not too deep underground. It is much closer to Valles Marineris and Tharsis.

Volcanic regions like the Tharsis Bulge, Olympus Mons or Elysium Mons have much to offer in the way of lava tubes and caves. But at least in the beginning these may be placed off limits for direct human contact in order to protect them from contamination.

The DRA 5 [24] suggested sites for a strategy aimed at exploring three different places. These included Centauri Montes, Nili Fossae and Arsia Mons which, between them, feature relics from Noachian, Hesperian and Amazonian epochs.

---

[9] NASA/JPL-Caltech/Arizona State University–JMARS, Public Domain

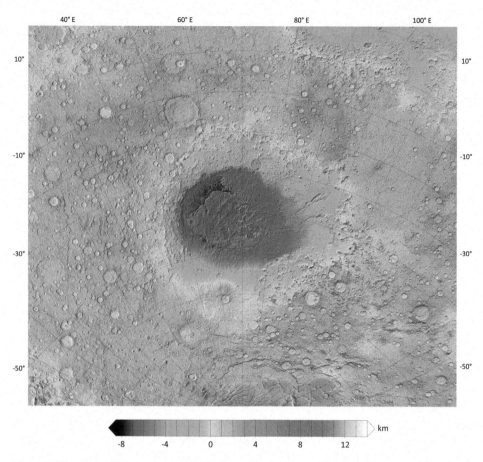

Figure 3.13 A map of the region around Hellas crater from the MOLA instrument of Mars Global Surveyor. (NASA/JPL-Caltech/Arizona State University-JMARS)

3.9   CONTAMINATION

3.9.1   Forward contamination

The term 'forward contamination' refers to the possible contamination (mainly biological) of a planetary environment by human and robotic exploration missions. This issue has been debated since the earliest Mars robotic probes, and guidelines for planetary protection have been issued by COSPAR and are included in the Outer Space Treaty.

Forward contamination is prevented primarily by sterilizing the spacecraft. While this is already difficult to achieve in the case of robotic missions, it is even more so in the case of human missions. In the simplest of circumstances, contamination may be due only to dormant micro-organisms that have not yet encountered conditions that permit them to metabolize and reproduce. In this case, contamination due to exploration may be biologically reversible. Also, if needed, it might be possible in the future to remove all the life we have brought to Mars and return the planet to its previous biologically pristine state.[10]

The difficulty with contamination is that there is a common perception that if Mars does not host indigenous life, then contamination is not an issue, and indeed could be pursued to terraform the planet. Contamination would be of relevance only if there is Martian life. If such life is discovered, strict anti-contamination practices (with the difficulties they would impose on human exploration) should be applied. But if a search fails to detect life, there will remain an uncertainty as to its possible existence. One suggestion for the case where no life is found is to divide the surface of the planet according to two possible situations:

- A zone in which we are sure there is no life, in which less strict precautions can be taken in order to make exploration easier.
- A zone, usually referred to as Special Zone, where life might be present and therefore stricter rules should apply.

In the latter case, a human presence might be totally banned and exploration carried out only using carefully sterilized robots and telemanipulators. If no life is found in a particular region, this could be reassigned to the first category. An interesting option is to combine this strategy of categorizing the planetary surface with that of landing all missions in the same place. In that case, the whole surface of the planet could be studied by teleoperators, thereby imposing direct human control over a single outpost. This would reduce contamination to a minimum while enabling the very presence of humans to make teleoperation practicable. An extreme version of this strategy would be to keep humans in orbit or on one of the moonlets of Mars, and allow only teleoperated robots to land on the planet. These issues will be further explored in Chapter 8.

As plans for future exploration are developed, a bioethics debate should start concerning whether it would be appropriate to import terrestrial life to Mars, and whether the process of terraforming might later be implemented.

3.9.2 Backward contamination

Backward contamination refers to contamination of Earth by biological material coming from space and in particular, for the case under consideration here, from Mars. This problem was addressed at the time of the Apollo project. At least for the first few missions, the astronauts returning from the Moon were quarantined. However, the Moon was soon proven to be a non-biological body, and these precautions appear with hindsight to have been excessive. In the case of Mars things are different. A heated debate began as soon as a sample-return mission was considered. Here we have two extreme positions: one stating that there is absolutely no danger because a parasitic organism evolves to infect a well-defined type of host. It was also pointed out that a number of meteorites that originated on Mars have reached Earth so that, if contamination were possible, it would have already occurred. The fact that these meteorites caused no contamination implies that there is no danger. The opposing view is that of the ICAMSR (International Committee Against Mars Sample Return), a body that, although not opposing sample return missions in principle,

---

[10] C. P. McKay, "Planetary Ecosynthesis on Mars: Restoration Ecology and Environmental Ethics," NASA Ames Research Center, 2007.

stipulates that the samples shouldn't be brought directly to Earth but instead be left for a long quarantine onboard some facility in Earth orbit, where they can be studied and perhaps stored forever. The committee holds that Mars is almost certainly inhabited by microorganisms which, like all bacteria, are potentially dangerous to any form of life. Exchanges of biological materials between planets, even if this happens naturally, are extremely dangerous. The committee further holds that many epidemic diseases may be caused by meteorites, comets and asteroids. Between these opposing positions, the majority of scientists and the NASA administration maintain that reasonable quarantine measures for all Martian specimens must be implemented and that even more strict quarantine procedures will be necessary when astronauts return to Earth from Mars.

These measures will obviously be required in particular on a human mission to Mars that discovers life. The very fact that the search for life is presently one of the scientific goals of Mars exploration guarantees that all Martian specimens will be studied in depth from this point of view, thereby eliminating any chance that life forms that can interact with terrestrial life will go undetected. It must be noted that the surface environment of Mars is so harsh that biological material eventually found in protected places is quite unlikely to survive when exposed to the surface.

Usually not included in backward contamination is non-biological matter. The surface of Mars is rich in chemicals that are hostile to all organic matter, and of fine dust which is harmful to both humans and machinery. These issues must be considered in designing the mission and all the machinery that must operate on the planet's surface.

Organic compounds such as prions, although not alive, may be pathogenic. On Earth they are biologically produced but we do not know whether they can be the result of non-biological reactions. Even if the surface of Mars proves to be free of such substances, care will have to be taken to prevent this type of back contamination.

3.10   MARS' SATELLITES

Phobos and Deimos are two small, irregularly shaped moons which orbit close to Mars. Their orbital data are given in Table 3.7. They might be similar to asteroid 5261 Eureka, which is a Martian Trojan, but their capture by an almost airless world is difficult to explain. The moons are commonly believed to be carbonaceous chondrites. Like other asteroids of the same type, they may be a target for mining operations. This may lead to synergies between human Mars exploration and industrial exploitation of its moons by

Table 3.7 Orbital data.

	Phobos	Deimos
Orbit semimajor axis (km)	9,378	23,459
Sidereal orbit period (days)	0.31891	1.26244
Orbital inclination (°)	1.08	1.79
Orbital eccentricity	0.0151	0.0005
Major axis radius (km)	13.4	7.5
Median axis radius (km)	11.2	6.1
Minor axis radius (km)	9.2	5.2
Mass (10^5 kg)	10.6	2.4
Mean density (kg/m³)	1,900	1,750

space agencies or (more probably) by private ventures. There have been proposals to base human exploration on Phobos to permit astronauts to teleoperate robots on the planet's surface, either as a precursor or an alternative to ground exploration of Mars. It is likely these bodies contain water and, if so, they may be used to produce propellants close to Mars but high up in its gravitational well.

3.10.1   Phobos

Phobos is the larger of the two moons and is the closest. Actually it is very close, its distance from Mars is 60 times smaller than the distance of the Moon from Earth. It is in an orbit lower than synchronous, hence it rises in the west, sets in the east and rises again just 11 hours later. Its orbit is decaying due to tidal forces and in about 50 million years it will either fall onto the planet or fragment to create a ring. In fact, there are traces on the surface of Phobos suggesting cracks that are probably due to tidal forces, so the process which will eventually result in its fragmentation has already started.

The large tidal forces on Phobos will make orbital maneuvering close to its surface tricky. And because it is so near the planet, to depart from it will require a larger velocity than its own mass would imply. Because Phobos spends so much time within the planet's shadow, using a SEP vehicle to access Phobos will be more difficult than for Deimos.

Figure 3.14 shows the largest crater on Phobos, called Stickney. It is on the Mars facing side. A portion of the floor of the crater is protected by its walls, by the mass of Phobos itself, and by the planet from about 90 percent of cosmic radiation, possibly more. Since it is one of the places in the solar system which is best protected from cosmic radiation, it is widely regarded as a good site to locate an outpost.

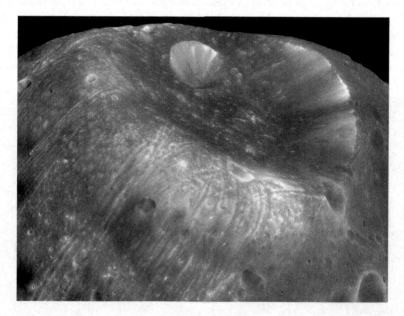

Figure 3.14 A close view of the crater Stickney on Phobos by Mars Reconnaissance Orbiter. (NASA image)

As a final remark, there have been suggestions that Phobos is either an artificial satellite built by Martians or a gigantic starship left in orbit around Mars by unknown aliens. These speculations, which were spread by Carl Sagan himself as provocative discussion starters, survived the intentions of the well-known scientist and will not be quashed until Phobos is investigated by direct exploration.

3.10.2   Deimos

Although in most studies of missions to a moon of Mars it is Phobos that is chosen, a case can be made in favor of Deimos (Figure 3.15).

Deimos is just above synchronous orbit and thus its apparent speed in the sky is low. Even though the period of its orbit is 30 hours, the interval from its rising in the east to its setting in the west is 2.7 days.

Its south pole is almost permanently in shadow, so it is among the coldest places in the solar system and is permanently protected from solar radiation. It is therefore a place to seek refuge from solar flares and to store cryogenic propellants for long times. The presence of water ice may also make it an ideal place to produce cryogenic propellant.

Because its orbit is almost synchronous, it will be able to maintain line of sight contact with any point on Mars for something like 60 hours. Moreover, because its rotation is tidally locked, most of the surface of Mars is visible from Deimos for 45 percent of the time; only 2.5 percent of the planet cannot be directly viewed (i.e. towards the poles). Hence a point on Deimos facing Mars would be an excellent site to install the antenna for teleoperating robots almost anywhere on the planet. However, even larger benefits could be obtained, perhaps at a lower cost, by deploying three telecommunication satellites into synchronous orbit to serve as relays to teleoperate robots from a base on the surface or from a spacecraft in orbit or perhaps stationed at one of the Lagrange points.

Even more than Phobos, Deimos will also be useful for mining, mostly because it is much easier to land on and to launch from. In 1997 there was a proposal by David Kuck to establish a Deimos Water Company to supply Earth orbit with water from this moonlet.

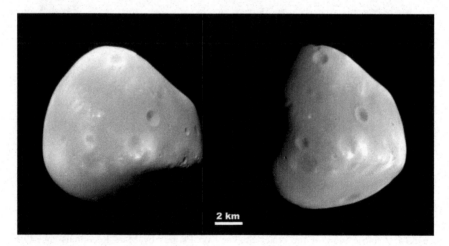

Figure 3.15   Color-enhanced views of Deimos by the High Resolution Imaging Science Experiment (HiRISE) camera on NASA's Mars Reconnaissance Orbiter. (NASA image)

4

Space environment and radiations

To travel to Mars, astronauts must spend several months in space. But space, and above all space beyond low Earth orbit, is a very harsh environment both for human beings and their machines, so it is essential to consider it in some detail. The danger of radiation is one of the factors that could cause us to postpone plans for a human Mars mission, or perhaps make it impossible. Whilst radiation is not necessarily a 'showstopper,' it must be considered very seriously. And if the vehicles for the mission are to be assembled in low Earth orbit, we must understand that environment too.

During a human mission to Mars the astronauts must endure three kinds of environments:

- Low Earth orbit.
- Interplanetary space.
- Low Mars orbit.

The first is particularly important if the vehicles are to be assembled in LEO or must anyway remain in LEO for some time. Interplanetary space is much more dangerous than space close to Earth. Low Mars Orbit (LMO) is quite similar to interplanetary space, and so usually is not considered separately.

In early efforts to design a human Mars mission, it was widely thought that being on the surface of Mars was much more dangerous than being in space, and the mission was devised in such a way that, in the event of an emergency, the astronauts would lift off from the planet and remain in orbit until an opportunity opened to return to Earth. Now this is considered to be wrong. Space is the most dangerous environment that will be encountered during a Mars mission, so astronauts are expected to spend the shortest possible time in space.

Fortunately, today's machines are sufficiently autonomous to be expected to work for long periods without maintenance and strict human supervision, so there will be no need for part of a crew to remain in orbit about Mars tending to the spacecraft while the others do their work on the planet. In this respect, a human Mars mission will differ from an Apollo lunar mission. In case of an emergency, Mars must be considered as a safe haven where the astronauts will stay until either the problems are solved or a rescue mission can be mounted from Earth, with the latter taking a very long time to arrive.

© Springer International Publishing Switzerland 2017
G. Genta, *Next Stop Mars*, Springer Praxis Books, DOI 10.1007/978-3-319-44311-9_4

4.1   THE LEO ENVIRONMENT

A low Earth orbit is one at an altitude of less than 1,000 km, hence this environment is relevant only for the early phase of a human Mars mission. The worst case is where the vehicle must be assembled in orbit. The Earth has a global magnetic field which forms a magnetosphere. This protects the planet and the environment in space beneath the Van Allen Belts from most solar and galactic radiation. As a result, the radiation in LEO is fairly moderate, even during solar flares. But there is an anomaly in the magnetic field. As shown in Figure 4.1a, the Van Allen Belts are symmetric about the Earth's magnetic axis, which is tilted about 11° with respect to the rotational axis. This angle, together with an offset of about 450 km, causes the inner Van Allen Belt to dip deeply off the coast of Brazil and produce a region of enhanced radiation, and again with lesser intensity over the North Pacific. The limits of the South Atlantic Anomaly (SAA) vary with altitude and its shape changes over time. At an altitude of 500 km it spans −90° to +40° in longitude and −50° to 0° in latitude, and its extent increases with increasing altitude (Figure 4.1b).

High energy protons, together with electrons, are trapped in the inner Van Allen Belt, with electrons being trapped in the outer one, both of which constitute a high radiation zone. Since their discovery at the start of the Space Age, various projects have been suggested to get rid of them, but it is likely we will have to live with them for the predictable future.

The inner belt goes from an altitude of about 1,000 km to 9,600 km, while the outer belt is located between 13,500 km and 58,000 km. The radiation dose that humans and machines can receive while crossing the belts is not large so long as they pass through

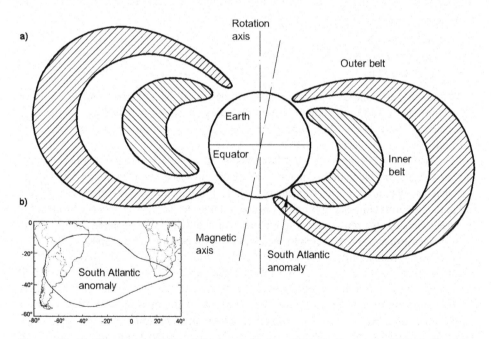

Figure 4.1   (a) The concentrations of charged particle radiation in the inner and outer Van Allen Belts. (b) The South Atlantic Anomaly.

rapidly, as is the case with impulsive propulsion. The radiation dose that an Apollo crew received when setting off on a lunar mission was between 0.16 and 1.14 rad, this low value being attributable to the fact that the trajectory did not cross the inner belt and went through the outer belt in a region where it was thinner. The return trajectory did not cause any problems either.

Problems can arise with a low thrust vehicle that reaches escape velocity by spiraling out slowly, and hence may remain for hours or days in the radiation zone. Also when re-entering, they spiral in and can spend a long time in this dangerous zone. Astronauts can be delivered using a 'taxi' a short time before the spacecraft reaches Trans-Mars Injection (TMI) and can leave it before entering the radiation belts on returning from Mars; this strategy has the added advantage of cutting more than a month from the total journey time, thereby further reducing the total radiation dose. In terms of hardware, the risk is to the components which are most sensitive to radiation, like solar panels, electronic circuits, sensors, etc. Solar electric vehicles may be more affected than nuclear electric ones, owing to their large solar panels and their slow climb out of the Earth's gravity well. The long dwell in the radiation belts must be accurately accounted for in the design of a spacecraft that will be powered by low thrust propulsion.

The characteristics of the upper atmosphere and of space above it are quite variable, both with altitude and time. At altitudes between 90 and 1,000 km we can use the average values for pressure, density, and temperature reported by the US Standard Atmosphere (which is an extension of the International Standard Atmosphere).[1] Because atoms in the upper atmosphere may be multiply ionized, the layer of atmosphere between 50 and 600 km altitude is known as the ionosphere. These are mostly oxygen ions. Above 300 km the composition progressively changes to mostly hydrogen ions. The conditions in that zone are well known, but are variable with space weather, a term commonly used for the ambient plasma, magnetic fields, radiation and matter in a region of space. The space weather close to the Earth is a consequence of the behavior of the Sun, the Earth's magnetic field, and our location in the solar system. Out in interplanetary space it is greatly influenced by the speed and density of the solar wind.

Data on the current solar wind speed and density and on solar flares are continuously available.

The solar wind density peaks at the maximum of the solar activity cycle about every 11 years. At such times, atmospheric drag on satellites increases and therefore the danger of losing altitude and possibly even inadvertently re-entering increases. The average duration of the solar cycle is 11.1 years and the maxima of the last four cycles (i.e. cycle 21 to cycle 24) occurred in the following periods:

Cycle	21	22	23	24
Maximum	December 1979	July 1989	March 2000	April 2014

---

[1] *US Extension to the ICAO Standard Atmosphere*, US Government Printing Office, Washington, DC, 1958; *US Standard Atmosphere, 1976*, US Government Printing Office, Washington, DC, 1976.

To remain in LEO between 200 and 400 km, satellites require a periodical reboost. This is particularly so if their ratio of surface area to mass is large. Owing to its large solar panels the ISS suffers high atmospheric drag. Reboost becomes even more important during periods of high solar activity. At higher altitudes, the pressure and density decrease fast. At about 1,000 km a satellite may well remain in orbit indefinitely, at least relative to normal service times for the machines that humans send up.

In the NASA DRA 5 study [24] appropriate allowance was made for the requirement to periodically reboost the spacecraft while being assembled in LEO. Reboosting has a non-negligible cost in terms of the total mass to be carried into LEO. For a mission powered by chemical propulsion, it was computed that the reboost modules would have a total mass of 145 t, which was more than 12 percent of the total IMLEO. This constraint argues for using higher orbits for assembling the interplanetary vehicle.

In addition to plasma, space is full of debris of various types – both natural and artificial. The natural type consists mainly of very small meteoroids, micrometeoroids, and dust grains that are drawn in by the Earth's gravity field and enter the atmosphere. The small ones burn up and the larger ones reach the surface as meteorites. However, most space debris in LEO is artificial. The large items of this type are accurately tracked by radar and telescopes, so their orbits are well known. Although new debris is continually being produced by new launches, explosions of spent rocket stages, and collisions, much of it is reclaimed by atmospheric drag and burns up. Only larger items reach the ground. As the density of the high atmosphere is much greater near solar maximum conditions than at solar minimum, the solar cycle has a strong influence on space debris. A periodic clean-up of debris in the lowest orbits thus occurs.

International treaties forbidding deliberate explosion of satellites are being prepared, and state that precautions must be taken against accidental events which may produce space debris.

The most critical orbits are those lying between 1,000 and 1,400 km altitude, where drag is insufficient to cause debris to re-enter. The probability that a large piece of debris will approach closer than 100 m from a satellite in LEO is about once in 100 years.

In the case of the ISS, very small objects do not cause damage and debris smaller than a few millimeters is stopped by external shielding. Dangerous debris is identified in advance by ground radar and the spacecraft maneuvers to avoid it. The most dangerous particles are those that cannot be detected in this way and are larger than a few millimeters. The probability that a particle exceeding 1 cm in size will pierce the ISS hull is about 1 percent during its 20 year life. The danger of such an accident occurring during the assembly of a spacecraft for a flight to Mars is small, but must be taken into account. There are well-consolidated design practices for commercial and scientific satellites in LEO, so this can be considered an environment that does not pose unexpected problems.

4.2   INTERPLANETARY SPACE

Space beyond the Van Allen Belts is crossed by the solar wind that pervades the entire solar system. This is mainly made of hydrogen ions (protons) flowing at high speed out of the Sun's corona. The temperature of the corona is so high that the coronal gases are accelerated to a velocity of about 400 km/s. This portion of the solar wind, which is actually the so-called slow solar wind, has a composition resembling the corona and a temperature in the range $1.4-1.6 \times 10^6$ K. Over coronal 'holes' the composition of the solar wind is more like that of the Sun's photosphere. This so-called fast solar wind can reach 800 km/s, and is about 8×10^5 K. The solar wind that emanates from the colder outer layers can be as slow as 300 km/s [16]. While the slow solar wind is mostly ejected from the equatorial region (up to 30–35° latitude either side of the equator) the fast solar wind originates from higher latitudes and the coronal holes near the magnetic poles. The rotation of the Sun and interactions between particles that have different velocities causes the solar wind to be dynamic and, consequently, the space weather across the entire solar system to be highly variable.

From time to time, large quantities of charged particles, mostly protons, are emitted by the Sun in a Solar Particle Event (SPE), also known as a solar flare. Associated with solar flares are the slower, but usually larger bursts of plasma known as Interplanetary Coronal Mass Ejections (ICME). They may disrupt the standard pattern of the solar wind, sending into the surrounding space electromagnetic waves and fast particles (mostly protons and electrons) to form showers of ionizing radiation. When these massive ejections impact a planetary magnetosphere, they temporarily deform the global magnetic field.

Owing to the motion of these charged particles, an Interplanetary Magnetic Field (IMF) pervades the entire solar system. On Earth this induces large electrical ground currents. The protons and electrons that penetrate the upper atmosphere in the polar regions form auroras.

Solar flares constitute a danger to spacecraft, both crewed and uncrewed. Solar flares are intense but relatively brief phenomena. Most of the activity is over in about 4–6 hours, but the flow can continue for some more hours. Figure 4.2a shows the time history of the proton radiation levels from a large solar flare that occurred on January 20, 2005.

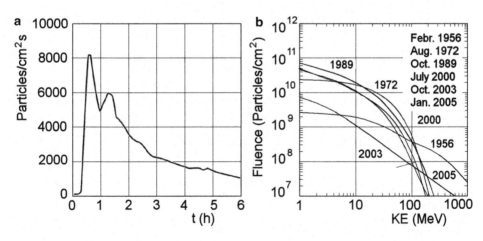

Figure 4.2 Solar particles events. (a) Time history of the fluence (in particles/cm²) of protons with an energy larger than 100 MeV in the first 6 hours after the beginning of the flare in the SPE of January 20, 2005. (b) Energy fluence spectra of some of the largest Solar Particle Events (SPE) events of the last 50 years.

Figure 4.2b plots the fluence (in particles/cm^2) of the most intense SPEs for the last 50 years as a function of the energy of the emitted particles. The envelope of the various lines yields a sort of 'worst case event' against which astronauts require to be shielded. However, it should be noted that only once, in February 1956, was a substantial contribution beyond 150 MeV.[2] But the relevant energy spectrum was only measured indirectly, and obviously not in space. It is considered reasonable to assume a maximum energy for an envelope of 250 MeV.

Since solar flares remain unpredictable, a crewed interplanetary spacecraft must carry a radiation shelter in which the astronauts will ride out dangerous solar events.

The interplanetary medium is also pervaded with Galactic Cosmic Radiation (GCR). The cosmic radiation that enters the Earth's atmosphere consists of 91 percent protons, 8 percent helium nuclei (α particles) and about 1 percent nuclei of heavy elements (which in this case means elements with atomic numbers exceeding 3) plus electrons (β particles), photons, and neutrinos. Although relatively few in number, the nuclei of the heavier elements contribute much to the radiation threat that astronauts must deal with in interplanetary space. This is because once atomic nuclei in interstellar space are stripped of all their electrons, the positively charged nuclei are accelerated by the galactic magnetic field to close to the speed of light. Their energy spectrum may reach to TeV or PeV, although at very low fluxes. As they travel through the very thin gas in interstellar space they emit gamma rays. Their composition is similar to that of Earth and the solar system.

Lower energy charged particles traveling through interstellar space are prevented from entering the heliosphere by the interplanetary magnetic field. Anomalous Cosmic Rays (ACR) are caused by neutral atoms from the interstellar medium. They can penetrate the heliosphere and travel at about 25 km/s. As they approach the Sun, these atoms lose one electron by either photo-ionization or charge exchange, at which time they are accelerated by the Sun's magnetic field and the solar wind. This population includes helium, oxygen, neon and other elements that have high ionization potentials.

The level of GCR can be 2–3 times higher at solar minimum than at solar maximum.[3] Hard spectrum SPEs pose a major radiation hazard to crews beyond LEO. They potentially occur at any point in the cycle and cannot presently be predicted,[4] but the danger posed by SPEs is lower at solar minimum and increase with the increase of solar activity.

Because all of these particles are deflected by the Earth's magnetic field or are stopped by the atmosphere they are not dangerous on the surface. In interplanetary space the radiation dose they give to human bodies depends on time of exposure. The average radiation in Mars orbit was measured by the Mars Radiation Environment Experiment (MARIE) carried by the Mars Odyssey spacecraft. Although it proved to be less than in

---

[2] P. Spillantini, "Manned Exploration and Exploitation of Solar System: Passive and Active Shielding for Protecting Astronauts from Ionizing Radiation, an Overview," 8th IAA Symposium on the Future of Space Exploration, Torino, Italy, July 2013.

[3] Adams, J.H. Jnr., Silberberg R. and C.H. Tsao, "Cosmic Ray Effects on Microelectronics Part I. The Near Earth Particle Environment," NRL Memo Rep., 4506 Pt. I. Publ. US Navy, August 1981.

[4] S. McKenna-Lawlor, "An overview of the energetic particle radiation hazard en route to and at Mars," in *The Particle Radiation Hazard en route to and at Mars*, Publ. Int. Acad. Astronautics, in press.

interplanetary space, it was still 2.5 times higher than at the ISS in LEO. This could be expected since Mars lacks a magnetosphere to protect it from cosmic radiation.

Apart from plasma, there is also a tiny amount of neutral hydrogen. At the distance of Earth's orbit from the Sun, this is present at about 10^4 atoms/m^3. Some of these atoms come from interstellar space.

All space is also pervaded by the cosmic microwave background radiation, consisting of very low energy photons (about 2.73 K) which are remnants from the time when the universe was only about 380,000 years old. Neutrinos, photons of different energies (produced by the Sun, other stars, quasars, black-hole accretion disks, gamma-ray bursts, and so on), electrons, muons, and other particles are also present. Most of them pose no danger.

There is a relatively small amount of dust in space, in the form of micrometeoroids. It is thought that much of it was produced in collisions between asteroids and in the shedding of material from comets while near perihelion. Around 30,000 t of interplanetary dust particles are estimated to enter Earth's upper atmosphere annually. The vacuum is much higher than in LEO, and hydrogen ions from the Sun substitute oxygen ions from Earth's atmosphere. Hence, while the environment in LEO is oxidizing, that in deep space, and also in Mars orbit, is reducing.

The light emitted by the Sun carries a lot of energy, with an average of 1.361 kW/m^2 at Earth's distance from the Sun. The energy per unit area received from the Sun is referred to as the Solar Constant even though the energy that reaches Earth varies, mostly owing to the eccentricity of its solar orbit, which causes the distance to vary over time by a small amount. The energy emitted by the Sun alternates slightly across to the 11 year solar cycle, although this variation is only about 0.1 percent. The energy in sunlight decreases with the square of the heliocentric distance, and at the distance of Mars the energy is about 590 W/m^2, i.e. 43.3 percent of that at Earth. And the variation there is larger owing to the greater eccentricity of Mars' orbit.

The radiation pressure of sunlight is about 9.08 µPa at Earth's distance from the Sun and 3.91 µPa at Mars. This effect can be ignored in the first approximation of a design for a Mars mission, but must be accommodated in a detailed analysis. For example, had it not been taken into account in planning the interplanetary trajectories of the Viking missions, those vehicles would have missed Mars orbit by about 15,000 km.

There have been studies of human Mars missions that would make use of solar sails for propulsion, but a sail for a heavy, crewed spacecraft would probably be prohibitively large. Electric sails that are propelled by the solar wind, rather than the radiation pressure of light, may be useful for human Mars missions, if not for directly propelling the crewed spacecraft then for propelling automatic vehicles that could extract water from asteroids, obtain liquid oxygen and hydrogen, and transport these to Mars to be used to power a return spacecraft without having to carry the propellant from Earth for that portion of a human mission.

4.3 PHYSIOLOGICAL ISSUES DUE TO RADIATION

Radiation in space is primarily high energy particles (i.e. electrons, protons, helium nuclei and some heavier nuclei, but very few neutrons) and a few electromagnetic x-rays and γ rays. For a mission which uses nuclear propulsion, there may also be neutrons and electromagnetic ionizing radiations coming from the reactor. Neutrons and x-rays can also be emitted by high energy particles hitting metal walls.

All of the above are ionizing radiations, and will cause acute and chronic health effects in biological systems in proportion to the magnitude of the radiation absorbed, the type of the radiation, the dose rate, the tissues affected, and the individual irradiated.

The damage caused by radiation is classified as either acute or chronic. Acute damage arises from a sudden dose that occurs over a few seconds to several minutes. It is due to raw energy which literally burns the internal organs and is usually referred to as non-stochastic or deterministic. The effects are generally related to the loss of a fraction of cells that exceeds the threshold for impairment of function in a tissue. These effects are deterministic, because the statistical fluctuations in the number of affected cells are very small if compared with the number of cells required to reach the threshold value. The appropriate assignment of dose limits can ensure that early effects will not be experienced by the crew during a particular mission. In space it may occur due to solar storms or, in nuclear spacecraft, to severe reactor accidents.

Chronic damage arises from a dose which occurs over a few days to several years. It is linked to damage to the cellular DNA, leading to cancer and genetic defects in the victim's future offspring. This is therefore referred to as stochastic. It may also cause skin ulceration and blindness due to cataracts scarring. Chronic damage can result from changes in a small number of cells, within which statistical fluctuations can be large, and some level of risk is incurred even at low doses. The usual models assume that there is no minimum dose below which no damage is incurred in. The normal background radiation in space, lengthy periods spent in the Van Allen Belts and, in the case of nuclear reactors, leakage through the reactor shielding can all cause chronic damage.

A distinction is made between the radiation absorbed dose, meaning the radiation energy that is absorbed by a unit mass of matter, and the radiation effective dose. The former is a physical quantity for which the SI unit is the gray (Gy), where 1 Gy of radiation is absorbed when a kilogram of material or tissue takes a joule of radiation energy. A non-SI unit is the rad, where 1 Gy is equivalent to 100 rads. Submultiples are the centigray and the milligray (and the millirad). By convention, deterministic effects are measured using the absorbed dose and not the effective dose.

The effective radiation dose measures the biological effect of radiation exposure. The SI unit is the sievert (Sv). It represents the equivalent biological effect of the deposit of a joule of radiation energy in a kilogram of human tissue. The equivalence to absorbed dose is denoted by a factor usually indicated by the quality factor Q such that

$$\text{effective dose} = \text{quality factor} \times \text{absorbed dose}.$$

A non-SI unit is the rem, where 1 Sv is equivalent to 100 rems. Factor Q depends on the type of radiation to which the body is subjected, and a weighted average is made where various radiations are acting simultaneously. Electrons, x-rays and γ rays have a quality factor of 1. For particles the quality factor depends on the particle's energy, but average values are $Q \approx 2$ for protons, $Q = 10$ for neutrons, and $Q = 20$ for α particles.

If only selected parts of the body are involved, which is usually not the case for radiation in space, the tissue that receives the radiation enters the computation of the effective dose.

The dose rate is the ratio between the effective dose received and the time that is needed to absorb it. This is measured in Sv/y (sievert/year). For instance, the average dose rate in the ISS has been evaluated at 160 mSv/y, although it varies with time.

Since the effective dose is used to measure the stochastic effects of exposure to radiation, there is an equivalence between the likelihood of eventually developing cancer (or rather, the increase of probability of getting cancer with respect to the probability that one already has as a result of the ambient radiation background, lifestyle and so on) and the effective dose that is absorbed. In this way, 1 Sv produces a 5.5 percent chance.

This is usually known as the linear, no-threshold model. If such a probability has to be limited to 3 percent, the effective dose must be limited to about 0.6 Sv. Individual studies, alternative models, and earlier versions of the industry consensus have produced other risk estimates which scatter around this consensus model. There is general agreement that the risk is much higher for infants and fetuses than it is for adults, higher for the middle-aged than for seniors, and higher for women than for men – although there is no quantitative consensus on this.

Due to latency effects, differences in sensitivity between tissue types, and differences in the average life spans between genders, the relationship between radiation exposure dose and risk is age- and gender-specific. Prior crew exposure is also a relevant factor, and cumulative Risk of Exposure Induced Death (REID) over several missions is considered in setting mission design requirements to ensure that the personal career Permissible Exposure Limits (PEL) of individual crew members are not exceeded. The progressive increase of the average life span can serve to increase the REID, whereas improvements in cancer therapy may decrease it.

Radiation exposure limits have not yet been defined for missions beyond LEO, but for LEO operation NASA has adopted the recommendations of the National Council on Radiation Protection and Measurements (NCRP) given in NCRP Report No. 98 (1989). This document contains monthly, annual, and career limits (see Table 4.1) that are based on equivalent doses on blood-forming organs – which are much affected by radiation – rather than on entire body exposure.

PEL figures for crew members on deep space missions are chosen to prevent the taking of in-flight radiation risks that are thought to be prejudicial to mission success, while also limiting chronic risks to acceptable levels based on legal, ethical, and financial considerations. Exposures are required to be As Low As Reasonably Achievable (ALARA) to ensure that astronauts do not approach their assigned radiation limits while in flight. This requires that measures be taken in the design and operational phases of the spacecraft to manage, and limit, the exposure of crew to ionizing radiation.

Table 4.1 Career limits for astronauts in LEO (expressed in Sv) based on National Council on Radiation Protection (NCRP) Report No. 98 (1989).

Age	25	35	45	55
Male	0.7	1	1.5	2.9
Female	0.4	0.6	0.9	1.6

Career exposure to radiation is estimated by NASA for individual missions and crews in order not to exceed a 3 percent of REID from fatal cancers. An ancillary requirement is that this risk limit is not exceeded at a 95 percent confidence level, using a statistical assessment of the uncertainties inherent in the risk projection calculations employed.[5]

At the present time, no space agency has assigned career dose limits for human personnel voyaging Beyond Low Earth Orbit (BLEO) and limits published thus far in the literature refer only to the LEO environment. NASA specifies short term as well as career dose limits for its astronauts. Its short term limits are set to prevent the occurrence of clinically significant non-cancer health effects. A probability of 10^{-3} is deemed in this regard to be a practical limit for the risks that occur above the selected threshold dose. Career dose limits are intended to constrain to an acceptable level the increased risk of contracting cancer incurred by members of the astronaut profession. The space agencies which participate in the ISS have agreed to a consensus limit in LEO for short term and annual exposure. However, they employ different constraints when defining career limits.[6]

Numerical modeling of a 400 day cruise to and from Mars during the minimum of solar cycle 23 have shown that the cumulative effect of incident, isotropic GCR could have posed a significant radiation hazard. Also, the occurrence of a hard spectrum SPE during this part of the mission can be catastrophic for the health of the crew. The effect of GCR at the Martian surface during a 30 day (short) stay at solar cycle minimum was estimated to be acceptable. The occurrence in this interval of a hard spectrum SPE was estimated to likely result in organ exposures in excess of NASA's current permitted exposure limits.[7]

A number of models for predicting SPEs at Mars are now available, but because they are based on different philosophies they require to be unified. Model predictions can meanwhile be validated using rovers and anthropomorphic phantoms, like the data recorded in space and at the Martian surface aboard the Curiosity rover.[8]

The effective dose measured by the Radiation Assessment Detector of the Mars Science Laboratory (MSL) during the cruise to Mars shows that astronauts would be exposed to a total of 1.8 µSv per day, while the readings taken on the surface show a dose of about 0.64 µSv per day. Assuming a 500 day surface stay and 360 days in space, the total radiation dose to which a crew would have been exposed on such a voyage is roughly 1.0 Sv across the whole mission. These measurements were obtained with no specific shielding, apart from that provided by the structure of the spacecraft. The MSL mission took place during a solar minimum, when GCRs were at a maximum and solar radiation was very low. In contrast, at a solar maximum GCRs are reduced significantly and only solar particles, which are more readily shielded against, reach their maximum.

---

[5] F.A. Cucinotta, W. Schimmerling et al, "Space radiation cancer risk projections for exploration missions: uncertainty reduction and mitigation," NASA/TP, JSC-29295, Publ. Johnson Space Center, USA, 2001; F.A. Cucinotta, W. Schimmerling et al., "Space radiation cancer risks and uncertainties for Mars missions," *Radiat. Meas.*, vol. 156, pp. 682–685, 200.

[6] S. McKenna-Lawlor, and the SG.3.19/1.10 team, "Feasibility study of astronaut standardized career dose limits in LEO and the outlook for BLEO," *Acta Astronautica* vol. 104, pp. 565–573, 2014.

[7] L.W. Townsend, M. PourArsalan, et al, "Estimates of Carrington-class solar particle event radiation exposures on Mars," *Acta Astronautica* vol. 69, pp. 397–405, 2011.

[8] D.M. Hassler, C. Zeitlin, et al., "Mars' Surface Radiation Environment Measured with the Mars Science Laboratory's Curiosity Rover," *Science*, doi: 10.1126/science.1244797.

The MSL data show that radiation in space and on Mars pose problems for human Mars exploration, but the problem is manageable. 1 Sv provides an increased probability of getting cancer of slightly over 5 percent; for instance increasing it from 20 percent to 25 percent. At that level of risk there would be no difficulty in recruiting volunteers. After all, people go on smoking even though they are aware of the cancer risks, and the risk linked to radiation on a Mars mission might be not much higher.

Other data from spacecraft in interplanetary space and on the surface of the planet will surely be available before any human mission is launched, and the problem will be much better known. Apart from the uncertainty due to the lack of data about the absorbed dose from GCR in interplanetary space and on Mars, there are uncertainties concerning the extrapolation of the results for x-rays and γ rays to the damage to be expected from GCR. A further difficulty is that the knowledge acquired on medical effects of radiation that is developed from studying subjects living on Earth's surface cannot be applied fully to people living in microgravity. In particular, microgravity itself might influence cellular damage induction and repair systems.

Owing to these uncertainties, instead of direct evaluation of REID using point values of the effective dose, nowadays there is a tendency to take into account confidence intervals. If a confidence interval of 30 percent, 60 percent, or even 95 percent is taken into account, then a much higher value is obtained than by simply using the most probable REID. For instance, in the case of a 95 percent confidence interval the REID is multiplied by about 3.5.

4.4 COUNTERMEASURES AGAINST RADIATION

In LEO the magnetosphere reduces the risks from space radiation, but in interplanetary space both solar radiation (mostly SPEs) and GCR are really dangerous. A human Mars mission will be in a high radiation environment not only while in space but also on the planet, since Mars lacks a magnetosphere and its thin atmosphere provides only a limited protection.

The simplest way to minimize radiation exposure is to reduce the time spent in space by the use of advanced propulsion systems. Any reduction of the travel time will deliver benefits not only in terms of radiation exposure, but also of the effects of microgravity and on possible psychological issues. But as will be seen, the time spent in travel and that spent on Mars are not independent variables, the shorter is the travel time, the longer the crew must stay on Mars. The advantage of fast travel remains because the radiation dose rate in space is larger than on the planet but if, for example, the travel time is halved, the total radiation dose is larger than half. At any rate, the goal is to explore Mars, so the faster the journey, the less time is spent soaking up radiation whilst doing little of value to that goal, and the more time is spent getting radiation (but less) while engaged in activities directly related to the objectives of the mission.

4.4.1 Passive shields

Passive screening is a common measure and is quite effective for many Earth applications, in particular those in the medical realm. All x-ray equipment has heavy lead screening and this is effective in protecting operators and patients. But in space things are rather different. In many Earth applications ionizing radiations are electromagnetic (e.g. x rays and γ rays) and the best screening is supplied by heavy metals such as lead or tungsten.

A completely different situation is posed by particle radiations like GCR or SPE. In this case screening is best achieved using light elements. Hydrogen is best. Next best are materials that contain a large amount of hydrogen, for example hydrogenated polyethylene or water or, even if it is less efficient, magnesium hydride. Some boron compounds, materials containing silicon (such as regolith) or, amongst the metals, magnesium or aluminum are also good.

However, passive shielding against particles has an intrinsic problem, in that the particles of GCR or SPE will strike the atoms within the shield and produce secondary particles (or in general secondary radiation) that may be as dangerous, or possibly even more dangerous than the original radiation. Hence a fairly thick shield will be required in order to stop both primary and secondary radiation. The amount of material for an effective shielding may then be well beyond what is practicable for most aerospace applications. The aluminum walls of the ISS, for example, are about 7 mm thick and are effective in reducing the radiation to which astronauts are subjected, but this is in LEO. It is unlikely that such shields would have the same effect in interplanetary space, where they might even increase the dose absorbed unless substantially thickened.

As noted, low density materials can improve things and hydrogen is the best choice. If there is liquid hydrogen on board, the fuel tank can be set around the habitat to protect it from radiation. However, this solution is problematic because most of the hydrogen will be used up in the initial part of the voyage. The hydrogen that remains on board for long enough to make it useful as a shield will pose severe boil-off problems. And in any case, it is difficult to think of placing a cryogenic liquid around a habitat.

Water and food are much easier to store around the habitat, and are effective, even if less so than hydrogen. Of course these are also used during the journey, but the water for the return trip will provide a good shield during the outward voyage. Another option is organic waste, which has the advantage of increasing during the journey. An interesting possibility is to place a number of tanks around the habitat which are initially filled with water or food and then, when they are empty, are filled with organic waste. In this manner, there will always be viable shielding in all stages of the journey.

The decrease of the absorbed dose as a function of the thickness of the shield is plotted in Figure 4.3 for different shield materials. The thickness is expressed in terms of mass per unit area, expressed in kg/m^2. By dividing this value with the density of the material in kg/m^3 the shield thickness in meters is directly obtained.

A NASA analysis found that a large space station required a shielding of 4 t/m^2. This is an enormous amount to achieve an effective dose rate of 2.5 mSv/y, which is a really small amount since it is lower than the background radiation in many sites on Earth. This might be achievable for space colonies where the shield mass is lunar regolith sent from the Moon by using mass drivers, as envisaged by Gerard O'Neill,[9] but it is impracticable for a craft which requires to be propelled.

The worst-case of fluence (envelope of the curves in Figure 4.2b) for SPEs in the last 50 years is repeated in Figure 4.4, together with the energy of the particles which are shielded by aluminum walls of various thickness. A 7 mm wall can thus shield out most of

---

[9] G.K. O'Neill, *The High Frontier: Human Colonies in Space*, Bantam Books, New York, 1977.

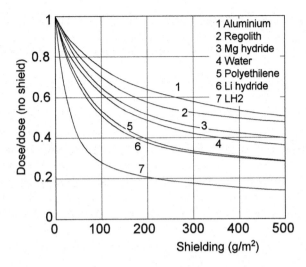

Figure 4.3 Absorbed dose as a function of the thickness of the shield for different shielding materials.

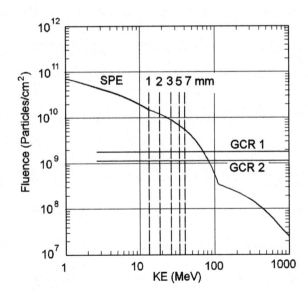

Figure 4.4 Worst-case fluence for the SPEs of the last 50 years versus particle energy, together with the energy of the particles which are shielded by aluminum walls of various thickness. The fluence due to GCR in one year during a solar minimum (GCR1) and a solar maximum (GCR2) are also shown.

the radiation due to SPE (up to 40 MeV). Also plotted is the fluence due to GCR in one year during a solar minimum (GCR1) and a solar maximum (GCR2). For the worst-case SPE to be as dangerous as one year of exposure to GCR, a shielding much heavier than 7 mm aluminum is required.[10]

4.4.2   Active shields

As an alternative to using solid or liquid material as radiation shield, it is possible to create a magnetic field about the spacecraft to act more or less as a magnetosphere by deflecting the charged incoming particles, or indeed to create an electric field in order to repel them.

The most common proposal envisages a shield produced by powerful superconducting magnets which create fields of several tesla. The possible health implications of exposure to such fields are still under investigation. It is possible to conceive the system in which the field is low (perhaps even nil) in the place where the crew is located, but achieving this would add complexity to the study.

It is likely that active shielding will require a fairly substantial amount of energy, even if a superconducting magnet requires much more power to charge it than to maintain its field. A nuclear power generator will likely be required, in which case it would make sense to have NEP. For NTP, a dual use engine could be employed which, when it is not providing thrust (which is most of the time) could produce the power for the shield.

Several research projects are under way, for instance the European Space Radiation Superconducting Shield (SR2S). A first step towards such a system is the development of superconducting magnesium diboride (MgB_2) cables. But it is not only the issue of critical temperature (which must obviously be as high as possible in order to reduce the problem of maintaining the coils in superconducting conditions) but also of critical field and critical current. Unfortunately, most of the so-called high temperature superconductors are not of much use for this application.

Active radiation shielding is essential not only for human Mars exploration but also, and perhaps more so, for placing space stations in the Lagrange points of the Earth-Moon system, for establishing lunar bases, and for the exploration and mining of asteroids. Developing active radiation shielding technology is one of the areas where a collaboration between the various programs can be more effective.

4.4.3   Storm shelter

As stated above, while GCR impose a continuous exposure to a low level radiation, solar flares are of short duration but very intense and there is no realistic way of shielding an entire habitat against these. The only defense is to provide the vehicle with a relatively small shelter in which the crew will take refuge for the short time that the spacecraft is subjected to such a radiation bombardment. The inconvenience of the very cramped

---

[10]P. Spillantini, "Manned Exploration and Exploitation of Solar System: Passive and Active Shielding for Protecting Astronauts from Ionizing Radiation, an Overview," 8th IAA Symposium on the Future of Space Exploration, Torino, Italy, July 2013

conditions may thus be tolerated and, owing to its small size, the total mass of the very thick shield will be acceptable.

This strategy was followed in the Apollo missions. In case of a solar flare, the crew had to move into the Command Module, which had thicker aluminum walls than the Lunar Module, and abort the mission to return to Earth as soon as possible. In addition, the cylindrical body of the Service Module would be faced towards the Sun as extra shielding. Almost all human Mars mission concepts dating all the way back to that of Stuhlinger in 1954 have been similar. In more recent times, the vehicle for the Inspiration Mars flyby was meant to have a 2m long tube surrounded by water in which the two astronauts could squeeze in the event of a SPE. The mass of the shielding water is reported in Figure 4.5 for different values of the tube diameter, relative to the kinetic energy of the protons to be stopped. By assuming an average kinetic energy of 150 MeV for protons arriving from various directions (see Figure 4.2b), a mass of 300 kg for the shelter is obtained. For a larger spacecraft with a crew of four or six, a heavier shelter is obtained but also the mass allowance is much larger.

A lighter shelter can be obtained if can be arranged for the spacecraft always to face its shield toward the impinging particles, so that the crew need not to be protected from the SPE coming from all directions. But it should be noted that because the protons in the solar wind spiral out away from the Sun, their direction of arrival is not exactly from sunward.

Having a storm shelter is practicable if the crew can receive sufficient warning of an impending solar flare. Whilst these are not really predictable with a substantial lead time, the spacecraft will surely carry sensors to warn the crew when the count of the particles starts to increase, allowing several minutes to enter the shelter before the radiation becomes intense. It would also be possible to position a probe with a sensor sunward of the human ship in order to report the rise in the particle flux with a longer lead time.

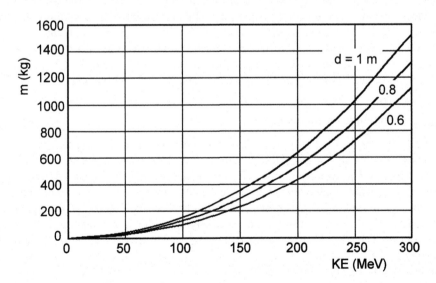

Figure 4.5 Mass of the water shield as a function of the kinetic energy of the protons to be stopped. The shelter is 2 m long and has different diameters d.

Suits that incorporate an inflatable outer layer which can be filled with water might offer another solution, even if a single shelter for several people may be lighter than a number of individual shelter-suits. An ESA study called PErsonal Radiation Shielding for intErplanetary missiOns (PERSEO) is developing wearable protection against SPE with selective shielding of the various organs. By concentrating the protection on the most sensitive organs, effective protection can be obtained with a smaller shield mass.

4.4.4   Protective therapy

Radiation effects can be reduced by putting the astronauts under a protective therapy. This is achieved by using drugs that improve the ability of the body to repair the damage caused by radiation. Some of the drugs being considered for this are retinoids, which are vitamins with antioxidant properties, and molecules that retard the process of cell division, thereby allowing the body time to fix damage before harmful mutations can be duplicated.

4.4.5   Hibernation

Studies of human hibernation have shown that this may be feasible and may be an alternative to decreasing the time spent in space. Depending on how hibernation is achieved, the sensitivity of human bodies to radiation may be decreased. In particular, hibernation slows the physiological functions such as cell division, and this is particularly beneficial in reducing radiation damage. Above all, a human body in hibernation is easily protected from harmful radiation by placing it in a shielded capsule similar to those considered as radiation shelters, but much smaller.

Hibernation does not require to achieve the extreme results common in science fiction, where it practically stops the vital functions of an astronaut traveling to the stars on a voyage lasting centuries. A much more realistic goal is to slow the vital functions for several months to render the astronauts unconscious, so that they can sleep in radiation-protected couches. Hopefully they would be free of the debilitating effects of microgravity and certainly they would be free of psychological problems. It is encouraging that mammals like squirrels and bears can hibernate to survive the harsh winter environment.

4.4.6   Radiation protection on Mars

Mars has no magnetic field and only a thin atmosphere but the radiation on its surface is much less dangerous than that in space. First, the mass of the planet blocks radiation coming from below, which eliminates more or less half of the radiation. Second, the air will stop at least some of the radiation. The column of air above the surface varies with the elevation of the site.

The shielding at the deepest point of the Hellas basin is about 50 g/cm^2, but for most of the planet it is 30–35 g/cm^2. On the summit of Olympus Mons the shielding can be as low as 2–3 g/cm^2. The atmosphere is mostly carbon dioxide. In terms of shielding, it should lie between the 'polyethylene' and 'water' lines in Figure 4.3b. This implies the atmosphere can remove 60–70 percent of the radiation arriving from space. But things will be significantly better, because radiation which does not arrive from directly overhead must pass through a greater column of air to reach the surface. This issue requires further study, in particular to assess the secondary radiation that will be produced by the primary radiation penetrating the thin atmosphere.

The levels of cosmic radiation to which an astronaut will be exposed on the surface of Mars should be 80–330 mSv per year, with these values being respectively for solar maximum and solar minimum. This compares with 110–380 mSv for the surface of the Moon. Figure 4.6 gives a map of the effective radiation dose on the surface as measured from orbit by the MARIE experiment on NASA's Mars 2001 Odyssey spacecraft.

Since, as stated, the average dose rate in the ISS is about 160 mSv/y, the radiation level on Mars (without shielding) is less than inside the space station in all the 'blue' and 'green' zones. The background radiation due to radioactive elements in the planet's crust should be lower than for Earth due to the lower trace quantities of uranium, thorium and potassium on Mars.

The habitat should at any rate be protected by regolith, either because it is built inside a cave or a lava tube, because regolith is spread on the top of the structure, or because it is built from regolith. Regolith is worse than hydrogenated material (see Figure 4.3b) but is abundant and increasing the mass of the habitat is not a major issue. The lower gravity will reduce the structural load placed on a habitat compared to a similar structure on Earth. About 1.5 m of regolith will be needed to reduce the radiation level inside the habitat to below 50 mSv per year in the worst conditions.

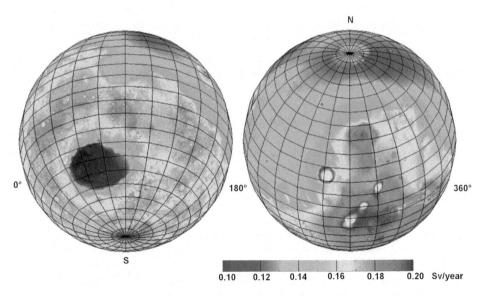

Figure 4.6 Effective radiation dose rate (in Sv/year) on the Mars surface. Map obtained from cosmic radiation data by the Mars radiation environment experiment (MARIE), on board NASA's Mars 2001 Odyssey spacecraft. The surface radiation was obtained using Mars altimetry data from the MOLA instrument aboard Mars Global Surveyor. (NASA/JPL/JSC image)

Another option is to position tanks all around the habitat and fill them with water from the permafrost. With the temperature on Mars being low, it would be possible to use ice to create high technology igloos as habitats!

If power is supplied to the habitat by a nuclear reactor, there will be sufficient power for active shielding, even if several tons of superconducting coils must be carried from Earth to achieve it.

A habitat on Mars should possess a shelter against solar storms. Such a facility could be used to store electronic equipment and materials which are sensitive to radiation. Also human-carrying rovers, in particular rovers designed for long distance travel, should have a roof for radiation protection, particularly if it is intended to travel to sites at high elevations.

4.4.7   Protection against radiation from nuclear devices

The first way to protect a crew from radioactivity released by a nuclear reactor is to locate it as far as possible from the crew compartment. Figure 1.8 depicts a very early design in which the largest component of the vehicle, the thermal radiator, is located between the reactor and the crew compartment, which are 150 m apart. The habitat will require shielding from all around. Even if the shielding against the x rays and γ rays from the reactor isn't the same as that used against CGR, screening out the reactor's neutrons will be very similar. The lightest way is to place a screen close to the reactor to block only the radiation that is emitted in the direction of the habitat. If there is sufficient separation between the reactor and the habitat, a directional screen will need to protect only a fairly narrow angle. The issue must therefore not be seen in terms of designing a screen to protect the crew, but of selecting a general architecture for the spacecraft to minimize the need for screening the reactor.

In the case of NEP, things are simplified by the low accelerations to which the vehicle will be subjected. A long, lightweight beam that separates the reactor from the crew (and the sensitive electronics) will not pose a problem. An NTP vehicle will accelerate more rapidly but it is likely that less screening will be required because the reactor will work at full power for only a brief time, and the power output from a dual use reactor will be low in the cruise.

Radiation screening for the reactor at the outpost on Mars will be easy, since the reactor will be located some hundred meters from the buildings and be buried under a thick layer of regolith. The task of burying the reactor could be assigned to robots prior to the arrival of the astronauts.

It is likely that a combination of the strategies seen above will be used. At any rate, all these solutions, namely active shielding, advanced propulsion, drug therapy, and hibernation require technologies that have yet to be developed and many people fear that to rely on them will postpone the first voyage to Mars by decades, perhaps indefinitely. Those who insist we must proceed as soon as possible say that the first missions should use only technology that is either currently on the shelf or could be developed within the target timescale, accepting the associated risks and dangers. They hold that the dangers posed by radiation are not so severe that we need to postpone going to Mars until we can proceed in a completely safe way, at least from this viewpoint. The statement that smoking is not less dangerous than going to Mars and nevertheless many people accept this risk despite the warnings on cigarette packets, has been made several times and is difficult to counter.

5

Human aspects

Radiation is not the only danger to human health posed by long space journeys. Microgravity is the other important stressor that threatens the health of people in space. Being isolated at a great distance from Earth is itself a source of danger in case of medical emergencies. Cognitive and psychological problems must also be faced, and they are important in designing the mission and the spacecraft.

During long distance, long duration expeditions in space involving humans, and in particular a mission to Mars, a number of issues need to be addressed to enhance crew safety and well-being, and to improve the chances of mission success. There are three principal stressors: low gravity, radiation, and psychological and cultural factors. All can negatively affect human physiology, cognition, and individual and crew psychosocial status.

The problems relating to radiation were dealt with in the previous chapter; the present chapter considers the other health risks posed by a mission to Mars.

A recent NASA study[1] defined the human health and performance risks linked with the space environment as:

- Radiation hazard:

    - Space radiation exposure.

- Reduced gravity hazards:

    - Vision impairments and intercranial pressure.
    - Renal stone formation.
    - Sensorimotor alterations.
    - Bone fracture.
    - Reduced muscle mass, strength and endurance.
    - Reduced aerobic capacity.

---

[1] NASA, "NASA's Efforts to Manage Health and Human Performance Risks for Space Exploration," report No. IG-16-003, 2015.

© Springer International Publishing Switzerland 2017
G. Genta, *Next Stop Mars*, Springer Praxis Books, DOI 10.1007/978-3-319-44311-9_5

- Host micro-organism interactions.
- Cardiac rhythm problems.
- Orthostatic intolerance.
- Intervertebral disc damage.
- Space adaptation back pain.
- Urinary retention.
- Pharmacokinetics (movement of drugs in the body).

• Hazards due to distance from Earth:

- Adverse outcomes due to in-flight medical conditions.
- Ineffective or unpredictable effects of medication due to storage.

• Hazards due to hostile/closed environment and to spacecraft design:

- Inadequate food and nutrition.
- Inadequate human-system interaction design.
- Injury from dynamic loads.
- Injury and compromised performance due to EVA operations.
- Celestial dust exposure.
- Altered immune response.
- Exploration atmospheres.
- Sleep loss, circadian desynchronization and work overload.
- Toxic exposure.
- Decompression sickness.
- Hearing loss related to spaceflight.
- Acute and chronic carbon dioxide exposure.
- Injury from sunlight exposure.
- Electrical shock.

• Hazards due to isolation:

- Adverse cognitive or behavioral conditions.
- Inadequate team performances.

These risks mostly arise from the fact that the environmental conditions experienced in spaceflight are very different from those in which humans evolved. Some of these risks are incurred even during short periods in space and so must be coped with. A spacecraft requires environmental control and life support systems to supply air, water and food, and to maintain temperature, pressure and humidity within acceptable limits. Other risks become severe only during longer periods spent in space. In the case of microgravity, the need to take remedial action becomes ever more important as mission duration increases.

The risks listed above are usually discussed in terms of being in space, but they are also present during a stay on the Moon or on Mars, even if in some circumstances they might be less severe. How the risks should be mitigated is the subject of space medicine.

Distance is in itself a cause of potential risks to people. Apart from the psychological problems, the fact that astronauts on a Mars mission are unable to rapidly return to Earth or receive medical supplies is an important factor. In addition, adequate medical expertise may not be available on site, particularly if the size of the crew is small and does not to include a professional doctor. The best available person may be a specialist in another field who got supplementary medical skills, for instance a biologist whose primary responsibility is to look for Martian life. Telemedicine may be possible but the communication time delay could be a factor. In a really serious emergency it will be impossible to hospitalize the patient, and this might cause anxiety. For sure, the astronauts will have to rely for long periods on their limited resources and medical advice from home.

Advances in the development of surgical robots and telemedicine, and the possibility of building on site medical equipment will undoubtedly help, but explorers have always been faced with having to deal with medical emergencies using inadequate facilities. Perhaps the point is that although most medical emergencies during the Age of Discoveries were in any case impossible to deal with, irrespective of where this situation occurred, nowadays we are used to receiving a level of health care that will not be available to an explorer in space or on another celestial body.

5.1   DIRECT EXPOSURE TO THE SPACE ENVIRONMENT

Apart from the presence of radiation, the thing that makes space such a lethal environment is a vacuum that is better than the highest vacuum available in laboratories on Earth. Animals that evolved on Earth cannot survive in vacuum due to lack of oxygen to breathe and lack of pressure.

The minimum partial pressure of oxygen that can be tolerated by humans is 160 mbar. At lower partial pressures of oxygen, the quantity of oxygen in the blood decreases and humans first become unconscious and then die from hypoxia. If a human is exposed to space vacuum, unconsciousness will occur in 9 to 12 seconds, when the blood lacking the necessary oxygen reaches the brain. Death follows in about 2 minutes.

When breathing pure oxygen from a tank, the maximum altitude at which it is possible to breathe is 13,000 m.

Too low an external pressure causes the body fluids to boil, a condition usually referred to as ebullism. Body fluids are mostly water, and following the state diagram of water shown in Figure 3.7 it follows that, at the temperature of human bodies, blood and other fluids boil at 63 mbar, corresponding to an altitude of 19,000 m on Earth. In fact, the pressure exerted by the blood vessels will prevent blood from boiling at that altitude. Nevertheless, despite this containment, in space blood does boil and the body swells to twice its normal size.

Space suits are worn over an elastic garment called a Crew Altitude Protection Suit (CAPS) to prevent ebullism at pressures as low as 20 mbar. Most space suits are pressurized with pure oxygen at 200 mbar, which is sufficient to sustain full consciousness. Even if at this pressure the blood does not boil, there is some evaporation and gases dissolved in the blood can cause decompression sickness.

Experiments on animals showed that a total exposure of 90 seconds to space vacuum is not lethal and is usually followed by a complete recovery. Some reports of accidents involving astronauts essentially confirm that survival after short exposures is possible and no permanent physical damage arises from an exposure of 30 seconds. An astronaut was exposed to vacuum due to a leaking space suit in a vacuum chamber at the Manned Spacecraft Center in Houston in 1965. He lost consciousness after about 14 seconds. The repressurization of the chamber was initiated 15 seconds after the accident. The subject regained consciousness and recovered. In the case of the Soyuz 11 accident in 1971, three cosmonauts were killed when the spacecraft inadvertently vented its atmosphere to space. The autopsies revealed that the cause of their death was the loss of pressure causing hemorrhaging of the blood vessels in the brain due to oxygen and nitrogen bubbles forming in the blood vessels.

It appears that rapid (or explosive, as it is usually described) decompression can be much more dangerous than exposure to vacuum. Even if the person exhales to prevent the difference of pressure from producing most of the damage, the venting may not be fast enough. A sudden pressure drop of only 130 mbar may be lethal while, if it occurs gradually, it does not produce symptoms. Partial exposure to space vacuum is much less severe if the subject can breathe.

In a vacuum there is no thermal conduction or convection, and the only way for thermal exchange to occur is radiation. The loss of heat between a human body at a temperature of 310 K and the heat sink at 3 K of space is quite slow, hence there is no danger of immediately freezing. Evaporative cooling of skin moisture may create frost, but this is not a significant problem. A space suit must have a thermal control system, in particular to deal with the heat received from the Sun, but this is off-the-shelf technology.

Exposure to the Martian atmosphere is only marginally different from exposure to space vacuum. Even the deepest point of the Hellas basin has an atmospheric pressure below that at which humans need a space suit (11 mbar versus 63 mbar), even if wearing a Crew Altitude Protection Suit (20 mbar).

Mars is quite a cold place, and picking up an object that has been left exposed on the ground may cause severe cold burns if the space suit is not well insulated. In addition, even though the low atmospheric density will limit it, there will be some convective heat loss. A space suit for Mars must be well insulated and must be supplied with heaters. Particularly challenging will be EVA activity during the night. This will probably be something that will occur only in an emergency. One difference between suits for use in space and for use on Mars may be that the former will provide mostly cooling, whereas the latter will provide mostly heating. Operating on Mars in full sunlight may cause strong heating on one side of the body and intense cooling on the opposite side, so it will be necessary to provide accurate thermal control.

Martian dust is potentially dangerous, particularly to the lungs, although less dangerous than lunar dust. If breathed in or ingested it will produce physical damage, and everything contaminated by dust should be introduced into the habitat only after careful cleaning. The Apollo astronauts found that when lunar dust was exposed to oxygen in the repressurized Lunar Module it had a smell reminiscent of gunpowder. Although little was done to prevent their coming into contact with the dust, they did not report serious problems. However, the Apollo crews spent only brief periods on the Moon. Precautions against dust contamination must be taken very seriously in planning long missions for a return to the Moon and to start our exploration of Mars.

5.2 LOW GRAVITY

During a mission to Mars, the crew is subjected to reduced gravity; almost zero gravity while traveling through interplanetary space and a period at 0.38 g while on the planet. Of course, there will be short bursts of higher acceleration during planetary entry and also, if chemical rocket propulsion is used, during impulsive maneuvers. If low thrust propulsion is used, the acceleration may be so low as to be barely perceptible. We know perfectly the effects of 1 g on human bodies, we have long experience of microgravity, and 12 men have experienced lunar gravity for periods of up to three days – which is not sufficient exposure to study the influence of this condition. But the effects of spending a lengthy period in the gravity of the Martian surface is completely unknown. On the one hand, 0.38 g might be only marginally better than microgravity. On the other hand it might be a substantial improvement on being weightless. Some people hold that what matters is that there is some gravity, and gravity on Mars will be sufficient to prevent most of the physiological damage that occurs when in the absence of a gravitational field.

The only way to address this problem is to carry out experiments in space using a rotating spacecraft in LEO. A facility of this kind will also be important in investigating the effects of lunar gravity, and is therefore an essential precursor to undertaking the colonization of either the Moon or Mars.

Although missions to the Mir space station and later to the ISS established that humans can survive lengthy periods in microgravity, there must also be a period of rehabilitation upon their return to Earth. The record for the longest time spent in microgravity on a single flight was set by Valeri Vladimirovich Polyakov, who spent 437 days aboard Mir. Because he had previously spent 240 days in space, his total time in orbit is 677 days (22.6 months).

Even if a stay on Mars will allow the astronauts' conditions to improve, the landing on the planet and the ensuing period while they prepare their habitat and other equipment will be critical phases of the mission and the prospect that the crew might be in poor conditions isn't pleasant.

5.2.1 Effects of low gravity on human physiology

Living in microgravity, or at least in low gravity, has several adverse effects on human beings. At the onset of microgravity, the astronauts are subject to Space Adaptation Syndrome (SAS). This is a temporary condition that affects different people (or even the same person at different times) in quite different ways. Other effects develop more slowly and hence become important only for medium or long periods spent in microgravity, but they are cumulative. These effects arise from the tendency of the body to adapt itself to the new environment.

The main effects of relevance to long duration missions are:

- Space Adaptation Syndrome (SAS) and balance disorders. Humans evolved a complex system to feel the direction of the gravitational acceleration vector as a consequence of their two-legged locomotion. In microgravity, the constant acceleration defining the 'down' direction is absent and this lack of signal from the equilibrium organs is matched by the absence of signals from the pressure organs in the feet and ankles and the absence of muscle contraction to maintain the posture. All of this results in visual-orientation illusions, as if the body is located upside-down. The disagreement between what the body perceives and what it knows to be the true situation originates the SAS and it symptoms. Over time, the brain adapts and most astronauts gain a perception of the down direction which is based more on their position in the environment. The SAS produces a nausea (similar to motion sickness) caused by derangement of the vestibular system, as well as vomiting, vertigo, headaches, lethargy, and a general malaise. It is limited to the first day or so spent in space. About 45 percent of all people who have flown in space have suffered this malaise.

- Deterioration of the skeleton (spaceflight osteopenia). Bones develop their strength mainly in the direction of mechanical stress, which in normal conditions is caused by weight. But in weightlessness they adapt to the lower loads and lose mass at a rate of approximately 1.5 percent per month. This is particularly evident for bones that are most stressed by the weight, such as the lower vertebrae, hip and femur. Bone density decreases rapidly and the bones become brittle, with symptoms similar to those of osteoporosis. In 1 g the effects of osteoblasts (producing bone tissue) and osteoclasts (destroying it) are balanced so that bone tissue is constantly repaired and renewed. In microgravity the increased osteoclast activity causes the bone to be reabsorbed by the body. Simultaneously, the activity of osteoblasts reduces. The calcium resulting from the decrease of bone tissue remains in the body, with resulting calcification of soft tissues and possible creation of kidney stones. The loss of bone is reversible after the end of microgravity conditions, but the recovery is slow and it can take years to make up for months of exposure to microgravity. On Mars it is expected that the recovery will be much slower because its gravity is less than that of Earth. But diet, exercise and medication may help to reduce bone loss and hasten recovery.

- Muscle atrophy. In a way similar to bone loss, muscles, which are less stressed by the weightless conditions, lose mass. This is particular evident for the muscles which are used to hold posture and to perform locomotion. Without regular exercise a decrease of up to 20 percent of the muscle mass can occur in 5 to 11 days.

- A slowing down of cardiovascular system functions. In microgravity the blood volume decreases by up to 22 percent. Since the heart is a muscle, when it is required to work less energetically it will atrophy. The blood pressure consequently decreases and the whole cardiovascular system reduces its ability to oxygenate tissues. One effect on the brain is dizziness. In general, fluids tend to collect in the upper body in weightlessness. Blood does too. When returning to Earth, the blood tends to stagnate in the lower parts of the body, inducing orthostatic hypotension which can cause blackouts. These effects on the circulatory system are accompanied by a decrease in the rate of production of red blood cells, but this trails the reduction of body fluid volume with a delay of weeks to months.

- Fluids redistribution. Gravity tends to force the fluids of the human body into the lower half of the body, and we have evolved to balance this condition. Lacking gravity, these compensation devices continue to function, causing a general redistribution of fluids into the upper half of the body. Apart from a general 'round faced' appearance, this condition causes balance disorders, distorted vision and, possibly, a loss of taste and smell. Eyesight difficulties are mostly linked with an increase of intracranial pressure which affects the shape of the eyeballs and slightly crushes the optic nerve. Distorted vision due to physiological problems can add to cognitive problems and constitute a danger in deep space and planetary missions. Some astronauts reported changes in their sense of taste. This is quite a subjective outcome. Some say that food loses its taste. Some develop a liking for food that they usually do not eat, or dislike food they usually eat. Many experience no change.

Since no specific cause was apparent, many explanations have been offered which range from the possibility of food degradation all the way to psychological changes such as boredom.

- Weakening of the immune system, mostly due to a reduced ability of lymphocytes to reproduce properly and to fight infection. This effect is synergic with the effects of radiation because both damage lymphocytes. Immunodeficiency prompts the viruses that are already present in the body to become active, and spreads infections among crew members. The confined environment in a spacecraft makes this problem worse. Decay of immune function may also be linked with the isolation of the astronauts within the closed spacecraft, and may decrease as the mission progresses.

- Sleep disturbance. Astronauts usually suffer problems with the amount and quality of sleep. This has been attributed to a number of reasons, such as variable light and dark cycles, poor illumination during daytime, and noise. Circadian rhythms are disturbed and this affects neurobehavioural responses and aggravates psychological stresses. On Mars, the day-night cycle is quite similar to that on Earth, so this may help to improve sleep quality in most of the mission. Sound levels in the spacecraft and Mars habitat may be high because in such an artificial environment there will always be machinery at work and fans in the life support system circulate air on a 24 hours a day basis.

It cannot be assumed that the weak gravity on Mars will significantly assist in reversing the losses of the sensorimotor, cardiovascular, and musculoskeletal capacity that occurred in transit. The astronauts' bodies will accommodate to these new working and living conditions homeostatically, but in the absence of definitive insight into the benefits of partial gravity, the best we might hope for is a pause in deterioration for the duration of the surface stay.

The combination of all the stressors (physical forces, weightlessness, isolation, radiation, and limited possibilities for direct human interactions) will affect the astronauts' behavioral health and capacity for performance in ways that we have not been able to explore during the relatively brief spaceflights thus far (relative to a Mars mission) and the human capacity for compensation. Hence we should expect biomedical surprises to arise during a Mars mission lasting a thousand days.

It is presently understood that biological results obtained under terrestrial conditions cannot be truly representative of what happens in space. In this regard, a variety of fluid redistribution effects and hormonal responses occur in microgravity which may influence cellular damage and repair processes, either directly or by controlling the state of oxygenation and the hydration of tissues. Indirect modification of circadian rhythms may also be involved.

Overall, there is evidence that the higher pro-oxidant state the human body adopts in microgravity may be part of a phase in which the deleterious action of ionizing radiation is mediated on a molecular, cellular, and tissue level.[2] Extensive studies in this regard are in progress, as are investigations into the potential use by crew of prophylactic

---

[2] G. Reitz, "Radiation Health impacts on the way to Mars," in *The Particle Radiation Hazard en route to and at Mars*, Publ. Int. Acad. Astronautics, 2015.

radioprotective drugs and the possibility of controlling phytochemical anti-oxidants in the body via dietary choices [19].

5.2.2   Suggested countermeasures

As assessed by the NASA Human Research Program (HRP), many human health risks currently are or might be unacceptable. Nevertheless, they are amenable to reduction through research. Operational risks associated with spacecraft mechanical safety may be greater than biological human health risks, but they typically occur during discrete, relatively brief mission phases. Human health risks may increase with exposure time, and then persist well beyond the end of the mission. Only a subset of all possible human health risks can be mitigated. Conscious programmatic decisions will determine which risks remain unmitigated.

As with radiation hazards, many risks are correlated with the time spent in space, so reducing the transit times between planets will dramatically reduce a crew's exposure to the risk environment. The best way to reduce most of the health risks encountered in a mission to Mars is therefore to use an advanced propulsion system.

At present, most of our countermeasures are focused on physical exercises to counteract muscular and skeletal deterioration in microgravity. The HRP developed the Digital Astronaut Project to investigate the exercise countermeasure regimes, based on musculo-skeletal models of humans exercising with the advanced Resistive Exercise Device (aRED) on board the ISS. This is essentially a weight lifting machine that has been adapted to microgravity, and it uses adjustable resistance piston-driven vacuum cylinders and a fly-wheel system to simulate free-weight exercises. Its primary purpose is to enable astronauts to maintain their muscle strength and mass during long periods in space but it is also effective in increasing their endurance for physically demanding tasks such as space walks. And of course, it provides biomedical data for studies.

A more radical risk mitigation strategy would be to eliminate weightlessness as a major contributor to the physiological changes observed during spaceflight. Rotating all or a large portion of the transit vehicle could produce 'artificial gravity.' Engineering considerations dominate this issue. For example, how would we design a vehicle which can withstand the mechanical stresses of rotating without exceeding the capability of the propulsion system to maneuver? Indeed, how would a rotating vehicle maneuver? And should there be duplicate internal systems (such as hygiene and life support systems) to allow the flight to continue in weightlessness if the rotation had to be terminated?

Although it is impossible with current technology to constantly accelerate a vehicle in space at a substantial fraction of normal gravity, it is possible to devise methods to provide artificial gravity discontinuously or continuously.

One method would be to install a centrifuge on the vehicle, in which the astronauts can spend short periods every day to ameliorate the effects of microgravity. Because they will not need to perform any task while in the centrifuge (it is even possible to think that they would sleep there) it will be possible to build a fairly small device to spin at a relatively high speed (a centrifuge 6 m diameter must spin at about 17 rpm). However, such a device would generate very strong gravitational gradients along the body, and the Coriolis forces would be large for any small movements of the body. We know very little about how much time would have to be spent in such a centrifuge, on the best frequency of these high-g periods, and the discomfort that will be caused. A program of experiments is needed. The discomfort could be studied in ground testing, but all other aspects would require experimentation in zero gravity. The 'space cycle' (Figure 5.1) is a human-powered centrifuge capable of creating artificial gravity ranging from 1–5 g. It was developed by the University of California at Irvine using a grant from the National Space Biomedical Research Institute and is meant to be used by two astronauts: one pedaling a sort of bicycle to power the device in order to subject a colleague to the artificial gravity.

As a second approach, there are essentially two ways of achieving artificial gravity: slow rotation with the humans located at a large distance from the center, and fast rotation with them nearer the center. Clearly any intermediate solution is possible.

The slow solution can be obtained by designing the vehicle to have two modules which can be attached to one another for the propulsive phases and trajectory corrections and then be detached and separated by a long tether. The required artificial gravity is obtained by spinning about the center of mass of the complex (Figure 5.2a). Another solution would set the whole spacecraft spinning about its center of mass (Figure 1.8). This is typical of nuclear electric spacecraft, where the counterweight is the nuclear reactor. This has the benefit of placing the reactor at the maximum possible distance from the crew. The electric thrusters can be set in any position and counter-rotated (i.e. 'despun') so that they are fixed in a non-rotating frame and can be operated during rotation, thereby eliminating the need to stop and restart rotation several times during the journey. For a NEP spacecraft the thrust is applied for all (or at least most) the time.

The fast solution where only the crew compartment is spun, is shown in Figure 5.2b. The vehicle is that depicted in the science fiction novel *Red Mars* by K.S. Robinson. The drawback is that the rotational speed is much higher and it is likely that gravity gradients and Coriolis accelerations will cause a degree of discomfort (how much is not yet understood), particularly if the vehicle is small. Here the choice is between one or more toroidal

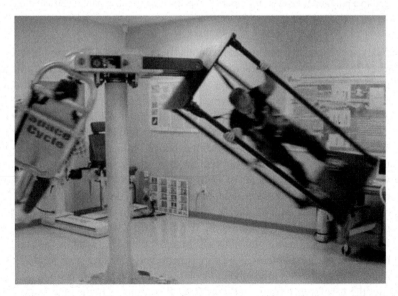

Figure 5.1 The 'space cycle' is a human-powered centrifuge built by the University of California Irvine, USA.

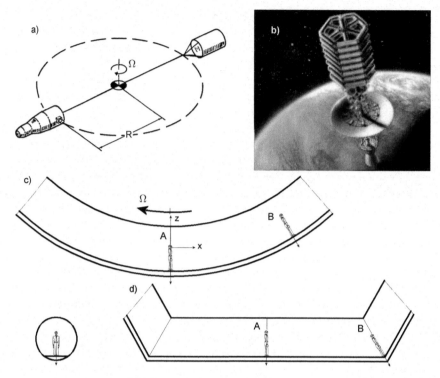

Figure 5.2 (a) Spacecraft made of two modules spinning about the center of mass to produce artificial gravity. (b) Spacecraft in which the crew compartment spins about the longitudinal axis. (c) Situation in a toroidal habitat. The floor is always perpendicular to the apparent vertical direction. (d) Situation in a polygonal habitat. The floor is perpendicular to the apparent vertical direction only at the center of each segment.

(Figure 5.2c) or polygonal (Figure 5.2d) habitats. In the former, a person will feel that the 'down' direction is always perpendicular to the ground, while in the latter it will be such only at the center of each segment (Point A). A person standing at Point B and looking outwards will sense the floor sloping down at an angle of $180/N$ with respect to the apparent horizontal, where N is the number of sides. In the case of Figure 5.2b, the habitat is hexagonal and the floor will seem to slope down at a 30° angle, which will make walking disconcerting. The situation will be somewhat better for an octagonal habitat with an apparent slope of 22.5°. The polygonal habitat also results in the value of the artificial gravity varying along each segment. In a toroidal habitat, a person standing at any point gains the visual impression of the floor sloping upwards, and while walking they feel that the floor is flat. In a polygonal habitat, a person standing in the center of a segment will have the visual impression of the floor being flat, but whilst walking they will feel it sloping downwards at an increasing rate. This disagreement between what is seen and what is felt might be disorienting, but there is no experimental evidence for how serious it will be. It is likely that both solutions are viable, at least if the radius is large enough.

Consider the xz reference frame in Figure 5.2c. It is centered in the inner ear of a person standing in Point A of the toroidal habitat. The centrifugal acceleration a_{ce} used to simulate the gravitational acceleration is directed along the z axis (downwards) and is proportional to the radius and to the square of the angular velocity

$$a_{ce} = -R\Omega^2 \tag{5.1}$$

where the radius is the distance between the axis of rotation of the habitat and the detector of the acceleration – in case of humans, the equilibrium organs of the inner ear. However, if the person moves, the centrifugal acceleration is not the only effect of rotation. There may also be a Coriolis acceleration that could cause discomfort. The value of the Coriolis acceleration is

$$\begin{cases} a_{coz} = -2\Omega v_x \\ a_{cox} = -2\Omega v_z \end{cases} \tag{5.2}$$

where v_x and v_z are the 'horizontal' and 'vertical' components of the velocity of the astronaut's head. This effect acts at right angles to the directions of both the angular velocity and the speed, so a vertical motion results in an acceleration in the horizontal direction, producing an apparent change of direction of the local vertical. The horizontal component produces a variation of the vertical acceleration that adds also to the change of the centrifugal acceleration due to the fact that it changes the peripheral velocity. A slower rotation and a larger radius can be used to reduce the Coriolis acceleration. It is generally believed that at rotation rates of under 2 rpm there will be no adverse effects from the Coriolis acceleration, but higher angular velocities are reckoned to be acceptable. The nausea-inducing effects of Coriolis acceleration can also be mitigated by restraining the movements of the head.

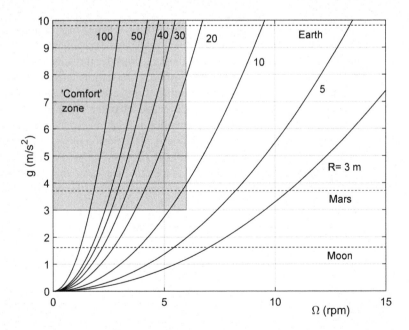

Figure 5.3 A plot of gravitational acceleration as a function of the spin speed for different radii.

A plot of the artificial gravity obtained as a function of the spin speed for different radii is given in Figure 5.3. The shaded zone labeled as the comfort zone is a tentative standard, but more research is required because it is limited to a zone in which the artificial gravity is larger than 0.3 g and the angular rate is less than 6 rpm. The habitat must thus have a radius of at least 10 m.

Consider for instance a toroidal habitat in which the floor has a radius of 10 m and it spins at 6 rpm. The artificial gravitational acceleration at the level of the feet is 3.94 m/s^2 = 0.4 g, which is slightly greater than the gravity on Mars. But if a person's inner ear is 1.7 m above the floor, the radius at which the gravitational acceleration must be computed is 8.3 m and the acceleration is 3.28 m/s^2 = 0.33 g. It is difficult to say which is the correct value: an astronaut will feel the acceleration at the head level, but the force felt under their feet is that corresponding the acceleration at the center of mass of their body, whereas the blood in the legs will be under an acceleration corresponding to that at the feet. Walking at 2 m/s (a fast pace that might not be easy in such a low gravity) they will feel an increase of the downward acceleration of 0.4 m/s^2 at their feet and 0.5 m/s^2 at their heads. This contribution is the centrifugal acceleration in the rotating frame and is always directed downwards, irrespective of the direction of walking.[3] Then there is the Coriolis acceleration of 2.51 m/s^2 that is independent of position. It will be downwards if the person walks to the left (the same direction as the rotation) and upwards in the opposite case. The total acceleration downwards at the level of their head is 5.04 m/s^2 when moving to the left and 2.52 m/s^2 when moving to the right. The effect of walking is a strong variation of the artificial gravity.

Now consider what happens when the astronaut moves their head vertically at 1 m/s in the downward direction. The Coriolis acceleration at the head level is 1.26 m/s^2 to the right. The sense is that the vertical is no longer perpendicular to the ground, but at an angle of 21° from it. This is a fairly substantial value and is likely to produce disorientation and possibly sickness.

Given these complications, it is obvious that the effects of achieving artificial gravity by rapidly spinning the crew compartment must be investigated experimentally.

To evaluate the importance of the radius of the habitat, consider the same torus but with a floor radius of 50 m, spinning at 2.68 rpm. The artificial gravitational acceleration at the level of the feet is the same 3.94 m/s^2 = 0.4 g and that at the level of the astronaut's inner ear is 3.81 m/s^2 = 039 g. The two values are now close to each other. If the astronaut walks at 2 m/s, then the increase of the downward acceleration is 0.08 m/s^2 at the feet and 0.084 m/s^2 at the head. The Coriolis acceleration is 1.12 m/s^2. If the person moves to the left, the total acceleration at the head level is 4.45 m/s^2 and it is 3.33 m/s^2 moving to the right. The effect of walking is much less disturbing. Consider again that the astronaut moves his head vertically at 1 m/s in the downward direction. The Coriolis acceleration at the head level

---

[3] If the computation is made in the inertial frame the total centrifugal acceleration is

$$a_{ce} = -\frac{1}{R}\left(R\Omega \mp v_x\right)^2 = -R\Omega^2 - \frac{v_x^2}{R} \pm 2\Omega v_x.$$

The second term is always negative (downwards) and the last term is seen in the rotating frame as the Coriolis acceleration.

is 0.56 m/s^2 to the right and the perceived divergence between the vertical direction and the perpendicular to ground is only 8.4°.

The prime biomedical issue to be determined is how much gravity would suffice? Even a half-century into the Space Age, definitive research has yet to been undertaken to identify the long term benefits of fractional gravity between 1 g of Earth's surface and zero g of orbital spaceflight. Any current postulations of adequate partial gravity levels are the result of other considerations such as engineering expediency rather than of rigorous biomedical inquiry.

5.3 COGNITIVE ISSUES

In addition to the effects of low gravity and radiation, extended periods spent on board a spacecraft or in the cramped quarters of an outpost on Mars may have detrimental effects on human health. The peculiar environmental characteristics of a spaceship and an outpost that could negatively influence cognitive performance, underlying neurological structures, and brain mechanisms include physical confinement, isolation, and noise.

Theoretically, the mental performance of astronauts can suffer from both direct effects on brain mechanisms by microgravity-induced neurophysiological changes and indirect effects imposed by stress-induced alterations to the attentional state.[4] A number of psychomotor and neurocognitive functions are degraded during spaceflight, including central postural functions, the speed and accuracy of aimed movements, internal timekeeping, attentional processes, limb position sensing, and the central management of concurrent tasks. Thus general research and cognitive activities such as clinical assessment and readiness to perform critical operations (e.g. transfer activities while in Mars orbit, space walks, landing, and activities which will later be conducted in reduced gravity on the surface, etc.) should be performed in space.

The crew must be prepared for any situation that may affect cognitive performance, and thereby mission success and perhaps their survival. Special environmental characteristics of long duration space missions may affect performance, and some of these are also present in Earth analogues and space simulations. Cognitive aspects that must be monitored during an expedition to Mars include astronaut attention, language, memory, learning, reasoning, and perception. Several cognitive tests and batteries such as MINICOG and AGARD have been used during both space missions and simulations, and the Spaceflight Cognitive Assessment Tool for Windows (WinSCAT) is the current standard for this type of assessment on the ISS.

Despite the utilization of self-administration assessment tools such as WinSCAT and the more recently developed Psychomotor Vigilance Test (PVT) to investigate risk factors in the cognitive and neurobehavioral domains, more integrated assessment and monitoring tools must be developed and incorporated into the on-board systems that can measure changes in psychological areas such as cognitive, psychosocial, and emotional/mood, in addition to a variety of psychophysiological areas such as pulse rate and EEG performance.

---

[4] A.B. Newberg, "Changes in the central nervous system and their clinical correlates during long-term spaceflight," *Aviat., Space, Environ. Med.*, vol. 66, pp. 562–572, 1994.

This type of integrated system could help to provide and apply the best countermeasures possible. Monitoring and detection measures can be based on pattern recognition techniques that record the facial expressions of individual crew members and analyze these in relation to surrounding conditions and actions of other members of the crew. The technology should be able to interpret the body language of individuals, their posture, their interactions, and their facial expressions. Through collation and analysis of facial and body language data, together with physiological markers such as pulse rate, it may be possible to detect both positive and negative interaction patterns between crew members. This technology will also be able to identify the location of each person, his or her proximity to colleagues, their frequency of interaction, and time spent at work and leisure, and correlate these factors with emotional state and performance. Furthermore, supplemented with data collected from facial and voice recognition, cognitive performance, and biosensors, it ought to be possible to gain insight into patterns of performance for each person and for the team itself. Two important aspects to be monitored in the cognitive realm during long term missions, are reaction time and accuracy. Reaction times vary significantly in different circumstances, such as neurological disease, brain injury, and under stress or fatigue. Computerized systems allow crew members to track reaction times and accuracy levels in the tests that they perform, in order to detect clinically significant variations in the indices. Such systems can also offer insight into the coping and resolving strategies of the crew.

As the time delay in communication increases during an interplanetary flight from Earth to Mars, the crew will experience increasing autonomy. Furthermore, there will be times when communication will be impossible, although these periods must be reduced to a minimum. Once in orbit around Mars, the spacecraft will spend a significant fraction of its time behind the planet as viewed from Earth. And for a crew on the surface, the rotation of the planet will break the line of sight for half of each sol. These gaps in communication are sure to be eliminated by placing a set of at least three relay satellites about Mars. This would not greatly increase the total cost of the expedition. Telecommunication satellites will facilitate continuous communication with an orbiting spacecraft or an outpost on the planet, and will also assist with teleoperation of rovers and other automatic exploration devices operating at large distances from the outpost. In any case, such satellites will be required to enable human-carrying rovers executing long range journeys to communicate with the outpost.

Another factor will influence a long stay mission. There is a time in which Mars is in conjunction, meaning that it is on the far side of the Sun from Earth. In this configuration, it may be difficult to maintain radio communication (it will depend on the apparatus used). A prolonged period during which communication with the expedition is impractical will be a dangerous point in the mission. It may have a strongly negative effect on the crew. The only way to solve this problem is to put a relay satellite in one of the two Lagrange points of the Sun-Mars system; namely L4 roughly 60° ahead of the planet in its solar orbit and L5 60° behind it. Although more costly than placing a set of telecommunication satellites in orbit around Mars, having a facility in interplanetary space would still be a small fraction of the overall cost of a human Mars mission and would be justified by the elimination of risk.

Another issue relates to the long months spent in the confined environment of a surface habitat on Mars. It is well known that the submariners lose their ability to estimate distances after spending a long time in a confined space. In fact they are forbidden to drive cars for the first few days after a mission. To perform EVAs and drive a rover in these condition could be dangerous. It is likely that this may be mitigated by providing the habitat with windows so that the crew can see outside (unlike in a military submarine). The experience of the ISS shows that astronauts enjoy just looking outside, and viewing the landscape of Mars is sure to be a favorite pastime for explorers. But providing large windows may be difficult; particularly if regolith is used as a protection against GCR or if the habitat is in a cave or a lava tube. And if the habitat is built from regolith using additive manufacturing, carrying large windows from Earth will represent a non-trivial mass penalty. Windows are likely to be substituted by large electronic screens showing the surrounding landscape. However, a screen with a 2D image will not address the issue of loss of depth perception. Cameras and screens that provide a high fidelity 3D image may improve the illusion of having a window sufficiently to prevent a loss of depth perception.

5.4 PSYCHOLOGICAL AND CULTURAL ISSUES

5.4.1 Experience from on-orbit missions

There have been several research studies involving astronauts and cosmonauts that provided information about important psychological, interpersonal, and cultural issues that affect space crew members. The Mir space station and then the ISS have proved to be very valuable assets in this regard. Nick Kanas and colleagues conducted two NASA-funded international studies of psychological and interpersonal issues during missions on board Mir and the ISS, and this section is based on their work [32].

A total of 30 astronauts and 186 mission control personnel were studied. The test subjects completed a weekly questionnaire that included items from several valid, well-known measures that assessed mood and group dynamics. There was significant evidence for the displacement of tension and negative emotions from the crew members to mission control personnel. The support role of the leader was significantly and positively related to group cohesion amongst crew members, and both the task and support roles of the leader were significantly related to cohesion amongst people in mission control.

Russians reported greater language flexibility than Americans. Americans scored higher on a measure of work pressure than Russians, but Russians reported higher levels of tension on the ISS than Americans. There were no significant changes in levels of emotion and group interpersonal climate over time, and there was no general indication of the so-called 'third-quarter' phenomenon, such as occurs when some crew members in isolated and confined environments experience depression and other negative emotions after the half-way point of their mission.

V.I. Gushin and his colleagues used an analysis of speech patterns and a measure of subjective attitudes and personal values to study both crews in space and people working in space analogue environments. They found that, with the passage of time, these isolated groups displayed decreases in the scope and content of their communications and a filtering of what they said to outside personnel, which was termed psychological closing. Members of crews interacted less with some mission control personnel than with others that they perceived as opponents. This tendency of some members of a crew to become more egocentric was called autonomization. This research team also found that crew members became more cohesive by spending time together, and that the presence of subgroups and outliers (e.g. scapegoats) negatively affected group cohesion. In one study of a dozen ISS cosmonauts, the investigators found that personal values generally remained stable, with those related to the fulfilment of professional activities and good social relationships being rated most highly.

L. Tomi and colleagues examined potentially disruptive cultural issues affecting space missions in a survey of 75 astronauts and cosmonauts and 106 mission control personnel. The test subjects said the biggest problem was difficulties in coordination the different space organizations involved with the missions. A.P. Nechaev surveyed 11 cosmonauts regarding their opinions of possible psychological and interpersonal problems that might occur during a Mars expedition. They found the following factors to be rated highly: isolation and monotony, distance-related communication delays, leadership issues, differences in management styles between space agencies, and cultural misunderstandings within an international crew.

G.M. Sandal and others surveyed 576 employees of the European Space Agency and found a link between cultural diversity and the ability of people to interact with one another. Especially important were factors related to leadership and decision-making. Finally, J. Stuster performed a content analysis of personal journals from ten ISS astronauts that were oriented around a number of issues which had behavioral implications. He found that 88 percent of the entries dealt with the following categories: work, outside communications, adjustment, group interaction, recreation/leisure, equipment, events, organization/management, sleep, and food. In general, crew members reported that their life in space wasn't as difficult as they expected prior to launch, despite a 20 percent increase in interpersonal problems during the second half of the missions.

5.4.2   The Mars 500 Program

A unique ground-based space analogue mission called the Mars 500 Program was carried out between June 2010 and November 2011.[5] It was designed to simulate a 520 day round trip to Mars, including periods where the crew functioned under high autonomy conditions and there were appropriate communication delays with outside monitoring personnel in mission control. Six men were confined in a simulator at the Institute for Biomedical Problems in Moscow.

During a 105 day pilot study in 2009 that preceded this mission, N. Kanas studied the mood and group interactions of a six man Russian-European crew and the relationships of this crew with outside mission control personnel, utilizing measures similar to those used in earlier studies of orbital crews. The high work autonomy (in which the crew members planned their own schedules) was well-received by the crew, all the mission goals were achieved, and there were no adverse effects (which echoed recent positive autonomy findings in other space analogue settings). During the high autonomy period, crew member mood and self-direction were reported as being improved, but mission control personnel reported more anxiety and work role confusion. Despite scoring lower in work pressure

---

[5] D. Urbina, R. Charles, "Enduring the isolation of interplanetary travel: A personal account of the Mars 500 mission," Paper # IAC-12-A1.1.1. International Astronautical Federation. Proceedings, 63th International Astronautical Congress, Naples, Italy, pp. 1–5, October 2012.

overall, the Russian crewmen reported a greater rise in work pressure from low to high autonomy than did their European colleagues.

In addition, several psychosocial studies were conducted during the actual 520 day mission.

V.I. Gushin found changes in time perception by crew members which was evidence for the displacement of crew tension to mission control, and decreases in crew member needs and requests during high autonomy which suggested that they had adapted to this condition. G.M. Sandal reported that the crew exhibited increased homogeneity in values and more reluctance to express negative interpersonal feelings over time, which suggested a tendency toward 'groupthink.' B. Van Baarsen et al. found the crew members experienced increased feelings of loneliness and perceived lower support from colleagues over time, which had a negative effect on cognitive adaptation. M. Basner et al. analyzed data from wrist actigraphy and the psychomotor vigilance test, as well as various subjective measures and identified a number of individual differences in terms of sleep pattern, mood, and conflicts with mission control. Finally, C. Tafforin assessed fixed video recordings of crew behavior at breakfast and found variations in personal actions, visual interactions, and facial expressions. One particular observation was a general decrease in group collective time from the outbound to the return phase of the simulated flight to Mars.

5.4.3   Space psychiatry and salutogenesis

A number of psychiatric problems have been reported during space missions in LEO. Most common are adjustment reactions to the novelty of space. These largely consist of transient anxiety or depression. Psychosomatic reactions have also been reported. Asthenization – a syndrome consisting of fatigue, irritability, emotional lability, and attention and concentration difficulties – has been reported to occur commonly in cosmonauts by Russian flight surgeons. However, problems related to major mood and thought disorders (e.g. manic depression and schizophrenia) have not occurred in space, probably because potential crew members were screened psychiatrically against any predisposition to such psychotic conditions prior to crew selection.

Post-mission personality changes and emotional problems have affected some returning space travelers. These include depression, anxiety, alcohol abuse, and difficulties in marital readjustment that in some cases have necessitated the use of psychotherapy and psychoactive medications [21].

Isolated and confined environments can also yield positive experiences. For example E.C. Ihle et al. surveyed 39 astronauts and cosmonauts and found that all of the respondents reported positive changes as a result of flying in space. One subscale especially stood out, namely their perceptions of Earth. One of the items in this subscale dealing with gaining a stronger appreciation of the Earth's beauty had the highest mean change score.

Extending pioneering research that was started in the early 1990s on the salutogenic (or growth-enhancing) aspects of space travel, P. Suedfeld and his colleagues analyzed the published memoirs of 125 space travelers. After returning from space, the subjects reported higher levels on categories of Universalism (i.e. a greater appreciation for other people and Nature), Spirituality, and Power. Russian space travelers scored higher in

Achievement and Universalism and lower in Enjoyment than Americans. Overall, these results suggest that spacefaring is a positive and growth-enhancing experience for many of its participants.

5.4.4   Suggested countermeasures

Despite their relevance for future orbital and lunar missions, the above findings may well have limited generalizability to long-distance, multi-year expeditions such as to Mars, where new stressors will occur which are related to autonomy, two-way communication delays of up to 44 minutes, and extreme loneliness due to perceiving the Earth as an insignificant dot in the heavens (the so-called Earth-out-of-view phenomenon).

There are several countermeasures that can be implemented to help ameliorate the impact of the above issues on the crew of a Mars expedition in terms of selection, pre-launch training, mission monitoring and support, and post-return adaptation. Crew members should be chosen who are comfortable working alone on a project, yet are also able to interact socially with their crewmates when appropriate. Commanders should possess both task-oriented and supportive leadership skills.

Pre-launch training should include both didactic and experiential sessions that deal with important psychological, interpersonal and cultural issues that may occur, as well as how to work effectively under conditions of high autonomy. Conjoint training involving both crew members and key mission control personnel should take place, and address issues related to displacement and possible crew-ground miscommunication [21]. Computer-based refresher courses echoing some of this training should be available in flight. Crews should reserve time during the mission to identify and deal with stressful personal and interpersonal issues before they fester and become problematic. Crew members should develop ways of communicating under time-delayed conditions with people on Earth, for example by appending anticipatory questions to e-mail messages in order to minimize repeated back-and-forth communications. The Earth-out-of-view phenomenon could be addressed simply by providing a telescope. The crew members should know their families are being supported during the mission by a mix of formal and informal group activities.

Re-adaptation debriefing and private time together need to be scheduled post-return in order to help the crew members and their families readjust to each other and to deal with the fame and glory issues that will inevitably result from a highly publicized space expedition.

6

Interplanetary journey to Mars

The interplanetary journey is just one of the phases involved in a mission to Mars, but at the present level of technology it is a highly critical one. Moreover, it must be undergone twice and because it requires a number of different space vehicles it greatly influences both the risk and cost of the whole enterprise.

6.1 TRAVELING FROM EARTH TO MARS

A voyage from the surface of Earth to the surface of Mars can be divided into a sequence of phases that strictly depend on the type of propulsion employed.

Chemical and nuclear thermal rockets apply large thrusts for very brief times, so they are defined as High Thrust (HT) systems. The simplest way to determine the trajectory of a spacecraft propelled by such a device is to assume that the application of the thrust causes an instantaneous change of the velocity of the vehicle, ΔV, after which it proceeds in an inertial manner along its trajectory under the influence of the gravitational attraction of a number of celestial bodies.

To simplify the computation of the trajectory, two assumptions are usually made:

- The celestial bodies influencing the spacecraft are spherical.
- At any given time only one body interacts with the vehicle.

The first assumption enables us to treat the celestial bodies and the spacecraft as point masses, whilst the second, which is known as the two-body problem, greatly simplifies the computation of the trajectory because it turns out to be made by a number of conical sections that are either ellipses, parabolas, or hyperbolas.

Interplanetary travel from the surface of Earth to the surface of Mars can be divided into the following phases:

1. Launch to LEO, followed by a certain time spent in orbit.
2. A powered maneuver to achieve escape velocity. This is followed by a parabolic (or better still a hyperbolic) trajectory in which the two bodies considered are the vehicle and the appropriate planet. This phase lasts until the limits of the sphere of influence of the starting planet are achieved.

© Springer International Publishing Switzerland 2017   119
G. Genta, *Next Stop Mars*, Springer Praxis Books, DOI 10.1007/978-3-319-44311-9_6

3. The Trans-Mars Injection (TMI) maneuver, followed by a long time spent tracing an elliptical interplanetary trajectory that connects the heliocentric orbits of the initial and the destination planets. The two bodies which are relevant to this phase of the mission are the vehicle and the Sun.
4. A powered maneuver to synchronize the velocity of the spacecraft in the heliocentric interplanetary trajectory with that of the target planet at the time of entering into its sphere of influence. This is followed by a parabolic (or hyperbolic) trajectory in which the two bodies of significance are the vehicle and the planet.
5. Mars Orbit Insertion (MOI), followed by a certain time spent in Mars orbit.
6. Deorbiting and descent to the planet, followed by landing.

Some of these phases can be grouped together. For instance, it is possible to launch the spacecraft directly from Earth into the interplanetary trajectory, thereby grouping phases 1, 2 and 3. Similarly, it is possible to directly enter the atmosphere of Mars and undertake the landing from the interplanetary trajectory, without entering Mars orbit. Even more important, phases 2 and 3 are usually executed as a single maneuver and the interplanetary trajectory is directly achieved from orbit around the departure planet. As will be seen later, the closer to the planet that the burn is applied, the less propellant it requires. Furthermore, phases 4 and 5 are usually performed as a single maneuver.

This process of splitting the whole trajectory into a number of elliptical, parabolic, or hyperbolic parts which are connected by propulsive phases, is known as the patched conic approximation.

Strictly speaking, only the first three phases require a rocket burn, because phases 4, 5 and 6 aimed at achieving Mars orbit, deorbiting, and then landing (or alternatively direct entry into the atmosphere from the interplanetary trajectory) can be performed by utilizing, in part or entirely, aerodynamic braking in Mars' atmosphere, because that will save large quantities of propellant. Whether this is expedient or not, depends on the trade-off between the mass of propellant that is saved and the mass of the necessary aerodynamic apparatus, such as heat shields and aerodynamic fairings.

However, HT devices are not the only ones which may be used for a Mars mission. If electric propulsion is used, the rocket will apply a low thrust for much of the time or perhaps even for the entire journey. Low Thrust (LT) devices cannot be used for all the phases of the interplanetary voyage. In particular, the launch to LEO must be at any rate performed using high thrust systems. Also the landing on Mars, if not performed using aerodynamic devices, will require a high thrust system.

For low thrust devices the phases of the interplanetary voyage are:

1. Launch to LEO (using high thrust rockets), followed by a certain time spent in orbit.
2. Raising the orbit by low thrust on a spiral trajectory until the escape conditions are achieved.
3. The interplanetary low thrust trajectory, which involves a period of acceleration and a period of deceleration, possibly with a period of coasting in between.
4. Entering Mars orbit by performing an inward spiral around that planet for a number of days until Low Mars Orbit (LMO) is achieved.
5. Deorbiting, descent and landing.

The first and final phases are the same as before. As stated, launching and landing are done in the same manner, irrespective of whether high thrust or low thrust rockets are used during the interplanetary voyage. The other phases, however, are completely different. To save propellant, phase 4 may be performed by aerobraking rather than by thrusting. However, it is unlikely that aerobraking will be proposed in conjunction with electric propulsion, whether this be SEP or NEP.

A large variety of trajectories can be devised, because it is possible to go beyond the simple approach stated above. In the last half-century astrodynamics has progressed such that nowadays it is possible to exploit the opportunities permitted by the complexities of the gravitational field of the solar system and design space trajectories that take account of the presence of many celestial bodies. In particular, maneuvers like a so-called 'gravity assist,' in which a spacecraft gets close to a celestial body in order to employ its gravitational field to gain speed at the expense of the latter, or to change the direction of its travel, have become a common trick.

In the preliminary study of a mission, the planets may be treated as being in circular and coplanar orbits, but to obtain realistic results their ellipticity must be taken into account; this is particularly so for Mars.

6.2 LAUNCH TO LEO

Like all space missions, a human mission to Mars must set off from the surface of Earth, at least until colonization has proceeded to the point that spacecraft can be built on the Moon, some asteroids, or in space itself. All material must thus be carried from the bottom of the Earth's gravitational well, which makes the cost of achieving LEO one of the primary costs of the entire operation, and in all likelihood the largest.

Hence any reduction in the cost of launching large payloads to LEO is of fundamental importance in making human Mars exploration affordable. In this, private organizations and companies will play a major role, even if space agencies will conduct the actual exploration.

Many Mars mission architectures assume that at least three interplanetary ships will be sent to Mars: two carrying cargo and one carrying the crew. These will incorporate the Mars lander/ascent vehicle, the habitats for the interplanetary phase and for use on the surface of the planet, and the propulsion system for the return voyage. It is usually assumed that the total IMLEO is in the range of 500 to 1,000 t, although some schemes are even heavier.

At 7,000–20,000 \$/kg, achieving LEO is the largest fraction of the recurring cost of a Mars mission. For the highest values for both the IMLEO and the cost, a total launch cost of \$20 billion is obtained. This is a not-inconsiderable sum, but well within the realm of possibility. Heavy lift launchers are usually defined as launchers with a LEO capability that exceeds 50 t. Superheavy lift launchers with a LEO capability of at least 100 t are not currently available, although they were mastered in the past (i.e. the Saturn V and the Energia) and will soon again become available in the form of the NASA SLS Block II, which will have a LEO capability of 130 t; the Long March 9 developed by the Chinese Space Agency will have a similar capacity. Another launcher in this class is the Falcon Heavy, which is being developed privately by the SpaceX company.

Some launchers that can lift more than 20t to LEO are listed in Table 6.1 and sketched in Figure 6.1. The list is far from complete, and includes launchers which were either never built or were never made operational.

Table 6.1 Some launchers able to carry more than 20 t to LEO, as sketched in Figure 6.1.

#	Launcher	Status	Payload to LEO	Launch mass (t)
1	ULA Delta 4 H	O	22.5	773
2	NASA Space Shuttle	R	24	2,030
3	NASA Ares I	C	24	—
4	SpaceX Falcon 9 Heavy	BB	53	1,400
5	NASA SLS Block I crew	BB	70	2,650
6	NASA Nova	C	—	—
7	NPOE Energia Buran	R	100	2,400
8	NASA SLS Block IA crew	BB	105	2,700
9	NASA SLS Block IA cargo	BB	105	2,700
10	OKB1 N1	C	110	2,735
11	NASA Saturn V	R	115	2,800
12	NASA SLS Block II crew	BB	130	2,950
13	NASA SLS Block II cargo	BB	130	2,950

(O: Operational, R: Retired; C: Canceled before becoming operational; BB: Being Built.)

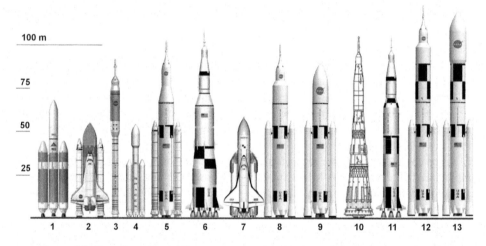

Figure 6.1 Some launchers able to carry more than 20 t into LEO, as listed in Table 6.1.

Launchers even larger than those in the superheavy class have been suggested. The Nova launcher proposed in the early 1960s was to have had a LEO capacity of 300 t in a form for human Mars missions. In that same era there was also a proposal for an even larger Sea Dragon launcher with a capacity of 550 t. Neither of these was developed, but recently SpaceX proposed a launcher, the Mars Colonial Transporter, designed to carry about 500 t to orbit.

Heavy lift launchers provide a major reduction in cost for large payloads in comparison to splitting the payload across several launches by smaller vehicles. They will be essential for a human Mars mission. Reusable rockets may further reduce the launch cost. For instance, the launch cost of the Falcon Heavy is predicted at 1,700 $/kg. Using a reusable first stage this may well be cut to around 600 $/kg. If the latter cost can be achieved, then launching to LEO the aforementioned IMLEO of 1,000 t would cost $600 million.

A major problem with heavy launchers is their launch rate. After the decision to launch is taken, the process of construction, assembly, testing, and final preparation of the vehicle will monopolize industrial assets for a year or so. In addition, in industry it is difficult to sustain a stop-and-go pace over a long period. Ideally, there would be one construction and launch per year (or very optimistically two) but it depends upon the type of mission. If the heavy launcher is only used for Mars missions, a more realistic production rate may be one every 1 or 2 years. At that pace, the industrial participants would require to take a very long term view.

Such constraints must be taken into account in planning the mission architecture and the preparatory phases. If an international mission requires more than one superheavy launcher and distributing the launch tasks amongst the participants, the fact that a smaller number of each launcher is built will only magnify the problems.

There are three options for starting a Mars mission:

- Direct launch.
- Single launch to LEO and then proceed to TMI after a short time.
- Multiple launches to LEO to assemble the interplanetary craft, and then proceed to TMI after a much longer time.

In the first two cases, the spacecraft must be launched in one piece, so this requires not only the capacity to orbit a large mass but also a bulky object. The construction of very large launchers (far bigger than superheavy ones) would make this possible.

In a project such as Mars Direct (Figure 1.10), the first option was chosen, but to make this possible it was necessary to design a very lightweight, small Mars spacecraft. Multiple launches may help along this way.

Most designs employ the third option, and include assembly of one or more spacecraft in orbit. On the one hand this may involve merely docking together the two parts of a vehicle. But it might be a lengthy task involving many elements carried into LEO using relatively small launchers, an operation that might take several months. And even in the former case there may be a long wait if a single launch pad has to be refurbished for the second launch.

In all cases, a basic design choice is the type, height, and inclination of the initial orbit. The inclination of the orbit relative to the equator is a very important parameter. On the one hand this may be a purely political decision on the basis of the location of the launch facility. The orbit of the ISS was chosen so that it was easily accessible from American, European and Russian bases, but it is unsuitable as a parking orbit for initiating a mission to Mars because the 51° inclination would make the TMI maneuver more costly to achieve. The starting orbit for a mission to Mars should really be governed by the characteristics of the interplanetary trajectory, which differ depending upon the launch opportunity.

If the launch base is very near the equator then all orbits are possible, albeit not at the same cost. In this regard, the European Space Agency base in Kourou (5° 11′ N) is a good choice. Next are the Indian Satish Dhawan Space Centre (13° 28′ N) and Cape Canaveral in America (28° 30′ N). The Chinese launch facilities are at latitudes of 28° N or more, and the Russian ones are beyond 45° N. The bases of other countries, such as Japan, are at latitudes which are not much lower. Another option is to use a platform at sea, such as the Italian San Marco platform or the one that is privately owned by the Sea Launch company.

The main decision to make for this first part of the mission is the altitude of the orbit. Usually the term LEO means an orbit between 160 and 2,000 km from the Earth's surface. Lower orbits impose sufficient aerodynamic drag to soon cause a spacecraft to fall into the atmosphere. Hence orbits below 200 km are seldom considered. Higher orbits are necessary for lengthy durations, such as to assemble an interplanetary spacecraft in space. Above 1,000 km, the spacecraft will spend at least a part of each orbit in the lowest part of the inner Van Allen Belt. As a result, the LEO choice is limited to orbits in the range 250–1,000 km.

The payload of a given rocket decreases with both increasing altitude and inclination. Despite the fact that orbital velocity decreases with increasing altitude, the ΔV required for the launch to that altitude increases (Figure 6.2). The figure is just a rough approximation because it is impossible to give general rules, as aerodynamic and gravity losses depend strictly on the launcher and on the stated trajectory, but the ΔV required for a 200 km LEO is about 9.4 km/s while the orbital velocity at that altitude is 7.79 km/s. A rough approximation of the ΔV to attain a circular orbit from the Earth's surface as a function of the altitude is plotted in Figure 6.2 (see Section B.1, where it is also explained that the values obtained in this way exceed the true ones. The result depends on the assumed ascent trajectory, and in particular the transition from the vertical path at lift off and the horizontal one at orbit insertion. The gravity losses that must be taken into account are mainly influenced by the duration of the burn. This is because the atmospheric drag losses are strongly determined by the exact velocity profile of the ascent. Their sum is something between 1.3 and 1.8 km/s for a 200 km orbit.)

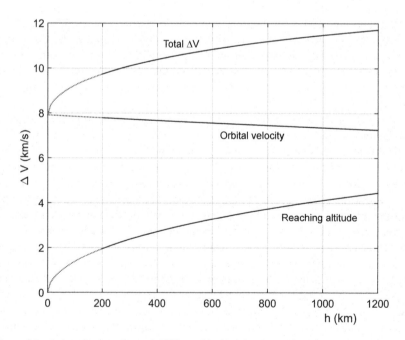

Figure 6.2 Approximate values of ΔV for achieving circular orbit as a function of altitude.

For reference, the capabilities of the SpaceX Falcon 9 launch vehicle in terms of payload mass in circular LEO at different altitudes and inclinations are given in Table 6.2.[1] This also gives the escape mass with different energies C_3 (see Section B.2.1). All data refer to a launch from Cape Canaveral.

In the case of a direct launch (the table on the right), the value of C_3 is about 8.7 $(km/s)^2$ for Hohmann trajectories between circular orbits (Figure 6.6). It uses actual values for the travel time and the launch opportunity; for instance, $C_3 = 20$ $(km/s)^2$ for the optimal 175 day trajectory of the 2042 launch opportunity (Figure 6.9).

The same data are plotted in non-dimensional form in Figure 6.3 for the masses in LEO. All masses are normalized with the maximum mass being to a 200 km orbit at an inclination at 28.5°.

A reference assembly orbit is a 400 km circular parking orbit. This already means a loss of about 5 percent of the IMLEO relative to an orbit at 200 km. The choice of altitude must take into account the propulsion used to make TMI and the time the spacecraft must wait in orbit. The first point is that a chemical rocket can start from a fairly low orbit, even at 200 km. But if nuclear propulsion is used then the starting orbit must be higher. This is particularly so for an NTP vehicle. The altitude at which a nuclear rocket can be started is still unclear. In proposals a figure of 1,000–1,200 km is often suggested. This would cause a loss of about 20 percent of the launcher's capacity (relative to 200 km), but provides a small advantage as a reduction of the energy required for the TMI maneuver.

There may be limitations for the reactor of a NEP system, although less severe than those for NTP, but this too is unclear. Probably it is safe to presume that the reactor will be switched on at an altitude no lower than 800–1,000 km.

In the case of SEP, there may be limitations because of the large drag that would result from the enormous solar panels. A fairly high orbit will be required, but this will depend on the launch opportunity chosen. At times of strong solar activity, the aerodynamic drag in the high atmosphere is larger and a higher altitude will be required. As an average, an altitude in the range 500–700 km would be sufficient.

If the time spent in LEO is long, orbits lower than 500 km will require reboosting, which may be achieved by chemical or even solar electric propulsion, so long as the altitude is not permitted to decay too severely. The main drawback of assembly in LEO is that, from time to time, a reboost of the various modules will be necessary in order to maintain the 400 km orbit, and these maneuvers will impose a significant mass penalty on the project. In the last NASA DRA5 architecture [24] the mass of the reboost modules exceeded 100 t.

An alternative would be to launch the payloads into a fairly low (but not too low) orbit and then use a space tug powered by SEP to raise them to an altitude at which there would be no requirement for reboosts, or even to the altitude at which the finished vehicle would start its nuclear engine.

Orbital assembly requires autonomous and automatic rendezvous and docking (just as the ATV does with the ISS) of several propulsion modules and payloads. The simplest way is to assemble in LEO the propulsive stage for the TMI maneuver with a 65 t payload module.

---

[1] *Falcon 9 launch vehicle payload user's guide - Rev. 1*; https://www.spaceflightnow.com/falcon9/001/f9guide.pdf

Table 6.2 Capabilities of the Falcon 9 launcher. Payload mass in circular LEO at different altitudes and inclinations and escape mass with different energies C_3. The data are for a launch from Cape Canaveral.

Orbital performance			
h (km)	Mass in LEO (kg)		
	Inclination (°)		
	28.5	38	51.6
200	10,454	10,221	9,823
300	10,202	9,975	9,586
400	9,953	9,737	9,358
500	9,727	9,508	9,138
600	9,503	9,289	8,924
700	9,287	9,076	8,719
800	9,080	8,872	8,522
900	8,879	8,676	8,331
1,000	8,687	8,486	8,148
1,100	8,500	8,303	7,970
1,200	8,320	8,127	7,799
1,300	8,147	7,957	7,635
1,400	7,979	7,792	7,475
1,500	7,817	7,633	7,320
1,600	7,662	7,480	7,172
1,700	7,510	7,330	7,028
1,800	7,364	7,187	6,888
1,900	7,221	7,048	6,753
2,000	7,085	6,913	6,622

Escape performance	
C_3 (km/s)^2	Mass (kg)
−16	3,823
−14	3,598
−11	3,373
−8	3,148
−6	2,923
−3	2,698
0	2,473
4	2,248
7	2,023
11	1,798
14	1,573
19	1,348
23	1,123
28	898
33	673
39	448
45	223

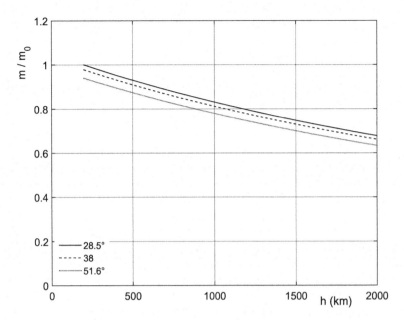

Figure 6.3 Non-dimensional values of the payload mass carried into circular LEO at different altitudes and inclinations by the Falcon 9 launcher. The data are for a launch from Cape Canaveral.

A 130 t chemical stage can typically boost a 65 t payload module to Mars, so the initial mass of this Mars transfer vehicle would be 195 t. The advantage of this strategy compared to direct TMI injection is a reduction of the launcher size and a possible increase of the departure rate (several payloads can be lifted into LEO before the departure window, then wait there for rendezvous). The NASA SLS Block II launcher is intended to have a capacity to LEO of 130 t. If zero-boil-off technologies aren't available, then just one payload per departure opportunity can be sent to Mars from a single launch pad. Eventually, if the waiting time in LEO is not excessive, the boil-off losses of cryogenic propellant might be acceptable but this remains an issue.

The advantages of the LEO assembly option are flexibility and relative independence of the launch date from the narrow window for the Mars departure opportunity, because the launches can be spread over time. The injected payload mass is not limited and, at a chosen window, a fleet of several vehicles can be sent to Mars simultaneously.

Zero-boil-off technologies will be mandatory with liquid hydrogen (LH2), including super insulation techniques and efficient cryocoolers systems, but the latter have not yet been mastered. Direct injection scenarios do not require zero-boil-off technology and they simplify the mission (no rendezvous and docking), but they are limited to a launch roughly every 2.1 years. Because the Mars window lasts only around four weeks, simultaneous construction and liftoff of several heavy launchers will be required in order to orbit more than 130 t. The effects on logistics, industrial capability, and facilities such as launch pads would be dramatic. The direct injection strategy is more appropriate in the case of 'split' architectures. This suggests, for instance, that the Mars ascent vehicle and the surface habitat should be sent to Mars (or at least placed into Mars orbit) separately and in advance of sending the crew.

A number of strategies can be employed in order to avoid the need to assemble a huge vehicle in LEO.

The first strategy is to choose a propulsion system with a high ISP in order to reduce the amount of propellant. In general, the same habitat and propulsion systems are also used for the outbound leg of the mission, which results in stronger requirements. The assembly of the vehicle in LEO is not avoided, but the different modules generally require no more than two or three launches of individual payloads of about 100 t.

In all chemical propulsion architectures, the assembly of a huge Earth Return Vehicle (ERV) in LEO is usually either avoided or significantly limited thanks to other options. With a LEO capability of 130 t for the launcher (maximum LEO payload of the NASA SLS Block II) the maximum payload for a direct TMI maneuver using chemical propulsion is 46 t and it is around 40 t for a fast interplanetary transit (May 2012). Crew size reduction and aerocapture allow some mass savings. The ERV can be split into two modules that are launched separately from LEO and assembled in Mars orbit. This may be straightforward if the modules are not too large. It is better to perform the assembly in Mars orbit rather than in LEO because it eliminates the need for a large propulsion system which would itself have to be assembled in orbit, then maintained by reboost modules. However, such strategies do not require to be implemented at the expense of the mass of the other vehicles, and especially the landers. The optimization of the mission suggests that all modules and materials that are needed for the return but are not required on the surface of Mars should remain in Mars orbit.

The privatization of the market for launching to LEO might become very important in reducing the total costs of a human Mars mission.

In a longer timeframe, revolutionary changes might radically alter the balance of trade-offs in planning human missions to Mars. One such change would be the introduction of a spaceplane such as the Skylon (Figure 6.4), since this would dramatically reduce the cost of putting payloads into LEO. This will be particularly true if the vehicle is capable of

Figure 6.4 The Skylon spaceplane, designed by the British Reaction Engines company to carry about 15 t to LEO.

operating with short turnaround times, so that the components of the spacecraft can be sent to orbit on a number of flights in a short interval. A fast turnaround will also be important because it is likely that the payload capacity of such a spaceplane will be much lower than that of heavy lift rockets, hence requiring a greater total number of launches for a given IMLEO. However, some people say that spaceplanes will be of little use for human Mars exploration precisely because of their limited payload. It is true that assembling a large number of small modules and refilling them with several tanker flights makes the overall scenario quite complex, and it may be that, even if spaceplanes become available, super-heavy non-reusable launchers will be preferred.

Another possibility is to refill the interplanetary vehicle with propellant produced on the Moon and transported to LEO. In practice this will be feasible only for LOX/LH2 propellants or, in the case of NTP, only for LH2. In the case of SEP or NEP, the propellant is likely to be argon, which cannot be produced on the Moon. However, the use of lunar propellants is still a controversial option (see Section 11.2.2).

A more drastic improvement would be the construction of a space elevator (see Section 12.3). Not only would a space elevator drastically reduce the cost of placing payloads in space, it would also release the assembled interplanetary vehicle beyond geosynchronous orbit with a speed higher than the orbital velocity. The Trans-Mars Injection (TMI) would therefore be much easier and would require a much lower amount of propellant. Very fast and economical trans-Mars trajectories would become feasible that would cut both the cost and duration of a human mission to Mars.

The final option would be to use NTP for achieving orbit. A few experts think that this is fully possible [23] but most are of the opposite opinion and, above all, current interna-tional treaties do not permit nuclear rockets to be fired in the Earth's atmosphere. There remains the possibility of launching from Earth with a chemical first stage, climbing to a higher altitude than usual, and then switching on the nuclear rocket in order to complete the ascent. As yet, however, this possibility has received only limited attention.

All of these advanced approaches would be effective in reducing the cost of human missions to Mars, reducing the time to reach Mars, or increasing the payload, but they are not absolutely necessary. Present Earth-to-orbit technologies are fully adequate, at least for the early Mars missions. The most pressing requirement is to reduce the cost of deliver-ing payload to orbit, and this is evidently more sensitive to organization and production issues than the need for new propulsion technologies.

6.3   IMPULSIVE INTERPLANETARY TRAJECTORIES

The most economical way to reach Mars is to pursue the semi-elliptical Hohmann trajec-tory in Figure 6.5 (see Section B.2.1). This takes about 258 days (8.6 months) to reach the planet. The Hohmann trajectory is, however, an abstraction that applies only to the ideal case of a transfer between circular and coplanar planetary orbits. Faster trajectories require more energy, which translates into more propellant. Trajectories like those that were defined as Type I in Figure 6.5 are usually faster than the Hohmann trajectory and the Type II are slower. Both types require more energy, at least within the assumption of circular, coplanar orbits.

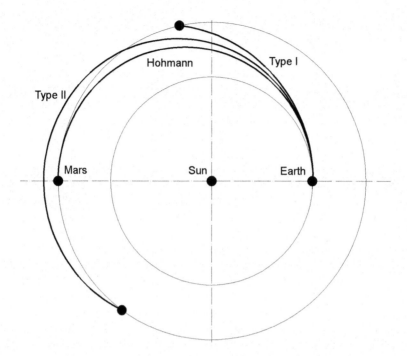

Figure 6.5 Earth-Mars interplanetary trajectories between circular and coplanar planetary orbits. The Hohmann trajectory is compared with a Type I fast trajectory and a Type II slow trajectory. Both of the latter trajectories shown in the figure start off tangent to Earth's orbit, but this is not necessary.

A spacecraft on the Hohmann trajectory must set off 96 days ahead of opposition, and will reach Mars about 163 days after it. The spacecraft will depart the gravitational sphere of influence of Earth with a hyperbolic excess speed of 2.945 km/s and after the cruise it will enter the sphere of influence of Mars with a hyperbolic excess speed of 2.649 km/s.

A parameter which specifies the energy required for the interplanetary trajectory is the square of the hyperbolic excess speed with which the vehicle leaves Earth. This is usually labeled C_3, and the plot which shows C_3 as a function of the departure and arrival dates is the so-called pork-chop plot. Usually such a plot is computed by taking into account the fact that planetary orbits are elliptical, and hence it depends on the particular launch opportunity. A general plot of this type that is calculated on the basis of circular orbits and hence applies in an approximate manner to all launch opportunities, is shown in Figure 6.6.

The pork-chop plot refers only to the energy of the Trans-Mars Injection (TMI), since the spacecraft is outside the sphere of influence of the departure planet. Actually, it is not convenient to achieve the escape conditions by applying a first ΔV in the departure orbit and then applying a second ΔV after escaping in order to achieve the desired hyperbolic excess speed; it is much more expedient to apply a single ΔV for the TMI as low as possible in the sphere of influence of the Earth. This single ΔV (hereinafter labeled ΔV_1), which includes the ΔV needed to achieve escape conditions, depends on the orbit from which the spacecraft starts.

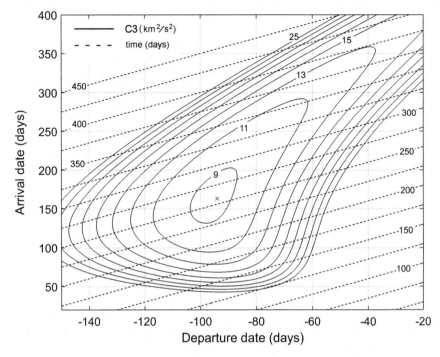

Figure 6.6 A general pork-chop plot for an Earth-Mars transfer computed assuming circular and coplanar orbits. Time is measured from the generic opposition. The minimum occurs for a Hohmann transfer ($C_3 = 8.673 \times 106 \; m^2/s^2$).

In a similar manner, it is possible to calculate the velocity increment ΔV_2 required to achieve the destination orbit about Mars. If the Mars Orbit Insertion (MOI) is performed by a propulsive maneuver, then a minimum $\Delta V_{tot} = \Delta V_1 + \Delta V_2$ can be chosen for a trajectory for a given transit time. This will be the most economical trajectory for reaching Mars in a given time.

Figure 6.7 shows ΔV_{tot}, ΔV_1 and ΔV_2 for such optimized trajectories in terms of the time spent cruising through interplanetary space, with the following specifications:

- The full lines refer to circular departure and arrival orbits which are at an altitude of 300 km around each planet.
- The dotted lines refer to an initial circular orbit of Earth at an altitude of 250 km and an elliptical arrival orbit which has a periareion at 250 km and an apoareion at 33,970 km above the surface of Mars.[2]
- The dashed line refers to a case in which the MOI is performed by aerocapture and/or aerobraking instead of by a propulsive maneuver. In this case $\Delta V_2 = 0$, ΔV_{tot} coincides with ΔV_1 and the optimization must be performed with reference to ΔV_1 only.

---

[2] The periapsis and apoapsis of an orbit around Mars are usually referred to as periareion and apoareion.

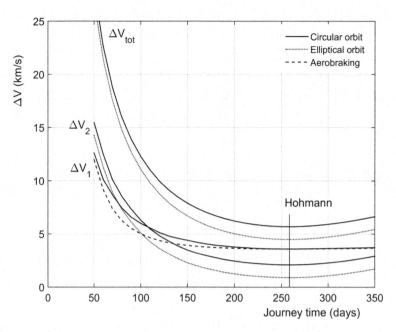

Figure 6.7 Values of ΔV_1 and ΔV_2 for fast trajectories as functions of the travel time T. For each value of time the trajectory leading to the lowest ΔV_{tot} was chosen. The plot has been computed assuming that the Earth and Mars orbits are circular and coplanar.

As can be seen, a moderate reduction of the travel time with respect to the Hohmann transfer can be attained with a reasonable extra expenditure of propellant but large amounts of energy are required for a really fast trajectory, and this can be accomplished only by NTP.

Figure 6.8 gives the departure and arrival dates for the same opposition used to calculate the trajectories in Figure 6.7. In this case the shape of the arrival orbit has no effect upon the departure and arrival dates.

Thus far, it has been assumed that the planetary orbits are circular and coplanar in order to make a rough approximation that will be valid for any launch opportunity. While for some planets, such as Venus, this assumption produces fairly good results, for Mars, whose orbit is significantly non-circular, it yields poor results and the Hohmann trajectory can be applied only to a point. As already stated, if the orbits of the planets are considered to be elliptical, the energy needed for the TMI maneuver varies from one launch opportunity to another. The pork-chop plots for the 3035, 2037, 2040 and 2042 launch opportunities are shown in Figure 6.9 with the departure and arrival dates relative to the day of the opposition.

The 2035 opposition (on September 25) is perihelic, meaning that it occurs when Mars is close to its perihelion. At closest approach, the distance between Earth and Mars will be near to the minimum possible distance, making conditions rather favorable. Consider the optimal Type I trajectory: the energy is quite low ($C_3 = 10.28$ km^2/s^2) and the duration of 194 days is at least 2 months shorter than the Hohmann trajectory. The optimal Type II trajectory is much longer (more than twice the duration of the Type I trajectory) and more demanding in terms of energy.

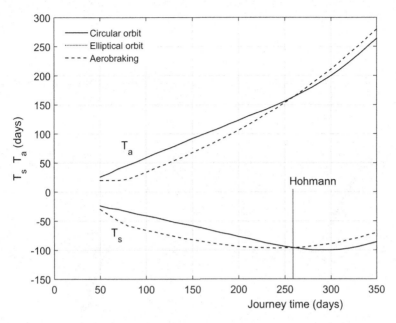

Figure 6.8 Departure and arrival dates relative to the opposition date for the same trajectories studied in Figure 6.7. The dates for the trajectories which start and end with circular orbits around each planet coincide with those which adopt a highly elliptical orbit upon arrival.

The worst case is the 2040 launch window, where the optimal Type I trajectory requires more than twice the energy of the Hohmann transfer. But it will be possible to use a slower Type II trajectory that takes much less energy than the optimal Type I trajectory. In 2042 the energy is low, particularly for the Type II trajectory, which is close to that of the Hohmann transfer, but the travel time is fairly lengthy. Very fast trajectories with travel times of around 120 days (or less) are possible at a perihelic opposition without large energy expenditure. The four plots in Figure 6.10 show the extent to which the energy and travel time depend on the selected launch window.

A comparison of the optimal Type I and Type II interplanetary transfers in terms of energy (C_3), travel time, and ΔV for departing and arriving in circular orbits at altitudes of 300 km during the launch opportunities of 2035, 2037, 2040 and 2042 are reported in Table 6.3, with the Hohmann trajectories included for comparison.

Figure 6.10 provides plots similar to those of Figure 6.7 and Figure 6.8, but for elliptical planetary orbits. The opposition of 2035 is perihelic, so the situation is very favorable. In these conditions the minimum ΔV is about the least that is possible and the transfer is fast. There is no advantage in using a particularly slow Type II transfer. As stated earlier, these very low minimums associated with perihelic oppositions occur every 15–17 years. They are not all equal, a pattern approximately repeats with a period of about 79 years. The next perihelic oppositions will occur in 2018, 2035 and 2050. The 2003 opposition provided the closest approach between the two planets in almost 60,000 years. There will be an even closer approach in 2287.

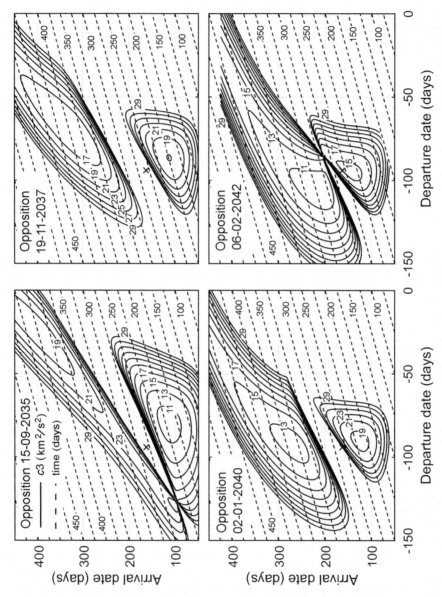

Figure 6.9 Pork-chop plots for the 3035, 2037, 2040 and 2042 launch opportunities. The departure and arrival dates are relative to the day of the opposition.

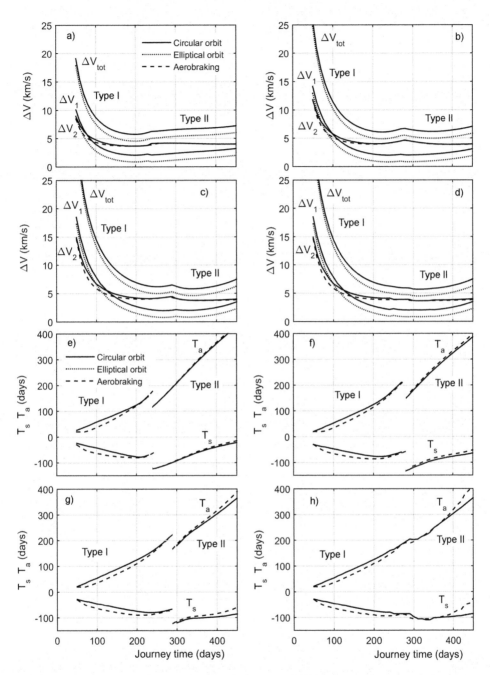

Figure 6.10 Optimal ΔV and dates for starting and ending interplanetary transfers between circular orbits at an altitude of 300 km and between a circular and a highly elliptical orbit. (a & e) 2035 launch opportunity. (b & f) 2037 launch opportunity. (c & g) 2040 launch opportunity. (d & h) 2042 launch opportunity.

Table 6.3 A comparison between the optimal Type I and Type II interplanetary transfers for the launch opportunities of 2035, 2037, 2040 and 2042, along with the Hohmann trajectories.

		2035	2037	2040	2042
Opposition date	Hohmann	Optimal type I			
C_3 (km²/s²)	8.67	10.28	16.96	18.33	12.28
T_s (days)	−96	−80	−86	−90	−95
T_a(days)	163	114	112	124	142
T (days)	259	194	198	214	237
V_∞(km/s)	2.945	3.206	4.118	4.281	3.504
ΔV_1 (km/s)	3.550	3.671	4.022	4.135	3.925
ΔV_2 (km/s)	2.070	2.076	2.215	2.457	2.425
		Optimal type II			
C_3 (km²/s²)	—	17.78	15.01	12.28	9.89
T_s (days)	—	−24	−69	−93	−106
T_a(days)	—	402	328	267	206
T (days)	—	426	397	360	321
V_∞(km/s)	—	4.217	3.874	3.504	3.145
ΔV_1 (km/s)	—	3.998	3.891	3.771	3.649
ΔV_2 (km/s)	—	3.021	2.485	2.112	2.020

For oppositions like 2042, the energy requirements of Type II trajectories are more convenient than Type I trajectories and the minimum energy is attained by using very slow transfers.

The trajectories discussed thus far can be generally defined as being direct trajectories. However, in the past many robotic planetary missions, and in particular those bound for the outer solar system, made ample use of gravity assists in order to gain speed and achieve a deflection of their trajectory. This requires great navigational precision, but nowadays such maneuvering is routine. The main reason that this is seldom considered for humans is that it implies long mission durations. The only reason for contemplating a gravity assist maneuver during a human Mars mission would be to facilitate departure outside of the normal launch window. This situation may arise for three different reasons:

- To achieve a flyby.
- To realize a short stay mission, where the time spent on the surface of the planet is much shorter than the 500–600 days of a long stay mission.
- To send materials to Mars outside of a launch window in an emergency.

In all cases the duration of the interplanetary travel time is longer than that of a typical direct mission and the energy expenditure is greater.

The third planet in a gravity assist trajectory for a mission to Mars is Venus. Its flyby may sometimes be made during the return journey (as with typical short stay missions) but very rarely during the outbound leg; an emergency launch of materials outside of a launch window requires an outbound flyby. A good reason to incorporate a flyby of Venus into the return leg of a short stay mission, rather than the outbound leg, is that a long period spent in microgravity traveling to Mars will commit the crew to a longer period of recovery on the planet, which will be worse than being compelled to remain inactive for a long time on Earth after the mission.

Another class of trajectories are the so-called free return trajectories, i.e. those which can lead back to Earth if the Mars orbit insertion maneuver fails or if an emergency situation rules out a landing on Mars. Something of this kind happened to the Apollo 13 crew during a lunar mission. In their case the spacecraft was approaching the point at which it would brake into lunar orbit, so instead it flew around the far side of the Moon on a trajectory that sent it back to Earth.

In the case of a mission to Mars, there are two free return options. One is a trajectory that is in resonance with Earth's orbit, for instance a heliocentric orbit whose period is two years and passes close to Mars; if it is not corrected in order to enter Mars orbit, it will lead back to Earth exactly 2 years after leaving it. The alternative is a trajectory which can be modified by a gravity assist at Mars that will deflect the spacecraft back to Earth. The Hohmann ellipse can be modified into a trajectory of this kind, but faster trajectories cannot. Free returns are also contemplated for Mars flyby missions (see Chapter 7) and often they can also include a Venus flyby.

A particular solution for Earth-Mars missions is a cycler. A cycler is essentially a space station which orbits continually between the two planets. People enter the cycler and goods are loaded aboard when it passes close to Earth, and they disembark when the cycler passes close to Mars – at which time the crew of a previous mission, who have been on the planet, enter the cycler for the return flight.

The main advantage of this scenario is that the mass of the cycler does not need to be accelerated at TMI and then again at TEI on every trip. As a long term facility, it will be able to offer a much more comfortable and safe voyage. A strong supporter of the cycler concept is astronaut Buzz Aldrin, who stated, "The Cycler system alters the philosophy behind a Mars program. It makes possible the dream of regular flights to the Red Planet and a permanent human presence there. That's the only way we'll ever succeed in taking mankind's next giant leap: a subway-in-the-sky between our planet and our future second home."[3] A more recent proposal, generally referred to as the Aldrin Mars Cycler, dispenses with a resonant orbit and instead uses a trajectory which utilizes multiple Earth and Mars gravity assist maneuvers to carry people back and forward between the two planets.[4]

6.4 MARS ORBIT INSERTION

At the end of the interplanetary trajectory, it is possible to either proceed with a direct entry into Mars' atmosphere or to enter into orbit around the planet. In some concepts, one part of the vehicle may plunge straight into the atmosphere while the other enters orbit. This was often done with robotic missions where the lander was released by the orbiter during the final phase of the interplanetary cruise. In some cases the orbiter then performed a deflection maneuver in order not to follow the probe into the atmosphere. In other cases the lander performed a maneuver to aim for the planet.

---

[3] B. Aldrin, *Popular Mechanics*, December 2005.

[4] D.F. Landau, J.M. Longuski and Buzz Aldrin, "Continuous Mars Habitation with a Limited Number of Cycler Vehicles" *JBIS*, vol. 60, pp. 122–128, 2007.

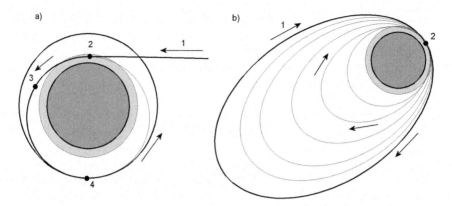

Figure 6.11 (a) Aerocapture. 1: Approaching hyperbolic trajectory. 2: Aerodynamic braking and insertion into a transfer orbit. 3: Jettisoning the heat shield. 4: Apoareion burn to circularize the orbit. (b) Aerobraking. 1: Initial elliptical orbit. 2: Subsequent atmospheric entry and braking to circularize the orbit. A final apoareion burn lifts the periareion out of the atmosphere.

Direct entry may be either passive or active. In the case of a passive entry the trajectory is adjusted prior to entering the atmosphere and then proceeds without further control. With an active entry, adjustments are performed in all phases of the flight. The entry maneuver begins at a range of 20,000–10,000 km from the surface, and then proceeds as follows:

- Deceleration starts at the outer edge of the atmosphere, at an altitude of about 131 km.
- After about 70 seconds, the heating rate reaches its maximum at 37 km.
- After about 10 seconds, the peak deceleration of 20–25 m/s^2 is reached.
- After about 80 seconds the parachutes start to deploy.
- The heat shield is jettisoned.
- After about 2 minutes the vehicle makes contact with the surface, either using air-bags or other devices to decrease the impact or, if the shock of an impact must be avoided, by using rockets to achieve a really soft landing.

Direct entry imposes a great deal of heating and very intense deceleration, so a heavy heat shield is needed in the front of the spacecraft and, in some cases, a smaller aft heat shield is also required in order to prevent the recirculating hot gases from reaching the spacecraft from that direction. The deceleration imposed by a direct entry may be too intense for a human crew, but would pose no problem for a cargo vehicle.

Rather than entering directly into the atmosphere, the spacecraft may enter orbit around Mars. If this is achieved by a propulsive maneuver then the ΔV is 1–3 km/s depending on the speed of the interplanetary trajectory and the parameters of the orbit.

There are three options for entering into orbit around Mars:

- All propulsive, where a propulsion system is used to brake the spacecraft.
- Aerocapture, where the vehicle enters Mars' atmosphere for intense atmospheric braking (see Figure 6.11a).
- Aerobraking, which is similar to aerocapture except that several passes inside the upper atmosphere complete the braking phase (see Figure 6.11b).

For an all-propulsive maneuver, the choice of the insertion orbit is essential. If a circular Low Mars Orbit (LMO) is chosen then the total value of ΔV_2 is about 2 km/s, although if a fast interplanetary trajectory is used it can be more. However, if an elongated orbit is employed, the value of ΔV_2 can be substantially reduced. One very interesting case is an elliptical orbit with a period of 1 sol. For an apoareion altitude of 33,970 km and a very low periareion of 250–300 km, the energy is quite high, and the ΔV of the maneuver to enter orbit is roughly half of the ΔV for LMO. However, the fact that the orbit has a high energy means the Mars Ascent Vehicle (MAV) must be able to launch from the surface with quite a high ΔV. Thus we see that the saving made when entering the initial orbit is spent when the lander subsequently lifts off from the planet's surface.

If the lift off from the surface is performed using propellant produced on Mars, then this has little effect on the IMLEO. Depending on how the In Situ Propellant Production (ISPP) system works, the quantity of materials to be transported from Earth may be the same as for a LMO, if the propellants are all locally produced. If something must be brought from Earth (usually hydrogen) there is a moderate increase in the quantity of materials to be sent from Earth. On the other hand, the decrease of the ΔV at arrival allows a significant saving in the IMLEO. There is also a matching advantage at TEI, since far less propellant is required to leave Mars orbit. The propellant for this maneuver must not only be brought from Earth, it must also be braked into Mars orbit. Therefore eliminating this can significantly reduce the IMLEO.

Adopting a highly elliptical Mars orbit makes planetary entry more difficult, because the vehicle has a higher energy. If the entry is by a propulsive maneuver, this will require more propellant. For an aerodynamic entry a heavier heat shield is required. A very elliptical Mars orbit therefore offers a good reduction of the IMLEO and therefore of the mission cost, so long as ISPP is used. In this case it may even be more convenient to perform a propulsive maneuver at arrival instead of aerobraking or aerocapture because it eliminates the need for the heat shield. The advantage of using a propulsive maneuver to enter into a highly elliptical orbit around Mars is shown in Figure 6.7 for the approximation of circular planetary orbits and in Figure 6.10 for the more realistic case of elliptical planetary orbits.

One possibility is to use several short burns rather than a single major one. First, a quite elongated orbit is obtained that has a periareion equal to the height of the final circular orbit; then, at successive periareions a small burn makes the orbit ever less elliptical until a circular LMO is achieved. This sequence is similar to that presented in Figure 6.11b for aerobraking.

In general, a propulsive capture and low orbit requires a great deal of propellant and is considered only for propulsion systems that provide a fairly high specific impulse. It can be achieved using NTP and is often considered for NEP or SEP, but in the latter two cases the vehicle will follow a sort of a spiral trajectory.

Aerocapture (Figure 6.11a) is undertaken by dipping deeply into the atmosphere of the planet, where the spacecraft is strongly braked by the drag and is inserted into an elliptical orbit that has its periareion inside the atmosphere. After emerging from the atmosphere the heat shield is jettisoned and remains in the elliptical capture orbit, where later braking passes either destroy it or cause it to fall onto the planet. Meanwhile, when the spacecraft reaches the apoareion of the capture orbit a small engine burn lifts the periareion out of the atmosphere, circularizing the orbit. The entire maneuver is completed in a relatively short time. Aerocapture requires a heat shield that can be more than half of the total mass at arrival. This method of capture demands very careful study of the aerodynamics of the vehicle and of the planetary atmosphere. At present, it seems to be suitable only for relatively small craft, with a maximum mass of about 30 t. Much investigation remains to be carried out for larger vehicles, and in particular for crewed ones.

Aerocapture is a risky maneuver because the vehicle must enter a narrow flight corridor. In addition, if the arrival velocity is high and the craft is either large or has a complex shape, then controlling its attitude during the atmospheric pass is difficult. Up to now, aerocapture has yet to be used, both because of the precision required and the stressing of the spacecraft.

Aerobraking has been used with small robotic spacecraft, but is generally not considered suitable for human missions because it is time consuming and the orbit must be adjusted after each pass. It starts immediately upon arrival with an engine burn that slows the vehicle just sufficiently to be captured by the planet. To reduce the amount of fuel required, a periareion burn is usually made at an altitude corresponding to the very thin upper layers of the planet's atmosphere. The highly elliptical orbit thereby achieved has its periareion close to the planet. Instead of a single burn, two burns may be made: the first in order to achieve capture and the second, at apoareion, to adjust the depth of penetration in the atmosphere and to make a final adjustment. At each periareion pass (usually referred to as an aeropass) the drag slows down the vehicle, progressively reducing the apoareion altitude (Figure 6.11b). When the correct apoareion altitude has been achieved, a burn lifts the periareion out of the atmosphere, in all likelihood to the same altitude as the apoareion in order to achieve a circular orbit.

This maneuver sequence may last several months, particularly if the elliptical capture orbit is so elongated that it has a period of a few sols and/or if the braking passes occur at a very high altitude in order not to subject the spacecraft to severe heating. In this situation, it may be possible to dispense with a heat shield and employ unprotected solar panels as drag devices. In the case of stronger braking, at each pass the solar panels may be secured to the vehicle's body to prevent problems. The Mars Reconnaissance Orbiter (MRO) was initially placed into a 300 × 45,000 km orbit and then gradually maneuvered into its final 250 × 316 km orbit over the next 6 months.

The density of the Martian atmosphere can vary, so the aerobraking timetable must be continually modified to account for fluctuations in atmospheric density. For instance, dust storms, which cause the atmosphere to swell, thereby increasing its thickness, can raise the density at the selected aerobraking altitude to a dangerous level. Once this is detected, the altitude of periareion must be increased to prevent damage.

As a general conclusion, the following recommendations may be made:

- Large vehicles generally have a high ballistic coefficient. Since atmospheric braking is easier for vehicles which have a low ballistic coefficient (see Section D.2), it may be best to use several small landers rather than a single large one. On the other hand, often the ratio of the mass of the heat shield to the payload mass increases for small vehicles. By decreasing the ballistic coefficient, the braking phase can be carried out at higher altitudes and impose a smaller deceleration peak.

- Integrate all modules into a simple shape which facilitates aerocapture (a conic shape for instance) and avoid modules attached to nodes that are located on the side of the vehicle (e.g. the manner in which the Orion spacecraft is attached to the interplanetary vehicle in the NASA reference mission).

These constraints seem to be manageable. As already noted, the selection of the orbit for Mars insertion is strictly dependent on the method of propulsion. If chemical propulsion is employed, then aerocapture must be chosen for all vehicles and the above recommendations must be taken into account. In addition, backup strategies must be considered if aerocapture fails, or if an abort becomes necessary (for instance if there is a problem with either attitude control or thermal protection).

At the very least, the backup strategy should be capable of aborting the aerocapture and adopting a free return trajectory. Alternatively it may be possible to utilize the propulsion of a lander to enter orbit around the planet. Other options depend on the availability of vehicles in orbit and on the surface. Whatever the situation, a reasonable recommendation is that a crewed vehicle should include a backup chemical propulsion stage for use in the event of a problem that occurs either prior to or during aerocapture or aerobraking. This propulsion stage would also serve as a means of avoiding hyperbolic trajectories that would cause a loss of the crew.

6.5   LOW THRUST INTERPLANETARY TRAJECTORIES

If either nuclear or solar electric propulsion is utilized, then thrust is applied during at least a good percentage of the entire journey. In contrast to the half-ellipse of a interplanetary transfer achieved by an impulsive maneuver, in this case the trajectory is no longer a conic, its shape more resembles an arc of a spiral. An important feature of low thrust propulsion is that the quantity of propellant required decreases monotonically as the duration of the interplanetary transfer increases. In contrast, an impulsive transfer requires a definite minimum amount of propellant and this requirement increases for longer travel times. This makes it possible to build slow cargo ships which can carry large payloads to Mars in a very economical way, albeit with a long travel time.

Whereas in the case of an impulsive interplanetary trajectory it is possible to identify a ΔV for each propulsive maneuver, and from the ΔV it is possible to compute the propellant required, the propellant consumed by a low thrust system depends on the precise trajectory and on the thrust profile.

It is possible to distinguish between systems that work with constant specific impulse or constant ejection velocity (CEV), and systems which work with variable specific impulse or variable ejection velocity (VEV). In the former, the thrust can be regulated by changing the power supplied by the generator to the thruster (to reduce the thrust, the power supplied to the thruster is reduced), while in the latter operations can be performed by maintaining a constant input power and regulating the specific impulse (to reduce the thrust the specific impulse is increased). When reducing the thrust, the propellant throughput q can be reduced more for a VEV system, so this will require a lower overall propellant mass than a CEV system. In both cases, the most important parameter to assess the performance of the propulsion system is the specific mass, which is the mass-to-power ratio α of the power generator.

Figure 6.12 illustrates the T-q (thrust-throughput) plot. On the left is the general case showing the two regulation modes: one for a constant ejection velocity (CEV) v_e and the other for a variable ejection velocity (VEV) at a constant power P. However, in real VEV thrusters the ejection velocity can be varied only between certain minimum and maximum

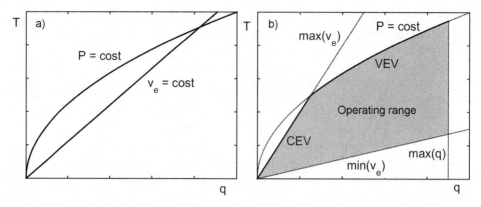

Figure 6.12 The thrust throughput plot. (a) The general case, with CEV (constant P) and VEV (constant ve) regulation. (b) A system in which the ejection velocity is limited between a minimum and a maximum value and the throughput is also limited.

values. This limits the actual variation in throughput. The regulation may be done at constant power only up to a certain value of the ejection velocity. If the ejection velocity cannot be increased further, to further decrease the thrust it is necessary to undertake the regulation at CEV by reducing the power (Figure 6.12b).

In both cases it is possible to define an objective function which must be minimized to optimize performances. In CEV systems this function is the characteristic velocity C (or J' in Eq. (C.25)) and is similar to ΔV; whereas for VEV the objective function is the performance index J [5] (Eq. (C.24)). The square root of J is directly proportional to the propellant that is required for the mission (see Section C).

Most scenarios for Mars missions based on NEP or SEP have assumed the employment of a CEV thruster, usually an ion thruster, and in most cases a value of the specific impulse of 10,000 seconds (i.e. an ejection velocity of about 100,000 m/s). A typical VEV thruster is the Variable Specific Impulse Magnetoplasma Rocket (VASIMR) that is now under development by the Ad Astra Corporation in the USA.[5]

The ideal VEV thruster will have no ejection velocity limitations. A plot of the square root of the performance index J for various Earth-Mars trajectories as a function of the time taken by the interplanetary voyage, between departing the sphere of influence of Earth and entering that of Mars, is shown in Figure 6.13. The figure refers to both SEP and NEP, and is based on the assumption of circular planetary orbits, so although it is a first approximation it applies to any launch opportunity. The basic parameter for all spacecraft propelled by electric thrusters is the specific mass of the generator α. In fact, if no limitation on the specific impulse is assumed then this is the only parameter that matters. Otherwise another important parameter is the maximum attainable value of the ejection velocity. The figure is based on the assumption that for NEP the power is held constant during the whole transfer (as already noted, in VEV thrusters the thrust is changed by varying the specific impulse and not the power) and that the specific mass of the power

---

[5] http://www.adastrarocket.com/aarc/VASIMR

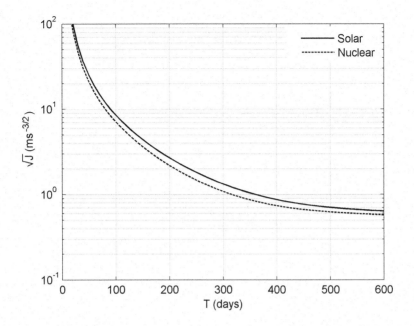

Figure 6.13 Values of $\sqrt{J}$ for an Earth-Mars transfer for both SEP and NEP as functions of the travel time. An ideal VEV has been assumed.

generator is constant. For solar power generation, the specific mass is assumed to be inversely proportional to the square of the distance from the Sun (because the energy in sunlight decreased with the square of the heliocentric distance). Its reference value is at 1 AU, which by definition is the average distance of Earth from the Sun.

The propellant consumption is moderately (less than 20 percent) higher for SEP because the specific mass of the generator increases with increasing distance from the Sun.

If α is low enough and the thruster is able to reach a specific impulse high enough, it is possible to make very short interplanetary transfers; much shorter than those obtainable using impulsive propulsion. With low values of α, transfer times of a few months are possible. With even lower values, which are not achievable today but may become possible in a more or less distant future, the interplanetary transfer may take less than a month.

Because the only important factor is the specific mass α, this must form the basis of the comparison between the SEP and NEP options.

Plots that can be named J-plots (similar to the pork-chop plot of Figure 6.6, i.e. those in which the performance index J is reported as a function of the departure and arrival dates) can also be obtained for VEV low thrust propulsion by either NEP or SEP. As in the case of the high thrust pork-chop plots, by assuming the planetary orbits to be circular it is possible to obtain the general case which applies to any launch opportunity. Alternatively taking into account the actual elliptical and non-coplanar planetary orbits will yield a plot for a specific launch opportunity.

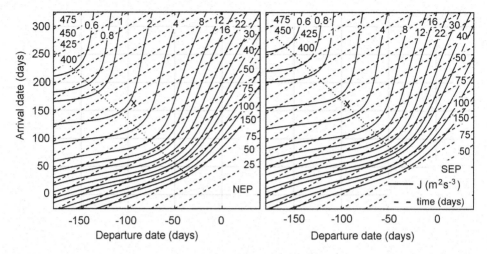

Figure 6.14 The J-plot is similar to the pork-chop plot of Figure 6.6, but for NEP and SEP. The Earth-Mars transfer assumes circular and coplanar planetary orbits. Dates are measured from the generic opposition. The × refers to the Hohmann transfer to provide a comparison with the low thrust trajectories.

Plots for circular planetary orbits are given in Figure 6.14. The lower right portion is for fast transfers with a duration as short as 75 days and a performance index J as high as 150 m²/s³. The upper left portion deals with slow transfer as long as 400 days with J as low as 0.6 m²/s³. Since the minimum is fairly flat for fast transfers, even a variation in the launch date of about 50 days makes little difference to J. The dotted line in the plot joins the points in which the minimum values of J for any value of the travel time occurs. If the scales of the axes were equal, the dashed lines would be parallel to the bisector of the first quadrant, whilst the line which connects the minima would cross the dashed lines at close to a right angle. If the surface is cut along this line, the square of the plot of Figure 6.13 is obtained.

J-plots similar to Figure 6.14 can be drawn for a specific launch opportunity by taking into account the actual elliptical and non-coplanar planetary orbits. Four NEP plots for the launch opportunities between 2035 and 2042 are shown in Figure 6.15. The difference between the cases is clear. A plot like Figure 6.13 but for the launch opportunities from 2035 to 2042 for the NEP case is presented in Figure 6.16. This was obtained by cutting the surfaces shown in Figure 6.15 along the dotted line.

If the planetocentric phases that depart from the orbit around the initial planet and achieve the final orbit around the destination planet are included, then a higher value of J is obtained. This depends on the radii of the departure and destination orbits. In this case the orbits are chosen with less freedom. For SEP, a moderate increase in the size of the orbits must be allowed for, in order to avoid the large drag on the solar panels that would make the orbit decay (or at least make it difficult to raise it). In the case of NEP, it may be necessary to increase the orbit radius in order to avoid switching on the reactor at low altitude.

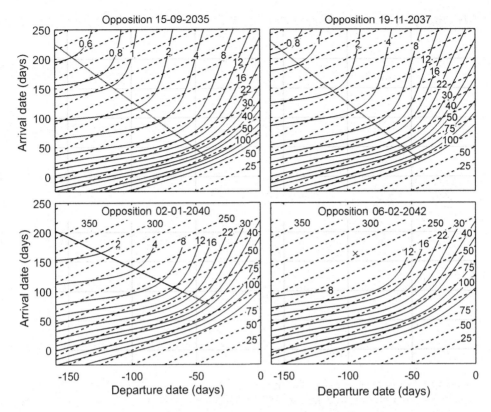

Figure 6.15 Figure similar to Figure 6.9 but for NEP.

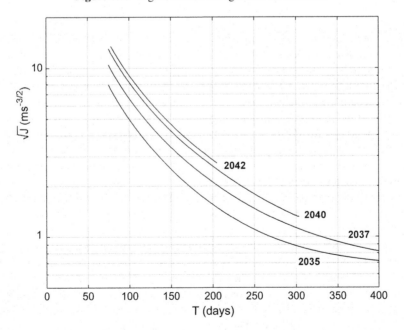

Figure 6.16 Values of $\sqrt{J}$ for an Earth-Mars transfer for NEP as a function of the travel time for the launch opportunities from 3035 to 2042. An ideal VEV has been assumed.

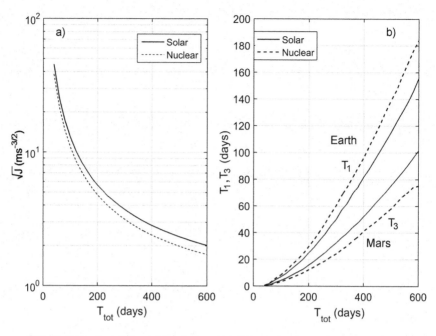

Figure 6.17 Optimized total Earth-Mars trajectories (with both planetary circular orbits at an altitude of 500 km) for SEP and NEP. It shows values of $\sqrt{J}_{tot}$ and the times required in order to escape from Earth and to achieve the final orbit at Mars in terms of the total travel time.

A fairly large circular or elliptical orbit about Mars may be chosen if the propellant that will be used to leave the surface of the planet is produced in situ. This would shorten the time spent spiraling out and reduce the propellant mass. The computation can be performed as in Section C.7. The plot in Figure 6.17a is similar to that of Figure 6.13, but takes into account the propellant that is spent in the planetocentric phases. In Figure 6.17b the duration of the planetocentric phases about Earth (T_1) and Mars (T_3) is shown in terms of the total journey time (T_{tot}). For SEP, the spacecraft must dwell more about Mars and less about Earth because it is expedient to decrease the thrust where there is less power and increase it when the power is more abundant (i.e. when closer to the Sun). Since this plot assumes the planetary orbits to be circular it is only a first approximation.

Consider a very fast spacecraft propelled by a NEP system which has a generator with a specific mass $\alpha = 1.2$ kg/kW $= 0.0012$ kg/W. At present this is unattainable but it would be a realistic expectation for a more or less far future. It also assumes a variable specific impulse thruster with an efficiency $\eta = 0.6$ and a maximum specific impulse $I_{smax} = 7{,}000$ seconds. Given the efficiency of the thruster, the effective specific mass $\alpha_e = 2$ kg/kW $= 0.002$ kg/W.

For the 2037 launch opportunity the vehicle travels to Mars in 105 days, with this time including the planetocentric phases for circular planetary orbits at an altitude of 500 km. By using the equations in Section C, the durations of the various phases are $T_1 = 9.9$ days, $T_2 = 91.1$ days, and $T_3 = 4.0$ days. The performance index is $J_{tot} = 101.059$ m²/s³ (subdivided in the various phases as $J_1 = 22.50$ m²/s³, $J_2 = 69.46$ m²/s³, and $J_3 = 9.10$ m²/s³). From

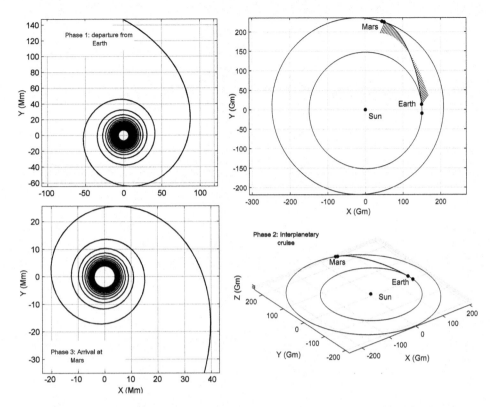

Figure 6.18 Trajectories for the three phases of the Earth-Mars journey. Note that each phase is with respect to a different frame.

this the propellant fraction of the spacecraft of $m_p/m_i = 0.450$ is easily obtained. The 2037 launch opportunity is a fairly favorable one. The dates of the outward leg are as follows. The vehicle initiates the maneuver in Earth orbit on September 18, 2037 and escapes Earth on September 28, 2037. It reaches the gravitational sphere of influence of Mars on December 28, 2037 and enters orbit on January 1, 2038. The trajectories of the three phases of the mission are given in Figure 6.18. The accelerations in the first and final phases are 7.27 mm/s^2 and 7.28 mm/s^2 respectively. Figure 6.19 displays the acceleration a, the specific impulse I_s, the propellant throughput $\dot{m}$, the power P, the thrust T, and the mass m of the spacecraft (the latter four all divided by the initial mass mi) as functions of time.

For an IMLEO m_i of 240 t, the payload is 72.6 t (minus the structural mass), the optimal generator mass is 59.4 t, and the total propellant mass is 108 t. The total power at the thruster output (since all losses are accounted for in αe) is 29.7 MW and the power of the generator is 49.5 MW. A single superheavy launch vehicle with a capacity of 130 t would be able to put the entire dry interplanetary spacecraft into LEO and a second such launcher could provide the propellant for the mission.

This example shows that if a value of the specific mass $\alpha_e = 2$ kg/kW can be attained, a journey to Mars in slightly more than 3 months is feasible. It is interesting that the

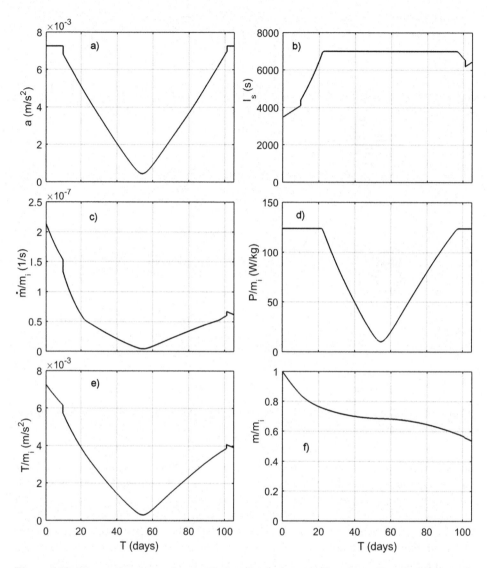

Figure 6.19 The acceleration a, the specific impulse I_s, the propellant throughput $\dot{m}$, the power P, the thrust T, and the mass m of the spacecraft (the latter four all divided by the initial mass) plotted as functions of time.

limitation of the specific impulse arising from the impossibility of achieving the very high values of the specific impulse that is theoretically required, doesn't significantly decrease the performance of the system, even though for most of its time the thruster will be working at constant specific impulse (CEV) instead of at constant power (VEV).

6.6 DESCENT VEHICLES AND EDL STRATEGIES

In a mission to the surface of Mars, the descent and the landing of goods and personnel is one of the most complex problems. As already stated, the entry, descent and landing (EDL) can be performed directly from the interplanetary cruise. Entry into the atmosphere at the high speeds of an interplanetary transfer is tricky and dangerous. It is manageable for lightweight robotic landers, but will be problematic for the larger landers envisaged for human missions.

The main issues derive from the low atmospheric density, which causes the velocity to remain high for a large portion of the entry phase and leaves most of the slowing down to be accomplished at low altitude. By leaving little time for the final part of the entry phase, this reduces the possibility of a safe touchdown at the required location. This is particularly true if the target is located at high elevation, where the atmosphere is even thinner. So if a landing on the floor of the Hellas basin is difficult, landing on the Tharsis Bulge or in the caldera at the summit of Olympus Mons may be impossible without using costly propulsive maneuvers in the final phase.

A descent can be made easier by starting in LMO, so as to perform atmospheric entry at the lower orbital speed. A small burn will be sufficient to deorbit the lander, which will enter the outer fringe of the atmosphere at hypersonic speed.[6] If the propulsive braking burn is able to achieve more than just deorbiting, the EDL becomes much simpler but at the cost of a non-negligible amount of propellant. If aerodynamic braking is employed, then a heat shield must be provided.

When the speed has reduced sufficiently, the first supersonic parachute is deployed. If the mass of this parachute is low then there will be an advantage to using it, otherwise it may be more convenient to use propulsion to perform all of the braking to subsonic speed. When the speed has been reduced below the speed of sound, the supersonic parachute is jettisoned (if there is more than one parachute) and a subsonic parachute may be deployed. This will slow the lander, but because the terminal velocity of a parachute is much larger in the thin Martian atmosphere it won't slow it as much as it would in Earth's atmosphere. And since the use of airbags or other similar devices is unsuitable for delicate objects such as those to be carried to Mars for human missions, and absolutely inadequate for crewed vehicles, there will be a final propulsive braking maneuver.

Some of the parameters to be considered are:

• The size and mass of the payloads to be landed: habitat, MAV, ISRU processing unit, surface power systems, surface vehicles, other items like workshops, greenhouses, etc.
• The ballistic coefficient of a landing vehicle.
• The initial parking orbit.
• The elevation of the landing zone.
• The guidance system to ensure that the cargo and habitat land close to each other and also close to the location assigned to the ensuing human expedition.

---

[6] The speed of sound in Mars' atmosphere is 244.2 m/s = 879.3 km/h close to the surface. It decreases to 212.9 m/s = 766.3 km/h at 30,000 m.

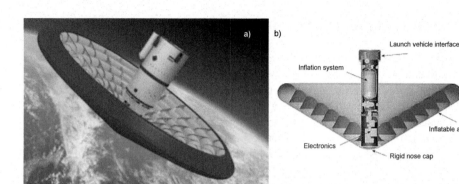

Figure 6.20 The Terrestrial HIAD Orbital Reentry (THOR) planned to be launched in September 2016. (NASA image)

There are several options for EDL. Rigid deployable heat shields or hypersonic inflatable atmospheric decelerators (HIAD) are promising systems.[7] At present, NASA is experimenting with small HIADs, like the Terrestrial HIAD Orbital Reentry (THOR) that is scheduled to be launched in 2016 (Figure 6.20). If the mass of the payload is of the order of 40 t, the HIAD appears to be the more mass-effective, but a large HIAD may present steering issues in the guided phase of the flight. Rigid heat shields (Figure 6.21) or smaller HIAD vehicles permit more control, but at the expense of size (the aerodynamic fairing of the launcher) or the mass of the payload. Because of the complexity of the physical models, the qualifications of EDL systems and procedures might be long and expensive, and impose a requirement to carry out several full scale tests in the Martian atmosphere. A solution may be found eventually, but at the expense of the complexity of the systems and high uncertainties of success in that phase of the mission.

A major problem with EDL is the variability and complexity of the physical models, making it difficult to estimate the probability of failure. The chances of success will typically increase with the number of successful full scale trials conducted in the Martian atmosphere. The risks can therefore be decreased beneath a desired threshold, but only slowly and with important implications for the cost and time spent on such testing. An important parameter of the maturity model is the decreasing rate of the probability of failure, which depends on the complexity of the EDL phase.

For simple EDL configurations, several tests at full scale might be sufficient, but more complex ones would require a long test campaign. If many heavy vehicles must be sent to Mars, the requirement to achieve an acceptable risk might render the project unsustainable and prompt its cancellation. As a consequence, EDL is the domain in which a high priority must be assigned to determining the simplest technologies and procedures. The results will impose strong constraints on other aspects of the mission.

---

[7] R.D. Braun, R.M. Manning, "Mars Entry, Descent and Landing Challenges," *Journal of Spacecraft and Rockets*, vol. 44 (2), 310–323, Mar-Apr, 2007; B. Steinfeldt, J. Theisinger, A. Korzun, I. Clark, M. Grant, and R. Braun, "High Mass Mars Entry, Descent, and Landing, Architecture Assessment," Proc. of the AIAA Space 2009 Conference and Exposition, AIAA 2009-6684, Pasadena, CA, pp. 14–17, September 2009.

Figure 6.21 Assembly of the biconical aeroshell of the Mars Science Laboratory. (NASA image)

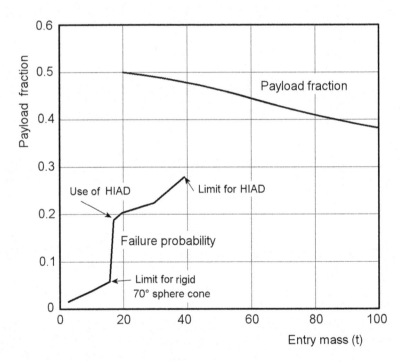

Figure 6.22 Upper line: Payload mass fraction as a function of the entry mass in tons. Lower line: Risk analysis for EDL systems before the first full scale test in the Martian atmosphere (estimated plot).

Parametric risk modeling may be crucial for EDL decision making. It will be essential to understand how complexity and risk increase with the mass, shape, and entry velocity of the landing vehicle.[8]

The problem is illustrated with an estimated result (the lower line) in Figure 6.22. Note that the figure is only qualitative. If the probability of failure is too high, the best strategy, if this were to be possible, might be to split the payload into several modules not heavier than 16 t and to use several small landers instead of a big one. For the same initial payload mass, several small landers might be lighter than a single heavy one (see the curve labeled payload fraction in Figure 6.22).[9]

This important issue is correlated with the Probability Of Failure (POF). If small landers are less risky, and overall lighter than a big one, there will be important implications for the architecture of the mission. Options in which important elements of the payload are sent down to the surface separately must be seriously considered and weighed against any complications that may arise.

---

[8] J.-M. Salotti and E. Suhir, "Manned missions to Mars: Minimizing risks of failure," *Acta Astronautica* vol. 93, pp. 148–161, 2014.

[9] J.A. Christian, K. Manyapu, G.W. Wells, J.M. Lafleur, A.M. Verges and R.D Braun, "Sizing of an Entry, Descent, and Landing System for Human Mars Exploration," *Journal of Spacecraft and Rockets*, vol. 45, No. 1, pp. 130–141, Jan–Feb, 2008; B. Steinfeldt, J. Theisinger, A. Korzun, I. Clark, M. Grant, R. Braun, "High Mass Mars Entry, Descent, and Landing Architecture Assessment," Proc. of the AIAA Space 2009 Conference and Exposition, AIAA 2009-6684, Pasadena, CA, 14-17 September 2009.

7

Mission design

A human mission to Mars is such a complex enterprise that there are countless possible design alternatives. In terms of a campaign involving a number of coordinated missions, the number of alternatives is even larger. As a further complication, the various design choices interact with each other and have an enormous effect on the affordability of the mission and on its possible timeframe.

7.1 MAIN DESIGN OPTIONS

A human mission to Mars can be envisaged in several different ways, and each of the possible design choices has deep implications for the goals, cost, complexity, safety, etc.

Some choices must be made right at the start. For instance, there is a general agreement that a one-off 'flag and footprint' expedition would be inappropriate. As a starting point, it is possible to conceive a set of three missions (with the prospect of follow-on missions) within a plan with various medium and long terms goals such as the establishment of an outpost on the planet as a precursor to colonization. The missions should be planned to maximize a mix of scientific returns and to apply the lessons learned to later and more complex missions. From this viewpoint a critical aspect is the planning of a timeframe and a roadmap, since to unduly stress long term goals might impose large delays, particularly if advanced technologies must first be created. On the other hand, relying on existing technology to start the campaign at the earliest possible time could disrupt the program by delaying the more advanced missions until the necessary technologies become available. The Apollo program provides a salutary lesson. Although the lunar missions were highly successful, the American public soon grew bored with seeing astronauts on the Moon, with the result that the program was terminated prematurely. And the "We've been there already" attitude became a severe obstacle to later efforts to initiate a return to the Moon.

During the past decade some mission choices have been resolved, but the list of those remaining open is still long. They involve the mission architecture as well as technical and operational issues.

© Springer International Publishing Switzerland 2017
G. Genta, *Next Stop Mars*, Springer Praxis Books, DOI 10.1007/978-3-319-44311-9_7

A sound proposal must aim to optimize not only the mission design itself, but also the program management principles and organization, with the clear goal of making the human Mars exploration program both appealing and affordable to the stakeholders and the political decision makers. In this context, thorough attention must be paid to all options, ranging from the most general to the most specific, and the influence of the corresponding choices upon the relevant program attributes and criteria, must be carefully considered.

Some of the options identified are:

- Type of mission, which is essentially a choice between conjunction or opposition (i.e. short or long stay).
- Crew size.
- Number of missions and landing sites.
- Interplanetary propulsion system.
- Mars orbit insertion.
- Descent vehicles and EDL strategy.
- ISRU options.
- Launcher to LEO strategy.
- Spacecraft architecture.
- Overall redundancy and multiple missions strategy.
- Preparatory missions and roadmap.

The two options that must be discussed first are the duration of the stay on the surface of the planet and the crew size. Amongst other points that will be discussed later in this chapter and in the following chapters, are the interplanetary propulsion system, the In Situ Propellant Production (ISPP), and the ascent strategy. All of these options not only influence the mission from many points of view that include (but are not limited to) human, technical and economic feasibility, they also deeply interact with each other.

The points mentioned above characterize the mission and the technologies that must be developed. A further set of decisions that must be taken are linked with the way a human Mars mission relates to other developments in the realm of space exploration and utilization. The creation of a space utilization market that ranges from space tourism to asteroid mining, from the creation of a lunar market to the development of orbital services, and so on, will have far reaching implications for how we plan Mars exploration in terms of its goals, its timeframe, and its affordability.

7.2 DURATION OF STAY

7.2.1 Flyby mission

The simplest flight plan would not include any stay on the surface of the planet, nor indeed entering orbit, the spacecraft would simple fly past it. Clearly, a Hohmann transfer cannot be used, because Earth would not be at the right place at the end of the return leg. Instead, a free return trajectory is required. A simple trajectory on a 2:1 orbital resonance (Sect. B) would give a total travel time equal to 2 years (730.5 days). However, much shorter journeys can be performed by using a gravity assist maneuver at Mars.

One such trajectory was first identified by Gaetano A. Crocco in 1956.[1] The basic idea is that a Mars flyby can cut the travel time to less than a year, at the cost of a high altitude pass (more than 1.5 million km) over the planet, but this range is not conducive to attaining good scientific returns. Crocco also found that a Mars-Venus mission that made a much closer pass by the Red Planet is practicable. He calculated that for the launch window of 1971 the flight times would have been 113 days from Earth to Mars, 154 from Mars to Venus, and 98 days from Venus back to Earth. Note that the total of 365 days is almost exactly one year. Clearly, this trajectory is not a minimum energy one, but the fact that no thrust was required at Mars in order to get back to Earth made it attractive.

The flyby mission designed by the Mars Inspiration Foundation was based on the 2018 launch window, which will be very favorable, and contemplated a total mission duration of 501 days (16.7 months).

A flyby human mission requires much less energy than a mission that includes landing on Mars, or on one of its satellites, or even a period in orbit. It can be undertaken using chemical propulsion without any difficulty, and present technology is fully adequate. However, it is of little interest and is seldom considered, except as the very first mission that is meant to test all the systems for a subsequent mission. After all, even in this case a flyby is not considered as a very interesting choice, because the descent, ascent, and Mars exploration devices are unable to be tested, and there is no real need to carry a crew to test the other machinery.

7.2.2 Missions to Mars orbit

Missions in which a crew spends time in orbit around Mars, or visits one of its satellites, have been proposed many times; even very recently. These plans are motivated by the presumed dangers at the surface of the planet, a consideration that is now considered as outdated. The advantages of robotic exploration controlled by human teleoperators who are in orbit of the planet, either in a spacecraft or on one of the moons, are:

- Reducing the dangers of forward (and backward) contamination.
- Reducing the risks and complexities of EDL, although aerobraking is needed anyway, particularly when using chemical propulsion.
- Reducing the risks and complexities of ascent from Mars.

The price to pay is:

- Reduced scientific efficiency and presumably the achievable yield.
- Almost no impact on future Mars colonization.
- Increased risks due to spending a longer time in space, including psychological risks and those due to radiation and microgravity.
- Additional cost in comparison to a flyby.

---

[1]G.A. Crocco, "Giro esplorativo di un anno Terra-Marte-Venere-Terra," Rendiconti del VII Congresso Internazionale Astronautico, Roma, Sept. 1956, pp. 201–225; English translation: "One-Year Exploration Trip Earth-Mars-Venus-Earth," Gaetano A. Crocco, Rendiconti pp. 227–252.

A Mars orbital mission without a landing might be made in order to test equipment for future missions on the surface (the EDL and MAV apparatus can be tested by sending down teleoperated robots to get samples from the surface). But there are better ways to perform a full mission rehearsal, without sending humans as far as Mars. Like missions that include landing on the planet, orbital missions can be of either the short stay or the long stay type. As an alternative to entering orbit around Mars, the mission could visit the L1 and L2 Lagrange points of the Sun-Mars system, but these are quite far away from the planet and are seldom considered. Nevertheless, as a first human mission to Mars a mission to enter orbit has been proposed for the launch opportunity of 2035 in the Humans Orbiting Mars Symposium [31].

The satellites are of genuine scientific interest and recently a mission was proposed to one of them, possibly Phobos (Figure 1.12).

7.2.3   Short stay missions (often called opposition missions)

Upon reaching Mars, it is possible to stay there (on the surface, on a satellite, or in orbit) for about 45 days. The return leg of the journey is much longer than the outgoing one owing to the positions of the planets; the situation differs for each launch opportunity. In any case, the return journey passes inside the radius of Earth's solar orbit, and for some trajectories there may be a flyby of Venus. Typical short stay missions are characterized by a ratio between the time spent in space and the time spent in the vicinity of, or on the surface of Mars of about 17 to 18, with a significant fraction of the time in space being spent within 1 AU, where the radiations from the Sun become progressively more intense, although during that time the GCR will be less dangerous.

In terms of science, a short stay mission provides little useful time to accomplish the mission objectives because during most of the short time spent on the planet the crew are recovering from the effects of the outbound journey. And, of course, they are exposed to a higher radiation dose and microgravity effects during the longer transit.

Most of the early studies were based on short stay missions and often, as noted, did not include a landing on the planet because the environment on Mars was considered to be more dangerous than space. It was commonly believed that it was better for the crew to remain in orbit or on one of the satellites and teleoperate rovers and other exploration apparatus on the planet.

Figure 7.1 is an example of a short stay mission. The trajectories for both the outbound and return legs are for the 2033 launch opportunity (see Sect. B.6) and there is a flyby of Venus.

7.2.4   Long stay missions (often called conjunction missions)

As the time spent on Mars on short stay missions is too brief to carry out much scientific and exploration work, and remembering that the crew must re-adapt to a gravitational field after spending months in microgravity, an alternative type of mission is to be preferred. The next return opportunity allowing a short travel time occurs after a time spent on the planet of 15–16 months. In comparison to the previous case the return journey is faster, requiring a duration similar to that of the outbound leg, and it never ventures within 1 AU.

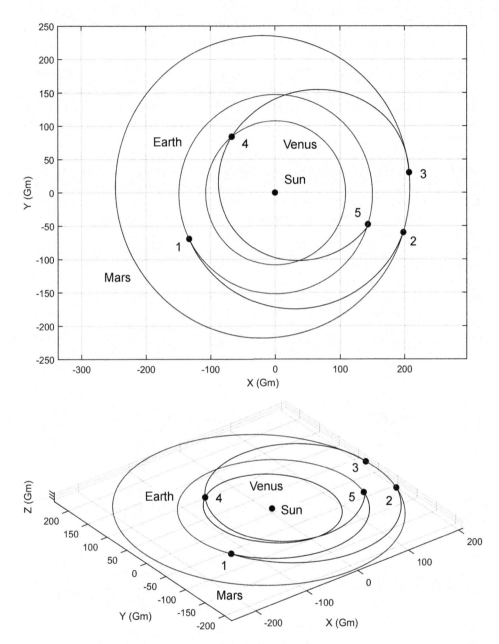

Figure 7.1 A short stay mission for the 2033 launch opportunity. 1: Departure from Earth (April 17, 2033). 2: Arrival at Mars (October 14, 2033). 3: Departure from Mars (November 23, 2033). 4: Venus flyby (May 12, 2034). 5: Arrival at Earth (September 4, 2034).

Furthermore, the time spent in space is comparable to the time spent on the planet. The propellant consumption is also lower because the trajectory of the return leg is closer to a Hohmann half-ellipse.

With the assumption that the planetary orbits are circular, making Hohmann transfers a possibility, Sect. B shows that a minimum energy mission must launch 96 days before Mars reaches opposition, the total time spent on Mars is 454.54 days, and the total mission duration is 972 days; i.e. 2 years, 7 months and 29 days. The ratio between the time spent on Mars and that spent in space is 0.878 (Figure 7.2).

If the outbound and return trajectories are faster, which is easily achieved with NEP, then the total mission time decreases somewhat but the time spent on the planet substantially increases (see Sect. B and Figure 7.3). The ratio between the time spent on Mars and that spent in space can increase substantially, easily reaching 2 and in some cases larger.

Average values for the time spent in space and on the planet on missions of the different types are given in Table 7.1; i.e. a short stay mission, a long stay minimum energy mission, and a long stay mission with a slightly larger energy. As usual, the actual numbers depend on the launch opportunity.

It is impossible to obtain an option in between short and long stay missions, since in order to decrease the total mission time at the expense of the stay on the planet requires a complicated trajectory, perhaps with a flyby of Venus and certainly a large expenditure of energy.

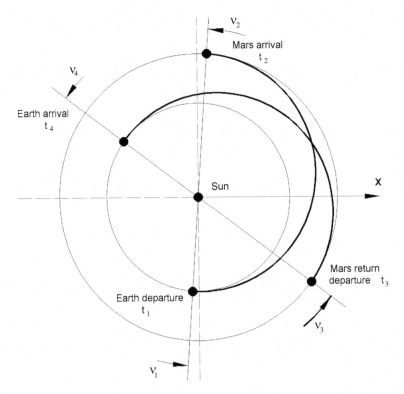

Figure 7.2 Departure and arrival positions of the planets for a mission performed using Hohmann trajectories. The x axis is defined by the position of the planets at the opposition.

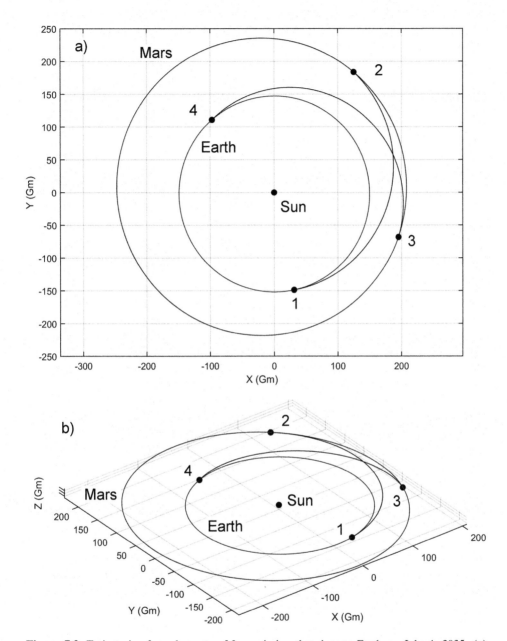

Figure 7.3 Trajectories for a long stay Mars mission that departs Earth on July 4, 2035. (**a**) Projection on the *xy* plane. (**b**) A three-dimensional plot. 1: Earth at departure. 2: Mars at arrival (December 31, 2035). 3: Mars at departure of the return leg (July 14, 2037). 4: Earth at arrival (end of mission, January 10, 2038).

Table 7.1 Time (in days and months) spent in space and on the surface of Mars for a flyby, a short stay, and two long stay missions (1: minimum energy; 2: fast)

	Flyby Days (months)	Short-stay Days (months)	Long-stay 1 Days (months)	Long-stay 2 Days (months)
Outbound	–	224 (7.5)	259 (8.6)	224 (7.5)
Stay	–	30 (1)	455 (15.2)	458 (15.3)
Inbound	–	291 (9.7)	259 (8.6)	237 (7.7)
Total	501 (16.7)	545 (18.2)	972 (32.4)	919 (30.6)
Tot. in space	501 (16.7)	515 (17.2)	518 (17.2)	461(15.4)
t*space*/t*Mars*	–	17.2	1.14	1

Only very high performance propulsion system (far beyond even the nuclear systems that can be envisaged within a medium timeframe) will permit us to circumvent the severe constraints imposed by the narrow windows of the launch opportunities.

7.2.5   Indefinite stay missions

Recently the realization that most of the difficulties involved in human missions to Mars are a result of the need for the crew to safely return to Earth, whilst colonization will usually mean one-way ventures, suggested the possibility of designing missions in which no return journey is planned, and the astronauts either spend their lives on the planet or wait for technological advancements that will allow cheap and fast interplanetary travel; for instance, when nuclear propulsion based on fusion is developed.

This option is usually presented as the only way to plan a human Mars mission for the near future, since traveling one-way to Mars is possible even using present technology. What is debatable is whether present technology could support an indefinitely long stay on the planet and whether this is feasible for a private non-profit organization such as Mars One, the only organization presently planning a mission of this type. Some of the criticism raised against the plans of Mars One was discussed in Sect. 1.7.

In fact, indefinite stay missions are not much easier than other scenarios, and require no less advanced technology than missions contemplating the return of astronauts. They just shift the difficulties from space travel to life support and planetary resources exploitation. It seems unlikely that a colony will be self-sufficient at the start, because the colonists will be heavily dependent on the supplies sent from Earth at each launch window, along with new colonists.

Whilst it is clear that sooner or later some people will decide to spend long periods, or even their lives, on Mars if true colonization of the planet is considered, it is questionable whether this will be able to be initiated with the first missions.

7.2.6   Low thrust missions

The above mission concepts are based on impulsive maneuvers typical of chemical or NTP, but NEP or even SEP low thrust trajectories (see Sect. C) will permit faster missions. If it proves possible to develop lightweight 100 MW NEP systems this will be particularly true.

A particularly interesting feature of low thrust missions is the prospect of devising very low energy missions which take a long time to reach their destination, even one year or more, because they will be able to be used to build true cargo ship, capable of ferrying to Mars the large quantity of supplies needed to establish an outpost or a true colony on the

Red Planet and also the propellants for the return journey of the fast ship used by the crew. These cargo ships do not need to return to Earth. They can be disposed of at Mars, particularly if they are low cost vehicles. With present technology, it is possible to build SEP spacecraft capable of performing this task, even without developing nuclear reactors.

On the other hand, a ship carrying only the crew together with the supplies required for the outbound (and later for the inbound) journey can cover the Earth-Mars distance in a short time, even without requiring a very advanced technology. An example of a two-way Earth-Mars journey is shown in Figure 7.4. This deals with an NEP spacecraft which in the 2035 launch opportunity can perform both the outbound and return trips in 175 days, of which 25 are spent in the Earth's gravitational sphere of influence and 10 in that of Mars. If the crew boards the spacecraft just before it leaves the Earth's sphere of influence and exits it in the same manner upon their return, the travel time is just 152 days. The spacecraft can leave a 500 km LEO on June 15, 2035, three months prior to the Mars opposition. It exits Earth's sphere of influence on July 9 to start the interplanetary cruise. It enters the Martian sphere of influence on November 26, then achieves a 500 km orbit around that planet on December 7. A suitable stay time is 620 days. The spacecraft starts to spiral out from Mars on August 28, 2037 and exits the planet's sphere of influence on September 7. After the cruise, it reaches the Earth's sphere of influence on January 25, 2038 and the crew disembarks. The ship then spirals in and achieves LEO on February 19. These trajectories are shown in Figure 7.4. A similar mission performed using a fast SEP is described in Sect. 13.5.

This split strategy performed using low thrust both permits carrying large quantities of cargo and sending people on a fairly fast trajectory in order to reduce radiation exposure. It is possible even with SEP, which is almost off-the-shelf technology.

7.2.7   Other mission configurations

As stated in the previous chapter, other mission schemes have been suggested, such as those with cyclers, large spacecraft traveling on a trajectory tangent to the solar orbits of both Earth and Mars and timed to reach those tangential points at the moment the planet is in the correct position for a rendezvous. With significant gravity assist maneuvers at both planets, a cycler can maintain such conditions for a very long time.

The outbound crew can readily reach the cycler when it is passing Earth, and remain on it until it is passing Mars, at which time a returning crew can enter the cycler to return to Earth, and so on. The cycler could also carry cargo, but this would mean accelerating the cargo to the velocity of the cycler and subsequently braking it into Mars orbit, which may require larger quantities of propellant. Cargo is therefore more likely to be assigned to NEP or SEP ships.

The shuttlecraft that ferries the outbound crew from Earth to the cycler and the inbound crew back to Earth must to be fueled on Earth, while the corresponding shuttlecraft carrying the homebound crew from Mars to the cycler and the Marsbound crew down to the surface of the planet can be fueled using propellant produced in situ. Of course, the two shuttlecraft will be entirely different. This approach requires carrying practically no fuel on the interplanetary journey, except for that needed to make the trajectory corrections of the cycler, and even this can be produced on Mars, whose gravitational well is shallower than that of Earth. And for a Mars shuttlecraft of the NTP kind that exploits NIMF, this architecture will be particularly efficient in terms of propellant.

However, although the long term prospects for a cycler are attractive, it is unlikely that they will be used for the initial human Mars missions.

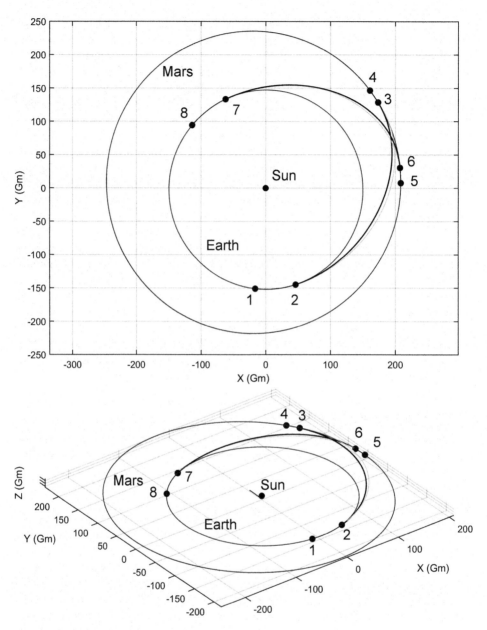

Figure 7.4 Two-way Mars mission performed by a passenger NEP ship in the 2035 launch opportunity.

7.3   NUMBER OF MISSIONS AND LANDING SITES

There is a consensus that the effort and the cost for sending a human mission to Mars is not justified in the context of a single 'flag and footprint' mission. Not even a single long stay scientific mission makes much sense. The very high development cost can be justified only in the context of a number of missions that share most of the technologies and the design effort. As already noted, we should not speak of a Mars mission, but of a well-coordinated campaign of several missions.

A good starting point may be to make a commitment for at least three initial missions with the implicit intention to proceed to other missions if everything goes well. The first missions to Mars must therefore open a new era of exploration with the long term aim of transforming humankind into a spacefaring species.

The choice of the landing site does not significantly influence the mission design, and can be left to a later stage of the design. In the context of a multiple mission program, the point that is worth discussing is whether or not the landing site should be the same for all missions.

The NASA DRM 5 study [24] decided that a single landing site is the best choice if the ultimate goal of Mars exploration is to prepare for colonization. If this is the purpose, each mission will contribute to the construction of a single well equipped and redundant outpost to last a long time. In this case, the site can be chosen more on the basis of its accessibility and the possibility of developing it into a true colony, rather than simply on its intrinsic scientific interest. Of course scientists, and especially biologists, will want to have access to intriguing sites, but sites which might harbor life must be protected against forward contamination and hence it would be wrong to construct the outpost in the immediate vicinity. Although humans might briefly visit such sites, they are best explored by sending robots that are teleoperated by people either in orbit or in the surface outpost. Another consideration is the requirement to be able to extract water ice from the permafrost, and this argues against a site near the equator.

As discussed in Chap. 3, there are possible sites in Argyre Planitia and Hellas Planitia, with the former being located closer to some interesting features and the latter being somewhat lower and therefore at a slightly higher atmospheric pressure.

If the main goal of the program is scientific, then the optimal strategy will be to choose a different landing site for each mission, because that will facilitate deeper understanding of the geology of the planet and maximize the chances of making discoveries, including finding life. Among the possible landing sites for this strategy are Centauri Montes, Nili Fossae, and Arsia Mons because they feature relics of the Noachian, Hesperian and Amazonian epochs.

The choice between these two strategies must be made very early in the mission design, since the design of the habitat, the rovers, and all the other equipment to be delivered to the surface of the planet is greatly influenced by the operational lifetimes incorporated into their designs. For instance, a habitat designed to last many years will cost more to develop and be heavier, but those factors will be justified if viewed as an investment in a long term effort.

A compromise between these two strategies is to land in three sites that are separated by moderate distances from each other and within the range of a pressurized rover which is left by the first mission. To expand their explorations, the follow-on crews would have to travel between sites with new equipment, so this option requires a relatively flat landscape across a the operating area. The downside of this strategy is the relatively small size of each outpost.

7.4   CREW SIZE

The number of people participating in the mission is one of the most important choices. The earliest concepts, which were developed without much regard to economic feasibility, assumed that there would be a large crew, as was the case for sea voyages in the age of sailing ships. It was expected that large spaceships would carry dozens (perhaps hundreds) of people to Mars. Later, it was realized that the cost, size, and complexity of such a mission are closely linked with the number of people on board and, at least for the first missions, this number should be kept to the minimum. Also, it was becoming evident that the increasing automation of many functions, both on the spacecraft and on the planet, would facilitate a reduction in the crew. Nevertheless, there is currently no agreement on what this minimum number is.

Crew size is a crucial factor because it affects the mass and volume of the spacecraft, the IMLEO, the capacity of the launch vehicles and the number of launches from Earth, the timescale for assembly in LEO, the interplanetary propulsion system, the EDL requirements (the options differ in TRL, complexity and risks), the feasibility of aerocapture, and the Earth return segment.

The mass of the deep space habitat is a function of crew size (see Table 7.2 [24]). If the crew is doubled from three to six, the habitat mass increases by a factor of 1.5, and this has a direct impact on many systems, complexity of the mission architecture, and above all the costs of development, launch and operation. The costs and risks suggest minimizing the number of astronauts, but the smaller the crew the narrower their range of skills. In order to compensate, candidates for a mission can cross-train to expand the cumulative know-how. And as noted, many tasks can be either automated or assisted by support teams on Earth. If these learning skills can be managed, then the task of optimizing crew size is a matter of minimizing risk.

For other mission systems, the '2-fault tolerance' that is the minimum requirement for human rated space systems, is recommended. In the 2004 ESA reference Mars mission, it is proposed to follow this principle and send a crew of three astronauts.[2] This number is seen as the absolute minimum. According to some physiology experts, three might be acceptable, provided they are properly selected and trained. However, from a human factors viewpoint three may be too few. A crew of six might be preferable to avoid psychological problems.

Table 7.2 Mass in tons of the habitable module as a function of crew size and mission duration

Duration	600 days	800 days	1,000 days
2 astronauts	19.7	22.2	24.9
3 astronauts	23.5	26.8	30.1
4 astronauts	27.1	31.1	35.2
6 astronauts	34.0	39.7	45.3

[2] L. Bessone and D. Vennemann, "CDF Study Report, Human Missions to Mars, Overall Architecture Assessment," CDF-20(A), February 2004.

A major reason for choosing a crew of three is that it would be very difficult to perform aerobraking and aerocapture with a vehicle carrying a larger number of people. In this case, it may be possible to send a pair of vehicles, each carrying three astronauts so that the two small crews would always be close to each other during the interplanetary voyage, and after landing would share the same habitat. This would also offer redundancy in a crisis, but it would increase the mass because (as shown in Table 7.2) the increase in mass of the habitat is less than linear with the crew size.

Whilst it is generally thought that the crew should include men and women, with a crew of three it will be impossible to have an even representation of the sexes.

7.5 INTERPLANETARY PROPULSION SYSTEMS

Over the years there has been a great deal of discussion about the optimum propulsion system for interplanetary travel. There are essentially four alternatives: chemical, nuclear (either NTP or NEP), and solar electric (SEP). Other alternatives such as solar or magnetic sails have been suggested, but they will not be considered here.

7.5.1 Chemical propulsion

Chemical propulsion is the simplest option, primarily because it does not require development of systems which currently have a low TRL. It is favored by those who advocate launching a human Mars mission in the near (or even very near) future. However, the low performance of chemical propulsion compels the use of systems based on low TRL technologies for specific situations, such as entering orbit around Mars at the end of the interplanetary cruise.

There have been many proposals. Some are overly complex. Others rely on optimistic estimates of the size and mass of modules, and indeed of the mission as a whole. In addition, architectures that save IMLEO might not be sufficiently robust, or provide sufficient margins in key mission phases.

An important issue with chemical propulsion is the choice of propellants. The physical characteristics of some liquid oxidizers and fuels are reported in Table 7.3. Liquid oxygen, methane, and hydrogen are cryogenic liquids that must be kept at very low temperature. But oxygen and methane are much easier to store than hydrogen, which has a much lower boiling point and a very low density requiring the use of very large tanks. A fully storable propellant combination is nitrogen tetroxide (NTO) and hydrazine; with the latter supplied as hydrazine, monomethylhydrazine (MMH) or unsymmetric dimethylhydrazine (UDMH), or possibly a mixture of them. Aerozine-50 is 50:50 mixture of UDMH and hydrazine. U25 is 25 % hydrazine hydrate and 75 % UDMH. The nitrogen tetroxide and hydrazine combination offers the advantage of being hypergolic, meaning they spontaneously ignite on coming into contact, thereby eliminating the need for an ignition system. Liquid nitrogen and liquid argon are also reported because they can serve as propellant for electric rockets, and carbon dioxide, ammonia and water because they can be used as propellant with NIMF (Nuclear rocket using Indigenous Martian Fuel).

Table 7.3 Characteristics of oxidizers and fuels. Kerosene is a mixture of hydrocarbons. The formula reported here is an average. NTO is nitrogen tetroxide. MMH is Monomethylhydrazine. UDMH is Unsymmetric Dimethylhydrazine. Nitrogen and argon are reported because they can be used as propellants in electric rockets. Water, carbon dioxide and ammonia are reported because they can be used for NTP

	Formula	M (g/mol)	ρ (kg/m³)	T
Liquid Oxygen	O_2	32.00	1141	−183.0 °C (90.19 K)
NTO	N_2O_4	92.01	1,450	21.15 °C
Liquid Hydrogen	H_2	2.016	70.85	−252.9 °C (20.28 K)
Liquid Methane	CH_4	16.04	423	−161.6 °C (111.7 K)
Kerosene (RP-1)	$C_nH_{1.953n}$	175	820	177–274 °C
Hydrazine	N_2H_4	32.05	1004	113.5 °C
MMH	CH_3NHNH_2	46.07	866	87.5 °C
UDMH	$(CH_3)_2NNH_2$	60.10	791	63.9 °C
Liquid nitrogen	N_2	28.01	807	−195.79 °C (77 K)
Liquid argon	Ar	39.95	1395	−185.8 °C (87.3 K)
Liquid CO_2	CO_2	44.01	1160	−78.5 °C (194.7 K)
Liquid ammonia	NH_3	17.03	730	−33.34 °C
Water	H_2O	18.02	1000	100 °C

Table 7.4 Mixture ratio, exhaust velocity, and tankage factor for some fuel/oxidizer combinations

Propellant	Mixture ratio	In vacuum		At see level	
		ve (m/s)	ve_{max} (m/s)	ve_{max} (m/s)	K
Monopropellant	–	2200			–
Solid	–	2850			–
Storable	1.08–2.10	3100	3375	2746	0.10
RP-1/LOX	2.29	3300	3512	2830	0.12
CH₄/LOX	2.77	3400	3660	2930	0.12
LH₂/LOX	5.00	4400	4600	3740	0.15

The performance of some combinations of oxidizer and fuel are reported in Table 7.4. Generally speaking, fuels contain either carbon or hydrogen (or indeed both). Using oxygen as oxidizer, the combustion products are carbon dioxide (CO_2) and water (H_2O). In the former, the stoichiometric fuel-oxidizer mass ratio is about 1/2.66 while in the latter it is 1/8. However, specific impulse is proportional to the square root of the ratio between the temperature of the exhaust gases and their molecular mass, and therefore a fuel rich mixture is often used which optimizes the specific impulse.

The table lists the maximum values of the ejection velocity for a number of propellant combinations, both in vacuum and at atmospheric pressure. However, realistic values are often smaller than the maximum values, so the values used for a first approximation evaluation of the performance of a Mars spacecraft are also specified. Note that only the values in vacuum are relevant, since most of the thrust is exerted in this condition and the only exception, liftoff from Mars, takes place in a very low pressure atmosphere anyway. Different values of the specific impulse can be found in other sources, because different values of the mixture ratio are used and also because the value is affected by the precise shape of the engine nozzle.

In the same table, approximate values of constant K, usually referred to as the tankage factor (see Sect. B.7) are listed. The tankage factor is the ratio of the mass of the tanks, the engine and all related hardware, and the mass of the propellant taking into account the ullage. These are only rough approximations, though. For example, the assumption that the overall mass of the various parts of the system is proportional to the mass of the propellant is fairly rough.

The highest ejection velocity for commonly used chemical propellants is obtained using liquid oxygen (LOX) and liquid hydrogen (LH2). It has a specific impulse in vacuum of 455 s and the practical value of the ejection velocity $v_e = 4400$ m/s. But storing cryogenic propellants (and particularly hydrogen) in space for long periods poses many problems. The low density of liquid hydrogen and the need to use thermal insulation to limit the boil-off rate, reduces its sheer performance advantages, in particular for all phases of a Mars mission subsequent to TMI. The next burn, which is to enter Mars orbit, occurs several months later, and the vehicle will have been in space for almost 2 years before it heads home to Earth. The boil-off rate for liquid hydrogen is approximately 3 % per month in LEO, 1 % per month in interplanetary space, and 0.5 % per month in Mars orbit. In order to use liquid hydrogen to enter orbit around Mars (or indeed to head home) much research is needed to reduce the boil-off rate. It is likely that active cooling of the tank will be required, which will require additional power to be generated. And installing cryocoolers will further increase the tankage factor. The low density of LH2 increases not only the size and mass of the tank, but also those of piping, pumps and various accessories.

Another option is the LOX/kerosene (RP-1) combination. It has a maximum specific impulse in vacuum of 358 s and a v_e of 3300 m/s. It delivers less performance but may be more convenient because kerosene has a much higher density which eliminates the requirement for very large tanks. Furthermore, there is no boil-off problem using kerosene. Liquid oxygen is cryogenic and therefore has boil-off issues, but these are much less severe than for liquid hydrogen.

A fully storable and hypergolic propellant combination is nitrogen tetroxide/hydrazine. This has a maximum specific impulse in vacuum of 344 s and a v_e of 3100 m/s. The mixture ratio depends on the exact type of hydrazine used (N_2H_4, MMH, UDMH, or indeed a mixture of these). However, their performance is fairly similar. In past mission designs for human missions to Mars storable propellants were usually considered for all burns occurring after TMI, but nowadays there is a preference for cryogenic propellants on the presumption that zero-boil-off technology becomes available.

Of particular interest for a human mission to Mars is the LOX/methane combination. It has a specific impulse of 370 s and a v_e of 3400 m/s. Both methane and oxygen can easily be produced on Mars by using atmospheric carbon dioxide and hydrogen carried from Earth, or even completely locally when using water extracted from the Martian surface. This option is attractive for the Mars Ascent Vehicle (MAV) when In Situ Resources Utilization (ISRU) is planned. If it is intended to carry the locally produced propellants into orbit, they can also be employed by the spacecraft which returns to Earth.

One problem might be the size of the tanks, particularly if carrying hydrogen from Earth, because by increasing the mass of the related heat shields and by making the craft harder to control, large tanks may complicate an aerobraking maneuver for entering orbit around Mars. Another possible issue is the ability of the turbopumps to start after a long time (in some cases more than 3 years) of inactivity on the surface of Mars or in orbit. To avoid the danger of the feed system malfunctioning, pressure-fed engines have been studied [24]. A pressure-fed engine is simpler than one fed by pumps. It is lighter and more reliable, but the tanks and the feed system hardware are heavier. And the pressure of the stored propellants is higher (about 17 bars compared to 3 bars) and therefore the tanks must be stronger, making them heavier. In addition, a larger mass of helium is required to pressurize the tanks. The specific impulse of a pressure-fed engine is usually lower too.

7.5.2   Nuclear thermal propulsion (NTP)

Nuclear thermal rockets were tested on the ground in the 1970s and shown to be practicable. Although this technology has a lower TRL than chemical rockets, its development is not so primitive as many people presume [23]. An issue facing their development is that testing is costly nowadays because direct testing in the atmosphere is no longer permitted. The only alternatives are testing in space (in a fairly high orbit; there is no agreement about how high but 800–1200 km seems to be a fair figure) or testing underground in a facility capable of capturing any fission product released by the rocket (Figure 10.4). Both will be quite costly. An underground facility of this type has been conceptualized by the Los Alamos Scientific Laboratory in the USA, and one has been declared available in Russia.

The simplest way forward would be to resume the studies that were discontinued in 1972, to produce a rocket that is a direct derivative of the Kiwi rocket developed during the NERVA project (Figure 7.5). It is often said that the specific impulse of that engine (about 850 s) was fairly low, and this is true in comparison to what might be achievable using more advanced nuclear thermal rockets, but it is almost twice the value from chemical rockets, and this would lead to quite a large saving of propellant (see Sect. B.7) which would provide a significant reduction of the IMLEO. A detailed comparison of nuclear and chemical propulsions systems was given in the report describing the 2009 NASA reference mission [24]. This demonstrates the clear advantage of an NTP system.

One factor playing against nuclear propulsion, is the fact that the use of a nuclear thruster will not be permitted in LEO. The starting orbit is therefore likely to be something like 800–1200 km and this will increase the cost of carrying all the required items into

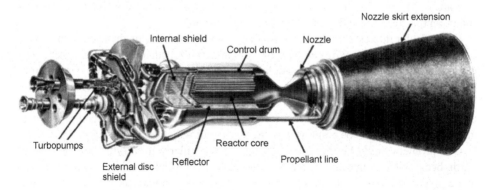

Figure 7.5 A nuclear thermal rocket was developed by the NERVA program. (NASA image)

Earth orbit; it may even erode much of what would otherwise be a reduction of the IMLEO. Another point is that the propellant for NTP is basically hydrogen, so it will face the same problems as the use of cryogenic chemicals.

Generally speaking, nuclear propulsion systems—whether thermal (NTP) or electric (NEP)—are essential technologies for the future of space exploration. Indeed, it can be said that if humankind intends to become a spacefaring civilization, then it must develop nuclear propulsion. In particular, if the first mission to Mars is considered as the start of a program, then the cost of developing NTP or NEP represents an investment which can be recovered during the follow-on missions to Mars and indeed to elsewhere in the solar system.

The main advantage of NTP, apart from a direct reduction of the IMLEO, is that it will eliminate dependence on technologies that do not yet have a sufficiently high TRL, such as aerocapture. One very important point is that nuclear propulsion may allow faster transit to and from Mars, thereby reducing the crew exposure time to radiation and microgravity. If the problem of radiation during the months spent in space proves to be a showstopper, and a fast transit is essential, then nuclear propulsion would become the only possible choice.

Nuclear propulsion allows designing a spacecraft capable of performing the round trip from Earth orbit out to Mars and back to Earth orbit, where (at least conceptually) it can be refurbished for another Mars mission. By executing multiple journeys, a single space-craft could act as a cost-effective interplanetary transportation system for so long as the nuclear reactor fuel lasted (and perhaps at that point the fuel could be replenished). Such a vehicle would not re-enter the Earth's atmosphere from a hyperbolic trajectory, it would maneuver into its high parking orbit. To reduce the dangers of back contamination, the crew will then spend a period in quarantine either on Earth after immediately returning there, or on a space station.

Hence even if chemical propulsion is affordable for a 'flag and footprint' mission, the much more complicated missions that will follow with the goal of constructing an outpost in preparation for eventual colonization will require nuclear propulsion. There is an important point to be stated here. Irrespective of whether chemical propulsion or nuclear propulsion is chosen, considerable research will be needed. In the first case, we need to develop and fully qualify aerobraking and/or aerocapture systems. But it should be noted that this investment is specific to Mars, and probably will be useful only for the early missions—those prior to the introduction of nuclear propulsion. The development of a nuclear engine is a true investment for the long term future of space exploration and exploitation.

As stated above, the specific impulse of nuclear engines based on the NERVA project is about 850 s. Although this is almost twice that of the best chemical propulsion, critics say it is not high enough to justify the research needed to develop this relatively new form of propulsion. But there is a flaw in this logic. Paradoxically, even if early nuclear and chemical engines had the same specific impulse there would be good reasons to develop NTP because nuclear engines of this kind have large margins for improvements whereas chemical rockets have attained their ultimate performance. In general, gaining experience of building nuclear systems and operating them in space is a major investment on the future.

Other points which must be considered in deciding whether to use chemical or nuclear propulsion are the timeframe for the mission to Mars (because the longer this timeframe, the more likely it is that nuclear propulsion will be available if it is recognized early on to be an indispensable technology) and the cost of launching payloads into Earth orbit (which might reduce by a factor of 2–2.5 if more private organizations enter the space business, developing space tourism or mining asteroids). The latter point will substantially reduce the cost of doing experiments in space to test advanced propulsion systems. Such considerations will not just permit a more informed choice between chemical and nuclear propulsion, but also between the NTP and NEP options.

A disadvantage of nuclear propulsion is that the best propellant is hydrogen, so nuclear rockets will have the same problems of long term storage of propellant as LOX/LH2 chemical rockets. In fact, because in this case all of the propellant will be hydrogen the issue could be rather worse owing to the greater size of the tanks. On the other hand, propulsive maneuvers will be used to enter and exit Mars orbit and there will not be any complications arising from coming into contact with the planet's atmosphere.

In the timeframe of Mars exploration, the concept of Nuclear rocket using Indigenous Martian Fuel (NIMF) is interesting. Any gas can be fed into a nuclear rocket, so it is possible to make direct use the Martian atmosphere, which is primarily carbon dioxide. But as already pointed out, in thermal rockets the specific impulse is inversely proportional to the molecular mass of the propellant. To use carbon dioxide rather than hydrogen will result in a significant decrease in the performance of the engine. However, other gases could be extracted from the atmosphere of Mars. There is argon but it is heavier than carbon dioxide. Both nitrogen and carbon monoxide are lighter. Even better would be to produce methane by mining water and then liberate the hydrogen from the water molecules in order to produce methane by using the carbon dioxide in the atmosphere. In fact, the water could be used directly as propellant in a nuclear thermal rocket. Although it is not easy to produce on Mars, ammonia could deliver a performance between that of water and methane.

The specific impulses computed by Zubrin[3] are given in Table 7.5 for a temperature of 2800 K and reduced by 5% for safety. The highest values are obtained using methane and then water. Note that methane, nitrogen and argon are all fully compatible with a reactor core designed for hydrogen, allowing both gases to be used. For instance, LH2 can be used for TMI and MOI and then in Mars orbit the tank can be filled by methane lifted from the planet to be used on the return journey. It is also true that hydrogen can be produced on Mars if local water is used, but lifting methane into orbit is much simpler than lifting cryogenic hydrogen.

Table 7.5 Ejection velocity characterizing nuclear thermal rockets of different types and using different propellants

Type	Propellant	ve (m/s)	Compatibility
NERVA	LH2	8300	–
NIMF	CO_2	2700	No
	H_2O	3500	No
	CH_4	5750	Yes
	N_2	2400	Yes
	Ar	1570	Yes
Project 242	LH2	27,000	–

---

[3] R. Zubrin, "Nuclear Rocket Using Indigenous Martian Fuel—NIMF," In NASA Lewis Research Center, *Nuclear Thermal Propulsion: A Joint NASA/DOE/DOD Workshop* pp. 197–216, 1991

Up to now it has been presumed that NTP means a system derived from the studies of the 1960s, and in particular a thruster derived from the Kiwi. Since those times, more advanced forms of NTP have been proposed, like liquid-core or gas-core reactors. These would yield a large improvement in performance, but they are in an early stage of development and there is no reasonable hope that they will become available for the early human Mars missions. It is also possible that perfecting them will be contingent upon obtaining substantial experience in operating simpler nuclear propulsion device in space.

The problems with fission NTR is that the propellant is heated by the core of the reactor, which must therefore be hotter than the propellant. This limits the maximum temperature of the latter and thus the specific impulse. As noted, designs where the core is liquid or even made by a gas have been conceived. Calculations suggest that by increasing the temperature of the core such reactors might yield values of specific impulse up to 10,000 s. However, these are speculative ideas and there is no serious research to develop the relevant technologies.

In the 1990s the Nobel laureate Carlo Rubbia proposed an innovative form of NTP. Some space agencies considered this proposal and the Italian Space Agency started Project 242 (the name comes from the Americium-242 that was suggested as a fuel). This idea is particularly well suited to human Mars missions,[4] and could be an alternative to the more classical NTP. The basic concept (which was proposed some decades ago) is based on the direct heating of a gas by fission fragments generated by a fissile material. As the ideal propellant is hydrogen, this design suffers from the aforementioned challenges of using a cryogenic propellant. But in this case the much higher specific impulse implies that for any given mission the quantity of stored propellant is smaller, which partially compensates for the size of the tanks needed to accommodate its very low density.

The initial design for a rocket based on this principle gave a specific impulse of about 2700 s and an ejection velocity of 27,000 m/s (roughly six times the best chemical propellants). The rocket was to supply a thrust of 3200 N, corresponding to a propulsion power of about 43 MW. Since the thermal power is 230 MW, the engine must dissipate a power of 190 MW and there are problems remaining to be solved concerning cooling and dealing with the high temperatures. The preliminary studies established that carbon-carbon composites are capable of doing the job.

A human mission to Mars with this engine was studied. It was a split mission with the two cargo vessels flying slow trajectories and the crew ship flying a fast one. Apart from the new engine, everything was off-the-shelf. For instance, medium launchers such as Ariane 5 and Delta IV were assumed for launch into LEO, there was no aerobraking and no ISRU. It was calculated that 20 launches from Earth would be required to lift a total IMLEO of 534 t. This could be reduced to five launches if a superheavy rocket became available. The mission analysis assumed average conditions. Circular, coplanar planetary orbits were used so that the solution did not rely on a specific launch opportunity. The cargo vehicles travel on separate Hohmann trajectories of the type shown in Figure 7.2.

---

[4]M. Augelli, G.F. Bignami, G. Genta, "Fission Fragments Direct Heating for Space Propulsion – Programme Synthesis and Applications to Space Exploration," *Acta Astronautica* vol. 82, pp. 153–158, 2013.

In this case, the advantage is a much larger payload than that for a conventional mission. The outbound voyage of the passenger ship takes 150 days. It spends 41 days on the surface. The return trip takes 178 days. Hence the total LEO-to-LEO duration is 369 days. After slightly more than a year of refurbishment, the ship is ready for its next mission.

The project is currently on hold, but it could be resumed by designing and building a new large ground test facility and by making a flightworthy rocket engine. Although further studies will be needed to verify this point, it is likely that the very low radioactivity from the engine will allow ground testing without the complex testing rig required for standard NTP. And if so, then the screening for the crew is also likely to be much lighter when operating a spacecraft with this kind of engine in space.

A revolutionary improvement would come from the use of fusion, rather than fission, in nuclear rockets. It may seem rather farfetched to advocate using nuclear fusion in this context when we have yet to attain controlled nuclear fusion for power generation on Earth, but it is actually easier to control nuclear fusion in a rocket engine than in a power station. This would produce such a large increase in the specific impulse that the resulting very fast interplanetary travel would make going to Mars seem rather pedestrian. Fusion thermal rockets could turn the solar system into the backyard of the human species. However, the impossibility of predicting whether and when such a breakthrough will occur means that human Mars exploration should not be predicated on the availability of such an engine.

Finally, it should be pointed out that the most significant obstacle to NTP is not technical but political. The rationale is the same as that which holds back a major use of nuclear power generation on our planet. Demonstrating that nuclear propulsion is a safe and viable way to travel in space may have the beneficial reverse effect of overcoming the reservations of public opinion against the nuclear industry in any form.

7.5.3   Nuclear electric propulsion (NEP)

Electric propulsion systems are attractive because of their high specific impulse. The power required can be drawn from either a nuclear reactor or a large set of photovoltaic solar cells. This section will consider the former.

In essence any electric propulsion system consists of two parts. These can be developed independently of each other, and hence can proceed along independent lines. In fact, electric propulsion for small space vehicles is already a reality and several space agencies have used electric thrusters for deep space missions, usually powered by photovoltaic cells.

In terms of electric thrusters, the challenge for human space missions is not to develop completely new ideas but to assemble very large propulsive units. Neglecting electrothermal thrusters that have a specific impulse not much higher than that for NTP, there are basically two types of electric thrusters: electrostatic (ion) and electromagnetic (plasma) as shown in Figure 7.6. In ion thrusters, the ionized propellant is accelerated by electric fields up to a very high ejection velocity. This yields a specific impulse much higher than that obtainable from a thermal rocket. Specific impulses of up to 10,000 s have been achieved with ejection velocities of about 100,000 m/s.

Small ion engines for robotic missions have a good TRL and are commonly used. They usually have an input power of 1–7 kW and yield a specific impulse of 2000–5000 s with a thrust of 25–250 mN and an efficiency of 65–80 %.

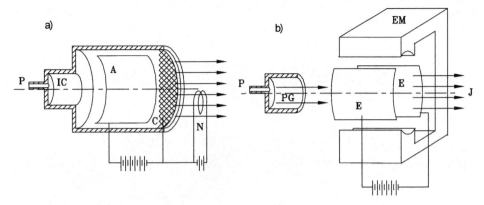

Figure 7.6 A sketch of electric thrusters for (**a**) an ion engine and (**b**) a plasma engine. Annotation A indicates the anode (focusing electrode), C the cathode (accelerating electrode), E an electrode, EM an electromagnet, IC the ionization chamber, J the jet, N a neutralizing electrode, P the propellant inlet, and PG a plasma generator.

It is a very different task to build a large engine suitable for human missions. Despite the fact that ion engines are often mentioned in proposals for Mars missions, usually delivering a specific impulse of 10,000 s, the TRL of these engines is still quite low. Above all, the very high specific impulse implies that the power to deliver the fairly large thrust required for a human mission is large. Although the propellant mass decreases as the specific impulse is increased, the mass of the power generator increases. There is therefore an optimum value for the specific impulse for any given mission. If generators with quite a low specific mass are not available, then the specific impulse of ion thrusters is generally too high for human missions

The commonly considered propellants for an ion thruster are xenon or cesium, but it is also possible to use iodine or bismuth. There are various types of ion thruster, in particular gridded electrostatic ion thrusters, Hall-effect thrusters, and field-emission electric thrusters.

A plasma thruster accelerates an ionized gas using magnetic fields and the Lorents force. The advantage of a plasma thruster is that it can deliver a large thrust and vary the specific impulse; e.g. the VASIMR (Variable Specific Impulse Magnetoplasma Rocket) engine of the Ad Astra Corporation (Figure 7.7). The development of such engines may make NEP a reality for large spacecraft. Units rated at 200 kW have been tested on the ground. They can deliver thrusts up to 5 N with a specific impulse which is variable between 1000 and 5000 s, and the efficiency between the DC supply and the jet is 60 %. These are quite attractive values. Future work may push the specific impulse to 30,000 s and also improve the efficiency.

If the planned experimental application of such an engine to reboost the orbit of the ISS proves successful, larger units might become available for human Mars missions. However, an interplanetary vehicle will draw a power in the MW range, and much larger units must be developed. An alternative is to use an array of small thrusters. For example, a cluster of thirty 200 kW engines will draw 6 MW, which is sufficient for a fairly fast transfer between Earth and Mars. Having a number of small units offers the advantage of enhancing

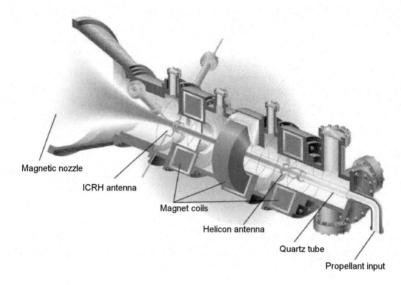

Magnetic nozzle

ICRH antenna

Magnet coils

Helicon antenna

Quartz tube

Propellant input

Figure 7.7 A section of a VASIMR engine.

redundancy and minimizing the impact of losing a single engine. But for a given power, large engines have a smaller mass-to-power ratio and an array of small engines may be much heavier than a single engine.

Many advantages derive from being able to vary the specific impulse. They allow the best specific impulse to be applied in each phase of the mission. It is also possible to control the thrust by changing the specific impulse at a constant power, rather than by regulating the power, thus enabling the power generator to operate under ideal conditions (see Sect. C).

As propellant, plasma thrusters can employ hydrogen, deuterium, helium, nitrogen, argon, and xenon. Hydrogen offers the bonus that its tank can provide very good radiation shielding for the crew compartment but it poses the same drawbacks as for chemical and nuclear thermal propulsion, namely the need for large tanks, its low storage temperature, and boil-off. In the experiments conducted on the ground, argon was mostly used since it is considerably easier to store, it has a much higher boiling temperature, and it requires much smaller tanks (see Table 7.3).

Argon and nitrogen can readily be produced from Mars' atmosphere. Once the Martian 'air' is compressed to produce oxygen and methane, extracting argon and nitrogen does not pose a problem. Interestingly, the percentage of argon in the atmosphere of Mars is higher than in Earth's atmosphere. The possibility of using propellant produced on Mars and lifted into orbit around the planet to enable the mission to return to Earth has cost implications. If both nitrogen and argon were taken from the atmosphere, this would increase the ISRU yield. The argon could serve as propellant and the nitrogen as an inert gas for the air that the crew breathe.

A problem with all types of low thrust engines is that they must operate for long times, possibly for the entire voyage, so their reliability may be problematic. As stated in Sect. C, the key parameter is the mass-to-power ratio of the generator, usually designated as α. It is even more important than the specific impulse. To obtain good performance, it is vital for α to be as low as possible, both in the case of constant specific impulse (CEV) and of constant power (VEV), as shown in Figure 6.12. In the latter case, when optimizing the thrust profile, the value of the thrust must be regulated by varying the specific impulse in order to allow the generator to operate at full power at all times; i.e. with the minimum value of α. In each phase of the voyage there is an optimal value of the specific impulse; it increases with decreasing α, hence only by decreasing the latter parameter it is possible to exploit the high specific impulse of electric thrusters. Moreover, with very low values of α, the thruster may be unable to reach a sufficiently high specific impulse. As a result, the more advanced is the generator, the more advanced must also be the thruster.

An alternative to generating electrical power using photovoltaic cells (as is done with many robotic interplanetary craft) the power can be drawn from a nuclear reactor. Nuclear generators to supply power in space have been built and tested in space, but not of sufficient size for interplanetary propulsion of a vehicle suitable for a human mission to Mars. Also, the units built thus far had a specific mass α much higher than would be appropriate. A research effort is needed in order to develop reactors that provide the desired performance.

Any achievement in this field will have major effects on space exploration since a nuclear reactor for a NEP Mars spacecraft is essentially similar to that which could be used for robotic exploration of the outer solar system on a scale far beyond that achieved to-date, and also for powering an outpost on the Moon or Mars.

The value of α of the earliest generators (e.g. the Russian Topaz tested in 1961) was of about 60 kg/kW (0.06 kg/W). A realistic value for systems that can be built at present lies in the range 20–30 kg/kW. Recent studies based on existing water-cooled reactors/steam-cycle technology point to a total specific mass (reactor, piping, turbine, and radiator) of 3 kg/kW.[5] Values down to 1 kg/kW have been suggested for systems based on graphene technology.[6] If such low values become practicable, then (as explained in Sect. C.9) very fast interplanetary spaceships that can reach Mars in only 3 or 4 months (perhaps less) will become possible.

7.5.4 Solar electric propulsion (SEP)

With electric propulsion, the duration of the voyage depends on a single parameter: the specific mass of the power generator. If a low value of α can be attained, low thrust interplanetary transfer times can be much shorter than for chemical or even nuclear thermal propulsion.

SEP may be achieved by using either photovoltaic cells or a thermal solar generator. The former currently seems to be the best choice, in particular for their specific mass, but in the future the latter may be best for very large plants.

For photovoltaic cells, the size and output of each array does not matter: what matters is the specific mass, in the sense that if the single array is quite large and has a low power, it is possible to use many arrays to obtain the required power. But there is a limit because a large number of arrays may result in a system that is difficult to deploy, is perhaps too delicate, or may require a heavy overall structure that takes the specific mass of the whole generator far above the specific mass of each individual array. The efficiency of solar

---

[5] J. Powell, G. Maise, J. Paniagua, "SUSEE - Ultra Light Nuclear Space Power Using the Steam Cycle," http://ieeexplore.ieee.org/stamp/stamp.jsp?arnumber=1235078&tag=1

[6] R. Lenard, "Impact of Advanced Technologies on Nuclear Power and Propulsion Systems," 62nd IAC, Cape Town, October 2011.

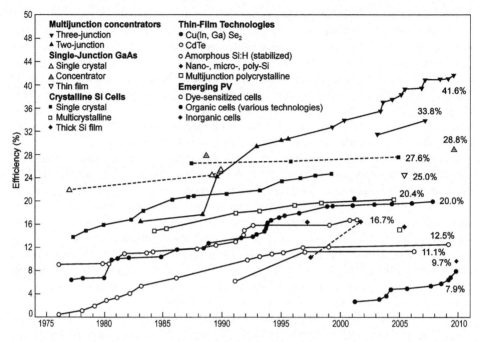

Figure 7.8 Efficiency of the best research photovoltaic cells of the various types in the 35 years between 1975 and 2010.

transducer cells has greatly increased in recent years, rising from about 10 % to almost 42 % in the best research cells and to at least 25 % for many commercially available cells, as detailed in Figure 7.8.[7] However, although the power per unit area is proportional to the efficiency of the cells, the power per unit mass is proportional to the efficiency divided by the mass per unit area of the cell (or more accurately of the array carrying the cells) since what matters is the specific mass of the entire generator. Although lighter cells, such as thin film cells, offer an efficiency of only 22 %, they may ultimately be more convenient than the rather more efficient triple junction cells.

Each of the eight solar arrays of the ISS was initially rated at 31 kW, with a decay in performance to 26 kW (16 %) in 15 years, and had an initial efficiency of 14 %. Each solar array module, including a part of the main truss, the panels, the radiator, and the Integrated Equipment Assembly (which is 7711 kg by itself) had an IMLEO of 15,824 kg. These figures give a very large value of the specific mass of $\alpha = 255$ kg/kW. However, this is for a space station power generator, where α has little importance. Much lower values can be expected today using more recent types of cells and designing the entire system for a minimal mass.

Assuming a solar constant of 1360 W/m² and arrays using thin film cells, then α_0 (the value of α at 1 AU) can be as low as 3.3 kg/kW.[8] This value is not only much smaller than

[7] http://sroeco.com/solar/images/PVeff-rev100414.png

[8] N.S. Fatemi, H.E. Pollard, H.Q. Hou, P.R. Sharps, "Solar Array Trades Between Very High-Efficiency Multi-Junction and Si Space Solar Cells," 28th IEEE PVSC, September 17–22, Anchorage, Alaska, 2000.

that mentioned above for the ISS, it is also quite a low value in comparison to values achievable at present using nuclear generators. There is another important parameter for a solar generator: the mass per unit area. A value of 0.0033 m^2/W (which is achievable today) translates to an efficiency of the array of 22 %.

Another point is that the efficiency of the cells increases as the temperature decreases, but it decreases with age. The former causes the power output to increase as a vehicle moves away from the Sun, such as from Earth to Mars (or more accurately, to decrease less than would be predicted by assuming constant efficiency). Also, the efficiency in Mars orbit is higher than in LEO. The second effect causes a decrease of the power available in time, particularly on long missions, such as one to Mars. If the spacecraft is reusable after the journey to Mars, it may be necessary to substitute the solar panels as part of refurbishing the ship for its next journey.

Research in this field is ongoing, and it is possible that lower values of the specific mass will be achieved in the near future.

An issue specific to SEP is that α is not constant, it increases as the vehicle moves away from the Sun since the power from the solar array or the thermal solar generator decreases. This decrease can be presumed to occur with the square of the heliocentric distance, even if this is not strictly correct because the temperature of the array reduces with increasing distance from the Sun and this increases its efficiency. The deceleration phase at Mars and the later departure to head home are therefore critical phases of the mission. However, it is possible to show that the effect is not so large across the whole journey, since the maneuvers around the Earth are more energy consuming than those at Mars. A solar generator requires a specific mass that is 15–20 % lower than a nuclear generator in order to provide the same performance.[9]

SEP, particularly if based on photovoltaic cells, is interesting for the following reasons:

- Solar panels are very cheap, simple, and reliable.
- SEP has already been used for small interplanetary probes (e.g. Deep Space 1 , Dawn, and Hayabusa). Its TRL is therefore higher than that for nuclear power.
- Photovoltaic technology has been improved since its first use in space. In terms of specific power, very lightweight and efficient solar panels can make it competitive against nuclear power.
- The use of high-power SEP makes a mission architecture significantly less sensitive to growth in mass and it improves flexibility.

It is often said that SEP may be used to build slow and efficient cargo ships to transfer large payloads from Earth to Mars with a relatively small IMLEO, but it is inadequate for fast ships carrying humans. Whilst it is true that slow SEP cargo ships could be built in a reasonable time and could greatly reduce the cost of preparing a human mission to Mars, the second clause of the statement does not stand up to close inspection. The same generator and array of thrusters used for the cargo could be used to build a fast ship by substituting propellant for the payload of the cargo ship. Moreover, although it may take up to a month to escape from Earth's gravitational sphere of influence, the crew could board shortly before this was achieved, thereby shortening their travel time by a significant fraction of the overall duration.

---

[9]G. Genta, P.F. Maffione, "Optimal Low-Thrust Trajectories for Nuclear and Solar Electric Propulsion," *Acta Astronautica* vol. 118, pp. 251–261, 2016.

Aerocapture may be impracticable for a ship equipped with giant and fragile solar panels. This may also be true for the large radiator of a NEP ship. But in both cases, the high specific impulse will make it unnecessary to resort to aerocapture; the ship will achieve low Mars orbit using its own engines.

Finally, at Earth the starting orbit of a SEP spacecraft may be lower than that of a NEP ship because in the latter case safety issues may require the use of a higher orbit. However, the large aerodynamic drag of the solar panels will tend to counter what at first appears to be an advantage in terms of IMLEO.

All in all, SEP may be a very interesting solution for early human Mars missions, before the introduction of lightweight nuclear generators, but in the long term its intrinsic limitations (arising from the fact that solar energy is very diluted) will make NEP much more expedient. Nevertheless, while SEP is facilitating missions to Mars, it will be thoroughly testing electric thrusters that will eventually be employed by NEP.

SEP can also be used to build space tugs to reboost the components of the interplanetary vehicle that are lifted into LEO during the assembly process, and subsequently to raise the orbit of a NEP or NTP spacecraft to the altitude mandated for firing nuclear propulsion. This may reduce the disadvantage of nuclear propulsion arising from the reservations against using it in LEO. The tug could be a low power device because the orbit-raising maneuvers can progress slowly. The crew could board their interplanetary ship shortly before it leaves Earth's sphere of influence.

A particularly expedient solution is to use a set of solar panels to power the thrusters of a NEP ship in LEO. Once a safe orbit has been attained, the solar array is jettisoned, together with an auxiliary tank that held the propellant for the orbit-raising process. At that point the nuclear reactor is started.

With solar power, it must be remembered that any maneuvering in LEO or in LMO will be hampered by the fact that the vehicle spends a substantial time within the shadow of the planet. Consequently, the orbit-raising and braking maneuvers will take longer than implied by the simple approach described in Sect. C. The actual trajectory must be computed by taking into account the details of the orbit, including its inclination.

7.6   INTERPLANETARY JOURNEY BACK TO EARTH

Basically, there are three alternatives for the return journey:

- In a vehicle that was placed into orbit around Mars, usually referred to as the Earth Return Vehicle (ERV).
- Returning in the same vehicle that was used for the outbound voyage.
- Launching directly from the surface of Mars for a direct return to Earth, possibly without an intermediate phase in LMO. This option is rarely considered, since it is unlikely that a vehicle large enough to accommodate the crew during the return journey could land on Mars and then lift off again.

There are also two alternatives for the final part of the return journey:

- Direct entry into the Earth's atmosphere, as in the case of Apollo.
- Entering orbit around Earth to allow the crew and their scientific samples to undergo a period of quarantine (possibly at a space station) prior to landing.

If chemical propulsion is used, the first option is usually considered for both situations but if high specific impulse propulsion is used, and particularly with electric propulsion, it is possible that a single spacecraft can do both journeys, with the returning interplanetary vehicle entering orbit around Earth to be refurbished for a new voyage.

For a mission which uses a special return vehicle, one of the main challenges, in terms of energy, is to send a habitat and a wet propulsion system from LEO to Mars orbit to be ready to bring the crew home. In addition, because re-entry in the Earth's atmosphere requires a heavy heat shield with a mass that is proportional to the size of the vehicle, it is usually assumed that there is a small capsule docked to the habitat and the crew will transfer to this on the final day of the mission in order to land. The total mass of the ERV cannot be less than 60 t (optimistic estimate). To transfer this from LEO to Mars orbit would require a heavy chemical propulsion system, and the total would greatly exceed the payload capability of a single heavy launcher.

In most chemical architectures the assembly of a large ERV in LEO is generally limited (or eliminated entirely) by other factors. Assuming that the launcher has a LEO capacity of 130 t (the maximum LEO payload of the NASA SLS Block II), then the largest payload for a direct TMI maneuver for chemical propulsion is 46 t and around 40 t for a fast interplanetary transit. Mass could be saved if the size of the crew was able to be reduced (bearing in mind the issues discussion in Sect. 7.4) and the use of aerocapture would help eliminate mass. In addition, two other strategies can contribute to reducing the IMLEO:

- Bringing more materials from the Martian surface and optimizing the list of elements that wait in Mars orbit. Zubrin has proposed a direct return from the Martian surface, but the mass of propellant to be produced on the planet would probably be too large.[10] Salotti suggested sending the main habitat back to Mars orbit and storing the resources for the voyage home in the small capsule.[11] Another option is to put a wet propulsion system into Mars orbit.
- Making the ERV as two modules that would leave LEO separately and be assembled in Mars orbit. If the modules are not too large this may not be complex. It is preferable to perform the assembly in Mars orbit rather than in LEO because it eliminates the need for a large propulsion system, which would also have to be assembled and kept in LEO with the help of reboost modules.

Of course, these strategies do not have to be implemented at the expense of the mass of the other vehicles, and particularly the landers. The optimization of the mission suggests that all modules and materials that are required for the return but are not needed on the surface of Mars should remain in orbit around the planet.

An interesting possibility for an architecture in which the same vehicle is to be used for both the outbound and the return legs and a slow cargo carrier is also sent to Mars, is to send the propellant for the return voyage aboard the cargo ship to enable the crewed vehicle to be refueled in Mars orbit. This solution is clearly best suited for NEP or SEP, but it

---

[10] R. Zubrin and D.A. Baker, "Mars Direct: Humans to the Red Planet by 1999," Proceedings of the 41st Congress of the International Astronautical Federation, 1990.

[11] J.M. Salotti, "Revised Scenario for Human Missions to Mars," *Acta Astronautica* vol. 81, pp. 273–287, 2012.

is also useful for mixed strategies such as operating a chemically propelled crew vehicle with a SEP cargo carrier. The same chemical (or nuclear thermal) rocket that was used for the outbound journey will be refueled for the return to Earth. The problem here is to store cryogenic propellants in space for up to several years. As noted earlier, this is a particular problem for liquid hydrogen. With standard storable chemical propellants there is no boil-off (although there may be other issues, such as corrosion), but if using NTP there is no alternative to the use of hydrogen. In the case of electric propulsion, argon can be used, and this is much easier to store for a long time than hydrogen.

7.7 ISRU OPTIONS

The early proposals for human Mars exploration were based on the strategy of carrying from Earth everything required during the entire mission, but in the 1990s studies began to address the concept of exploiting the resources of Mars. This was summarized in the slogan "live off the land." Particularly noteworthy in this regard is the Mars Direct concept of Zubrin [10, 11].

The two resources that can be most easily exploited on Mars are carbon dioxide from the atmosphere and water from the ground. In terms of mass savings, the most important thing that can be gained from local resources is propellant, at least to enable the ascent vehicle to reach Mars orbit and possibly to send the main spacecraft home to Earth. This process is known as In Situ Propellant Production (ISPP).

Most studies envisage the use of methane and oxygen as propellants for the Mars ascent vehicle. The main options are as follows:

- There is no ISPP, and both CH_4 and O_2 are ferried from Earth. The transportation of these chemicals is straightforward, and the mission is independent of local resources and complex ISRU systems, but the downside is a very heavy cargo lander and that in turn means a high IMLEO.
- The CH_4 is ferried from Earth and the O_2 is obtained from CO_2 in the atmosphere of Mars. This is fairly simple and robust. Since in the fuel-oxidant combination the mass of the oxidant is larger than that of the fuel (with methane, the oxygen/methane ratio for best specific impulse is 2.77:1) there are some mass savings. There is still a heavy cargo lander.
- The H_2 is ferried from Earth, but the atmospheric CO_2 is processed to obtain oxygen and also to turn H_2 into CH_4 (for each 1 kg of hydrogen, 4 kg of methane can be obtained). The mass savings are more substantial, but carrying liquid hydrogen to Mars requires either complex cryocoolers or some other means of inhibiting boil-off. In addition, the plant for producing methane adds to that for producing oxygen, increasing the mass of the ISPP plant.
- There are no carry-on resources. Both CH_4 and O_2 are produced on Mars from water from the ground and CO_2 from atmosphere. This minimizes the mass of the landers, but water extraction requires complex robotic processes. Automation may be difficult and human presence will probably be needed. In the latter case, there is no possibility of preparing the return fuel prior to the crew landing on Mars, and that will add to the risk of mission failure.

Some sizing considerations are provided in the NASA reference mission [24]. In the final case, according to some calculations, the mass of the ISRU systems (power included) may be less than 25 % of the mass of propellant needed by the MAV. That option was discarded in the NASA report, and the presence of humans for the deployment of ISRU systems and the supervision of the work was not considered.

A detailed analysis is required to determine the feasibility and risks associated with the extraction of H_2O and CO_2 to produce propellants. If it is deemed necessary to have the MAV ready and fueled before the crew even leaves Earth, to be certain they will be able to make the return trip, the best trade-off is probably to bring the methane from Earth and to produce only oxygen in situ, as recommended in the NASA design reference architecture.

If methane is also to be produced in situ, then once the plant for producing propellant is operating, it will be possible to produce additional methane and oxygen to serve as a fuel for a wide variety of other applications, such as rovers which are powered by internal combustion engines and/or rocket hoppers for exploration activities. Methane is also a good choice for NIMF, since it would enable an interplanetary spacecraft using NTP to make its outbound journey with LH2 and then refuel in Mars orbit with methane produced on the planet for the return trip. Moreover, the production of methane can yield oxygen as a by-product for use on Mars.

The decision to employ ISPP greatly influences the entire mission design, because if the propellant needed to leave Mars is to be produced on the surface, the vehicle that arrives from Earth can enter a highly elliptical orbit around Mars in order to reduce the propellant that must be carried from Earth for the TEI maneuver. The MAV fueled with in situ propellant will be able to rendezvous with the return ship in high orbit. Actually, the MAV could even make a direct launch toward Earth. If that vehicle was deemed too small to accommodate the crew on such a long journey, then the MAV could rendezvous with the ERV, launched from Mars orbit along the same trajectory.

Oxygen for the life support system can be produced in the same manner, thereby reducing the need for complete recycling of the atmosphere within the habitat. The trade-off between regenerative life support systems and in situ production of water and oxygen are dealt with in Chap. 8. Any excess water extracted from the ground that is not required by the crew could be used to start a greenhouse to grow plants for food. There have been many studies in this area.

Another resource that can be readily used is regolith. For example, simply by piling it on the habitat it will provide protection against radiation. In spite of the fact that this idea is fairly old, the configurations of the habitats in the NASA reference mission documentation [24] don't easily lend themselves to being covered with regolith. If this option is adopted, then the habitat must be designed accordingly. In fact, regolith can be used as a construction material to build the habitat from scratch, and this may substantially reduce the mass that must be delivered to the planet.

Experiments in constructing houses on Earth using machines derived from 3D printers (additive manufacturing) have been very successful. If habitats on Mars are built using such techniques, it will permit much larger habitable spaces to be created for the crew. There will obviously be some mass savings derived from not having to take a fully outfitted habitat to the planet, merely some construction tools and essential items such as airlocks and internal fittings, but these savings have yet to be defined. See Chap. 8 for further discussion of this topic.

Other materials can be extracted in situ on Mars, but this will occur much later, when colonization of the planet is underway.

ISRU technologies must be tested prior to sending a human mission to Mars which will rely on them, and the Moon seems to be the ideal test site because it is our nearest celestial neighbor and getting there takes only a few days. As stated in Chap. 2, the Moon has no atmosphere and it lacks a general permafrost, although there is ice in some of the craters in the polar regions. The regolith is not too dissimilar to that on Mars, and it will be possible to extract oxygen from its minerals. Hence the technologies that process regolith can be deeply tested on the Moon. However, the processing of lunar ice to provide hydrogen and oxygen to serve as propellant for a Mars mission is controversial.

7.8   SPACECRAFT ARCHITECTURE

The first choice is whether to design a split mission or to send everything and everyone in a single spacecraft. The advantages of a split mission are many, but the main one is that it allows the cargo to be sent on a slow, more economical trajectory and the crew on a fast trajectory. This is best exploited when low thrust trajectories are used, because the quantity of propellant decreases monotonically with the increase in travel time. In the case of impulsive trajectories, on the contrary, there is a minimum propellant requirement for a given duration and if a slower trajectory is used then the propellant consumption increases; this minimum strongly depends on the actual launch opportunity.

Another advantage of a split mission is that the items whose absence on the surface of Mars may result in the loss of the mission (or worse, the loss of the crew) can be delivered well in advance. Furthermore, the crew might not set off until it is confirmed that the MAV is waiting for their arrival and is ready for lift off. This is particularly important if ISRU (and in particular ISPP) is included in the mission design. If, however, the MAV goes to Mars with the crew (or independently in the same launch opportunity), then several months may elapse before the propellant to leave the planet can be produced, and during that time the crew will either have to bide their time in orbit or land immediately and hope that everything goes to plan.

The drawback of a split mission is that the cargo ship must travel unattended, and many of its components must be guaranteed to operate for a long time in this condition. If the vehicle uses cryogenic propellants, it will also require the development of zero-boil-off technologies.

Pre-deployment may, however, be quite difficult without human intervention (some ISRU options might be penalized) and, according to NASA, the long waiting periods cause other risk issues.

A split mission is impractical unless a precise landing can be assured, as the crew vehicle must set down very close to the cargo ship. Unless the outbound spacecraft provided a form of artificial gravity, the crew will have been exposed to weightlessness for a lengthy period and will be in fairly poor physical condition. Nevertheless, immediately after landing on Mars they must be able to walk to the cargo ship, deploy the rover, and then reach the habitat and all the other devices that are already on the surface. This requires all the landers to set down very close together. If it turned out that the crew landed off target, the rover could be teleoperated or proceed autonomously to collect them. The requirement for precision guidance suggests that a Mars Guidance, Navigation & Control (GNC) satellite be placed into orbit around the planet very early in the execution of the mission plan.

A few examples of the possible alternatives are:

- The return vehicle may be one of the devices that are sent in advance and waits in Mars orbit for the arrival of the crew, or alternatively the same manned interplanetary vehicle could be used for the outbound and return voyages through interplanetary space (this option is usually considered only for NEP or SEP missions).
- The manned interplanetary vehicle could land directly on Mars. The alternative is to provide a separate descent vehicle that will later dock in Mars orbit with the return spacecraft.
- The MAV may be integrated with the vehicle in which the crew lands on the planet and the same propulsion systems are used for both the landing and ascent. The other option is to send the MAV as part of the cargo to be pre-deployed.
- The MAV may include the main surface habitat, or alternatively this may be the ERV for a direct return.
- A pressurized rover may land as a part of the surface habitat, or as cargo aboard the crew lander. The surface habitat itself might be a pressurized vehicle.

These are only a few of the many possible configurations. The ERV can be split into two smaller vehicles that proceed through aerocapture independently, and then join in Mars orbit. The first one would include the main propulsion stage for the TEI maneuver and the Earth re-entry capsule. The second would have the interplanetary habitat and a small service module that would be jettisoned after the vehicles had docked in orbit. If the surface habitat is small enough, it might even be possible to integrate it into the MAV.

The options are endless.

7.9 OVERALL REDUNDANCY AND MULTIPLE MISSIONS STRATEGY

Ideally, there must be at least one backup habitat module at every point of the mission, and one viable backup strategy ready before committing to each successive phase of the mission. But in addition to complexity and cost, the provision of many backup strategies might add risks. A strategy to increase redundancy is to split the payload into several modules.

One idea might be to use very small modules with only two astronauts per habitat and to duplicate the entire mission. This would provide many backup scenarios. Another idea would be to exploit all the modules as follows:

- For the period spent on the surface of Mars, even if there is just one cargo module and one habitable module, a backup life support system can be added to the equipment of the cargo to provide redundancy.
- For the outbound trip, the ERV or the cargo ship can play the role of the backup vehicle.
- For the return voyage, the small re-entry capsule can be the backup habitable module.

For a strategy in which a number of missions will be sent to the same location on Mars, starting with the second mission it will be possible to accumulate materials on the planet. For instance, it is possible to have redundant power plants, rovers, and environmental control and life support systems at no increased cost compared to missions sent to different sites. If the habitat is constructed in situ by additive manufacturing with local materials (i.e. regolith), the machinery can continue to work during the time that there is nobody on Mars, to enable the second expedition to live in a much more comfortable environment. Furthermore, when the apparatus delivered by the first two missions starts to age and is replaced, it will constitute a stock of spare parts for maintaining other machinery.

For this strategy to be really effective, it must be incorporated well in advance, so that every single piece of machinery sent to Mars is designed accordingly. Each must contain the maximum of standardized components and be designed for maximum maintainability. All of the critical parts must be readily fabricated using the technologies that will be available (for instance, using additive manufacturing). Just as Apollo changed forever the way in which we design machinery (i.e. substantial use of numerical modeling, diffusion of Finite Elements and multibody codes, concurrent design, and a systems approach to design), sending humans to Mars could change completely the way we approach small scale manufacturing of complex, delicate, and critical machinery.

With this in mind, it is possible to plan a Mars campaign that unfolds over a period of at least a decade, some missions being robotic and others carrying humans, and all targeted at a single site.

- A preparatory mission (if needed) would deliver to Mars the return vehicle required for the next mission, if the design for that mission requires the launch of the return vehicle during the previous launch opportunity.
- A Heavy Mars Sample Return Mission. The vehicles are the same as intended for the human mission but there is no crew aboard. It functions as a final rehearsal, testing all the systems in their intended operating conditions. The scientific objective is several hundred kilograms of samples from Mars. In this launch opportunity, some material starts to be left on Mars and construction is started.
- Mission 1 sees the first human contact with Mars. This is a short stay mission with a small crew of perhaps three astronauts. Little actual work is undertaken on the planet, but all of the systems are tested in actual conditions and the reactions of the crew are tested without requiring them to commit to a lengthy stay on the surface.
- Mission 2 is the first application mission. A full crew of perhaps six astronauts have a long stay on Mars and start actual exploration of the area surrounding the outpost.
- Mission 3 is a consolidation mission to enlarge the outpost. Again a full crew spend a long stay on the planet during which they continue to explore the local area and start to extend their range.
- Mission 4 initiates colonization, possibly with an indefinite stay being made possible by the large quantity of machinery and spare parts that are on site.
- Subsequent indefinite stay missions.

8

The outpost on Mars

Once on Mars, a human mission must rely on a set of facilities, including at least a habitat, a power system, and one or more rovers to allow the explorers to travel to the most interesting zones of the planet. Since it is difficult to imagine that the astronauts would carry with them all that is necessary for their stay and their return trip, an In Situ Resources Utilization (ISRU) plant is often planned. Also workshops, greenhouses, and other auxiliary equipment such as a landing pad and landing guidance systems may be required.

8.1 THE HABITAT

8.1.1 The building

The habitat on Mars must provide sufficient volume in a shirt-sleeve environment. The long experience gained with life support systems on space stations will be precious, but it must be remembered that, in the case of Mars, the system must be able to operate autonomously for long periods of time with little or no resupplying and only the minor maintenance that can be done by the astronauts. The longer the stay, the greater the emphasis that must be put on regenerative systems. Sophisticated Environment Control and Life Support Systems (ECLSS) may permit recycling of resources; for example, recovering oxygen from carbon dioxide for respiration and water from urine and cabin humidity. A further level of 'loop closure' can be achieved by the production of food and the recycling of solid waste. The appropriate level of closure for the ECLSS will depend upon the crew size, the duration of the mission, the power consumption of the systems, and the strategy for optimizing mass.

If a number of missions are to be sent to the same site on the planet, then being able to run the habitat for many years is essential. In this case, a larger mass of the ECLSS can be tolerated and the incentive to employ regenerative systems increases.

To sustain a long term human presence on Mars, the habitat must not only protect the crew from the environment, but also be constructed in such a manner as to encourage a small society to flourish. Long stay missions will spend 450–550 days on Mars, but for a number of landings in the same place the useful life of the facility must be much longer.

© Springer International Publishing Switzerland 2017

G. Genta, *Next Stop Mars*, Springer Praxis Books, DOI 10.1007/978-3-319-44311-9_8

The surface habitat will support astronauts' living and working conditions by providing:

- Sufficient volume per crew member.
- A shirt-sleeve environment and a breathable atmosphere.
- Protection against the Martian environment.
- Food and water.
- The means to undertake scientific and other tasks.

Essentially four approaches are possible:

- A 'metal' habitat (one built in the same manner as the modules of space stations) will be built on Earth, sent to LEO, carried to Mars orbit, and finally landed on the surface.
- An inflatable habitat (such as are built on Earth) will be sent to LEO, carried to Mars orbit, and finally landed on the surface. Since it will not be inflated until it is in place, its stowed configuration will take up much less volume aboard the lander (or perhaps the same volume will be allocated in order to carry a larger habitat). It will probably be much lighter, easier to land, and easier to deploy.
- A 'masonry' habitat will be built using regolith and other materials found on the site. All that will need to be delivered from Earth are the construction tools, specialized units such as airlocks, and internal apparatus.
- A 'cave' habitat would exploit either a cave or a lava tube. This solution may increase the habitable space while minimizing the mass to be brought from Earth. It would also offer protection against radiation.

Metal habitats, often dubbed 'tin can' habitats, can be sent in a ready-to-use condition, and may also be used as space habitats during the outbound journey. The astronauts may even land in one, if it is designed as a lander. If the surface habitat serves as a space habitat then it will be necessary to have a similar habitat for the return voyage – unless it is also to serve as the ascent vehicle, in which case the habitat would be too small for a long stay mission and probably marginal for a short stay.

Usually metal habitats are cylindrical in shape, and can be mounted on the ground in either a horizontal or a vertical position, like the two habitats shown in Figure 8.1. Due to atmospheric entry constraints, this configuration has a limited internal space and the volume allocated to each person could be much less than ideal. Above all, it is almost impossible to protect the crew by covering the habitat with regolith.

Owing to their cost, the mass and volume of Mars habitats should be minimized. Volume is constrained by the aerodynamic fairing of the launch vehicle. Basically, the components of the outpost must fit inside a payload shroud that has a diameter of approximately 6 m.

An artist's impression of a habitat for NASA's Mars Semi-Direct mission (as outlined in DRA 1.0) is shown in Figure 8.1. The unit that landed along with the crew is joined to another one that had arrived in the previous launch window. The two units are provided with wheels in order to make their 'docking' easier.

To increase the internal space, the habitat may be deployable, inflatable (Figure 8.2), or assembled from several sections on Mars, ideally in an automated, self-deployable manner. The simplest option is an inflatable habitat. We have preliminary experience with such

Figure 8.1 An artist's impression of two joined habitats for NASA's Mars Semi-Direct mission, as described in DRA 1.0. Two rovers are also depicted. (NASA image)

Figure 8.2 An artist's impression of an inflatable habitat on Mars. (NASA image)

habitats. In 2016 a Bigelow Aerospace module was attached to the ISS and inflated for test purposes. By the time of a Mars mission, this technology will be mature. Actually, it is easier to produce an inflatable module for Mars than for space. The substantial experience gained with large air-supported structures on Earth may be applied in building inflatable habitats on Mars. In this case, the greater pressure differential between the inside and the outside may require using a stronger membrane, but it will also permit placing regolith on at least a part of the habitat to protect the interior from radiation.

Masonry habitats have being studied for about half a century. Indeed, when the Apollo missions brought samples of the lunar regolith to Earth there were several successful tests to produce cement out of it. It is not linked with any specific technology. In the context of long stay missions, the astronauts could live in their lander for some time and devote their efforts to construction work, but this is hardly a practical solution. The most precious resource on Mars is the time available to the crew, and to devote a substantial portion of it to work of this kind is surely not justified, particularly because undertaking building work while wearing a pressure suit may be difficult and dangerous. Hence the construction of masonry habitats has always been scheduled for after human activity on Mars switches from scientific exploration to colonization.

Today things are changing rapidly. In the last years, some experience has being gained of house building using very large 3D printers. Two examples are shown in Figure 8.3a and 8.3b. In the former, the printer is based on a huge portal, so that a Cartesian (or orthogonal) robot is obtained, while in the second the robot is based on a spherical arm. Both are able to complete the whole structure of the building, and with a suitable design the roof can also be made in one piece with the walls. In some experiments, an entire small sized building has been completed in one day. The second configuration is probably more suitable for a planetary outpost, because the device needing to be carried to Mars may be much lighter and less bulky. Both NASA and ESA are working in this direction. The machine may be mounted on a rover, so that, once the building is completed, either another one can be started nearby or a bigger one can be made by using the motion of the supporting rover for additional degrees of freedom. It may be sent out to Mars to construct the outpost using regolith prior to the astronauts' arrival. A further advantage is that such a habitat will provide much better shielding against cosmic rays than an inflatable or a pre-built habitat that must be installed in a trench and partially buried to offer a similar degree of shielding.

In a paper by R.P. Mueller et al.,[1] it was proposed that basalt regolith fines be used to build a variety of structures which may greatly simplify the exploration of the Moon or of Mars:

- Landing and launch pads.
- Blast protection walls and structures.
- Consolidated and stabilized slopes to prevent slides that could damage habitats.
- Paved access roads.
- Hangars for spacecraft and equipment storage.
- Sheltered habitats for humans.
- Radiation storm shelters at waypoints.

---

[1] R.P. Mueller et al., "Additive Construction using Basalt Regolith Fines," http://ntrs.nasa.gov/search.jsp?R=20150000305

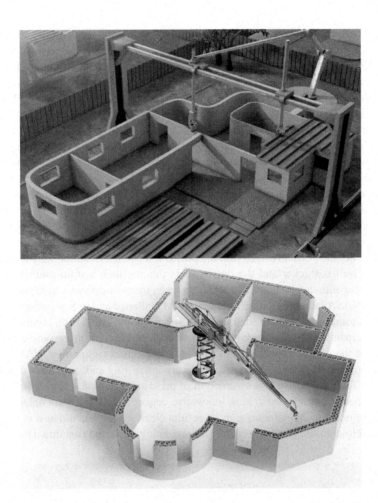

Figure 8.3 Additive manufacturing of buildings. (a) Cartesian machine. (b) Spherical machine.

- Radiation shielding panels for spacecraft.
- Dust free surface stabilized zones (paved areas).
- Thermal inertia pads to capture heat in the day and release it during the night with associated shade structures.
- Roof structures for subsurface shelters.
- Ablative heat shields for planetary EDL.
- Radiation shielding for a fission power plant.
- Antenna towers for line of sight communication.

The dust can be sintered using a laser or a solar concentrator so that a binding material is not required. In this way the only material needed for building the habitat is the regolith fines and nothing has to be transported from Earth except the machine itself, its power system, and various fittings such as airlocks, windows, etc. The production of the walls and other structural elements can readily be automated, but it is difficult to add the fittings in an automatic manner. For instance, the installation of the electrical network might have to wait until the astronauts arrive. Windows are a typical example of an add-on that requires particular care to be taken in their design. As pointed out in Chapter 5, windows are likely to be absolutely essential to the well-being of the crew but large screens (probably thin-film screens) could be installed on the walls as 'virtual windows' which display feeds from external cameras. This would eliminate any discontinuity in the regolith structure and not compromise either the thermal insulation or the protection from cosmic radiation. This solution would further reduce the materials to be carried from Earth, and may be sufficient to solve the psychological stresses of the crew, but may fall short of preventing the loss of the distance perception that occurs when people spend long periods in closed environments. Losing the ability to judge distances during EVA and/or when driving rovers will place astronauts at risk. Prior to committing to 'virtual windows', it will be necessary to conduct appropriate experiments. Double walls can be built and the space between filled with water from the permafrost, so that the protection against radiation may be improved. If the inner wall is thicker and the outer one is thinner, then a solid multilayer wall of regolith-ice-regolith can be created that will be even more impervious to radiation.

In 2015, NASA launched a 3D Printed Habitat Challenge and in September of that year it awarded a total of $40,000 to the three best submissions. The design competition challenged participants to develop architectural concepts based on in situ resources and to take advantage of the unique capabilities of this new fabrication technology. NASA received more than 165 submissions, and the winners were interesting and innovative designs. The proposal by Team LavaHive was awarded third prize and Figure 8.4 shows their design.

The competition was more to seek innovative concepts than to obtain detailed proposals that were ready to implement, and the results clearly established that it is possible, at least in principle, to use Martian regolith in fabricating large and functional habitats.

Figure 8.4 The habitat designed by Team LavaHive that won third prize in the 3D Printed Habitat Challenge undertaken by NASA.

Of course, the most difficult aspect of this plan will be to enable the machine to operate autonomously on Mars to build a habitat ready for the first astronauts. A possible alternative, particularly if a series of missions are to land at the same site, is to send to Mars in advance a small habitat such as one of those shown in Figure 8.1, to accommodate the astronauts when they land. The construction machinery, including the 3D printer, can land along with the crew and then can start working to build a large habitat under their supervision. When that has been achieved, the astronauts will move in. The construction machinery will then proceed to build other buildings, either to be used during the same mission or the following ones. Over time, a large outpost incorporating comfortable housing, laboratories, workshops, greenhouses, and other facilities can be built.

An interesting point is that we already have a lot of experience in building and operating large structures in locations with extreme climates. The large shopping centers and hotels that are built in hot deserts are completely isolated from the external environment. If they were also built to accommodate an inside/outside pressure differential and had airlocks and life support systems, then in principle they could be transported to the Moon or to Mars.

Finally, dwellings in Martian caves or lava tubes are an interesting possibility, so long as underground environments are not all classified as protected environments into which humans cannot enter, lest they forward contaminate possible abodes of indigenous life. Here too, there are various possibilities. For example, a standard metal or masonry habitat might be located inside a large cave or lava tube for enhanced protection against radiation. A small cave may be made airtight simply by inflating a habitat until it comes into full contact with the rock walls. The fabric of such an inflatable can be much thinner, since it does not have to withstand the pressure differential. And of course, the drawbacks of living in a cave may be mitigated by 'virtual windows' that display the landscape on the surface.

8.1.2 The interior space

The comfort of the crew undertaking a long Mars mission should equal that on Earth because their psychology is a key issue. In all likelihood the criteria for the selection of the candidates for a Mars mission will be rather more sophisticated than in the case of ISS crews.

The habitat will serve as home for each astronaut for a long period of time, at least on long stay missions. Moreover, the surface conditions on Mars are such that going outside will not be just a matter of spending some time away from home, but an EVA and hence something which should be minimized because it is tiring and dangerous.

In such an alien and unforgiving environment, it is essential that the crew feel 'at home' in their habitat. Each person requires some private space for sleeping, reading, or spending their free time when they wish to be alone.

'Hot bunking,' a common practice on naval submarines, where several people share the same bed in turn, following their work shifts, is not considered a suitable solution for a Mars mission. For psychological reasons, it is extremely important that every astronaut possesses an individual compartment into which he/she can retreat. There must also be a common space in which meetings can be held, and a common space where conventional common recreational activities can be pursued. The private and public places must be clearly separated. There is a need for semi-private or semi-public spaces to enable small impromptu groups to relax, chat and rest between assignments.

On the ISS the favorite off-duty activity for astronauts is simply looking at their home planet, and they spend much time in the cupola which offers panoramic views. In a similar way, as stated, in all likelihood a habitat on Mars will have 'virtual windows' to enable the crew to view the landscape. As far as possible, these should convey the impression of being real windows. For instance, the cameras for windows at the same level should be located at the same level too, and they should look in the appropriate direction, so that the mental picture the crew develop of the external environment corresponds to reality; if not, they face becoming disoriented during EVA activities. Even if real windows are present, having additional direct transmission screens could enhance the sense of contact with the outside world. If possible, there should be an 'observation tower,' an elevated point of view that provides views of the widest possible area to allow dust devils, sunsets, and other occasional events to be viewed to enhance the crew's sense of being in control of their new world. Every airlock hatch ought to have either a window or a screen to see out, as this will enhance safety when people outside are close to the habitat. Of course, view screens will be essential if the habitat is located in a cave or a lava tube where real windows will be inappropriate.

A telescope to view the Earth may help to reduce a crew's sense of isolation, even if the small size of its disk emphasize the tremendous interplanetary gulf.

Exercise is important to maintain physical health, particularly when in a reduced gravity environment, hence appropriate apparatus will be required and, if possible, an area should be set aside as a gymnasium.

In addition to providing a large pressurized space, a masonry habitat could offer a varied and pleasant architecture with curved walls, arches, etc. Some of the interior fittings may also be made of regolith material along with the building, and these should also be designed to be pleasant and comfortable.

To function efficiently in a confined environment, astronauts will need recreation time, therefore there must be facilities for reading e-books, watching movies, and catching up with news from Earth.

Summarizing, the required areas or rooms are:

- A galley or kitchen.
- Individual crew quarters.
- Meeting and communal spaces for social gatherings, dinner, and recreational activities.
- One or more laboratories.
- Work spaces (computer and bench with 3D printer).
- Greenhouse.
- Hygiene facility.
- Medical facility.
- Storage spaces.
- Space for the life support system.
- Airlocks, and/or hatches.
- A solar storm shelter.

Because the number of people and amount of materials that can be delivered to Mars is limited, it is sensible to plan a modular base that can be expanded by adding modules. It may be preferable to separate the main habitat from some of these areas, such as the greenhouse, laboratories, or workshops. These may be compartments in a structure that also serves as a garage for rovers or other forms of transportation.

Flexible, multifunctional spaces, and spaces that can be re-purposed for multiple uses, are essential because astronauts should be able to personalize their individual spaces. Crew members might want to adapt and add simple changes to the habitat, both in the private and common spaces. Volumes that do not serve a predefined purpose can be used to create more or less open spaces to allow the visual spatial impression to change, perhaps entirely altering the character of that part of the habitat.

Lighting, air conditioning, and even colors and textures ought to be changeable to suit requirements. Sliding or foldable walls might allow the architecture of non-fixed spaces to be varied. There should be sufficient flexibility to allow the astronauts to personalize the spaces that do not serve predefined purposes.

The habitat must have kitchen implements, like a freezer, oven, sink, dispenser, cooking and eating supplies; health equipment like hygiene facilities, exercise equipment (e.g. treadmill and rowing machine); and recreational devices (perhaps one for each person) like computers, televisions, video players, e-books, and music players. If virtual windows are used, these will also be able to be used as screens to watch videos or turned into 'virtual instrument panels' to operate apparatus like the life support system or the power generator.

Figure 8.5 gives an example of the interior of a two-level 'tin can' habitat of the style of Figure 8.1. It is actually the habitat of the Mars Desert Research Station (MDRS) which The Mars Society built in the San Rafael Swell near Hanksville, Utah, USA. It is designed to give crews of six people a simulated Mars exploration experience. Owing to its small size, it can be considered as the minimum required for a long stay mission.

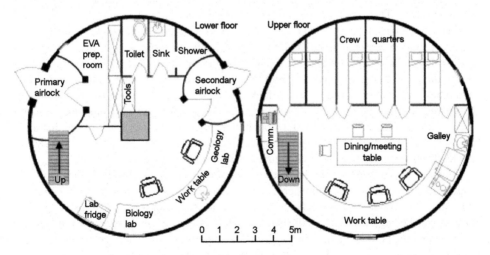

Figure 8.5 Internal plan of the two floors of The Mars Society's MDRS.

8.1.3   Accessibility

To enter and exit the habitat there must be a general purpose airlock in which astronauts can don their space suits prior to exiting for an EVA and then doff them upon finishing. The door must be large enough to permit easy transfer in and out of specimens to be analyzed and bits of apparatus that need to be fixed.

On the Moon, the Apollo astronauts could put on and take off their EVA suits inside the cramped cabin of the Lunar Module (LM). This was an adequate strategy for such brief stays, but contamination by lunar dust became a problem. The astronauts tried to brush the dust off their suits before heading up the ladder but everything that came into contact with it became coated. As extremely abrasive finely pulverized rock, it would rapidly have clogged up zips and ring seals if the stay times had been much longer than they were (the longest performed three moonwalks). The airlock of a future lunar habitat or one on Mars must therefore have the equipment necessary to remove dust from suits and apparatus to prevent contamination of the interior. Ideally, electrostatic dust precipitators for a Mars mission will first be proven in lunar conditions.

As the critical point for avoiding contamination both backwards and forwards on Mars, the airlock must be very carefully designed. Having a lot of space in the airlock will help to simplify decontamination procedures, so an inflatable airlock may be attractive. Also, there should be the minimum number of entries and exits.

An access hatch to which a pressurized rover can dock will enhance safety by removing the need for astronauts to make an EVA to enter or exit the rover, and it will also save time. The rover may be left docked to the habitat, as shown in Figure 8.4. If there is more than one pressurized rover, then more than one access hatch will be needed. As an alternative, a large airlock can be provided that acts as a garage in which it is possible not only to store the rover but also to undertake maintenance without wearing space suits. However, this will be possible only if large pressurized volumes are available, as in the case of 3D printed habitats. On the other hand, this may make it difficult to apply strict contamination controls. However, the alternative of carrying out maintenance outside while wearing a space suit is not much better. This aspect of operating on the Martian surface will require careful thought.

The garage must be connected to the habitat by way of a further airlock, because it will be impossible to decontaminate the rovers and the other equipment that must function in the outside environment. As astronauts pass from the garage to the habitat, even if the two places are pressurized, they need to pause in an intermediate room in order to clean off dust and other contaminants. And when a suited astronaut emerges from the personnel airlock, or a rover exits the garage, there will inevitably be a degree of forward contamination from the inside environment. Robotic or teleoperated rovers sent to zones of the planet that must be protected from contamination should therefore never enter the garage, and they should avoid places that may already have been contaminated by the human presence.

Suitports or suitlocks are an interesting alternative. A space suit such as that shown in Figure 8.9a has a device located where, in a normal suit, there is a backpack. This engages with an external hatch of the habitat so that the operator can transfer. The suit remains always attached to the exterior of the habitat (in that of Figure 8.4, there are two suits ready for use) and the person remains in the pressurized environment whilst entering or exiting the suit. In this manner, the internal environment is always isolated from the external one, eliminating the risk of contamination in either direction. With this strategy, each space suit needs to have a suitport and there must be at least as many hatches as there are people. But this system is not very flexible, particularly if each crewmember requires a tailored space suit. If each person requires to be able to exit both from the habitat and from the pressurized rover, there must be twice as many suits and suitports as people.

Of course, even if the pressurized rovers dock at hatches and the suits have suitports, a simple airlock will still be needed to transfer items in and out, and indeed for use during an emergency, such as to recover an astronaut who has suffered an injury that precludes their using the suitport.

Figure 8.6 reports an interesting concept in which a suitport is integrated with a more or less traditional airlock. The airlock is collapsible, and takes very little space when in stowed configuration. When deployed, it permits easy entry and exit. The airlock is subdivided into two parts and the separating wall has two suitports. In normal use, it is possible to wear the space suit through a suitport. However, the separating wall can be opened (as shown in the lower part of Figure 8.6b) so that the system becomes a conventional airlock with space to wear conventional space suits.[2] In this way, it is also possible to easily bring the space suits inside the habitat for maintenance.

All things considered, sophisticated concepts like docking hatches and, above all, suitports, may be an unnecessary complication, and the simplest and best solution may prove

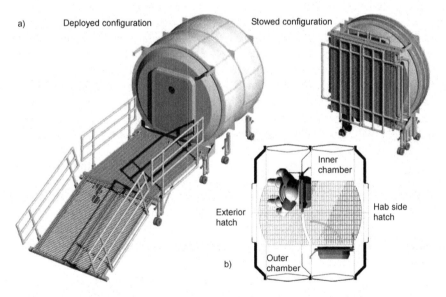

Figure 8.6 Collapsible suitport. (a) Sketch in stowed and deployed configurations. (b) Cross section of the airlock.

[2] A.S. Howe, K. Kennedy, P. Guirgis, R. Boyle, "A Dual-Chamber Hybrid Inflatable Suitlock (DCIS) for Planetary Surfaces or Deep Space," 41st International Conference on Environmental Systems, July 2011, Portland.

to be two simple airlocks: a small one for just one person and a larger one for more people and objects. Before committing to a specific design for a Mars habitat, the many alternatives must be thoroughly tested in a simulated environment on Earth and then in more realistic conditions on the Moon.

A storm shelter is essential, because the astronauts must be able to take refuge in a solar storm. This facility may also be used to store food or drugs (for instance some vitamins) or equipment (like some electronic components) that could be impaired by radiation. The shelter will require the strongest shielding. In the case of a masonry habitat, or one in a cave or lava tube, there may be no need to construct a specific shelter because the habitat will already be well protected. An exercise of design of an outpost built by sintering Mars regolith using a 3-D printing machine is shown in Fig. 8.7. The outpost can host 6 people in relative comfort and is provided with a solar storm shelter and a greenhouse. All windows, except those in the observation tower, are virtual, so that the structural continuity and radiation protection is not interrupted.

8.2   LIFE SUPPORT SYSTEM

8.2.1   Air

The air quality inside a habitat or any enclosed space built on Mars must be comparable with that on Earth, and will be regulated by the Environmental Control and Life Support System (ECLSS). The parameters will include oxygen pressure and humidity. As already stated, the atmospheric pressure on Mars is about 0.006 bar, which is less than 1 percent of the Earth's. The temperature at ground level on Mars ranges between −63°C and +20°C in the southern hemisphere in mid-summer, and it can fall to −100°C. Due to the thin atmosphere, thermal insulation is not a problem and conventional materials should be sufficient. Maintaining the same Standard Temperature and Pressure (STP) conditions as occur on Earth will impose a significant physical load on the habitat.

The air we normally breathe consists of about 21 percent oxygen, 78 percent nitrogen, and approximately 1 percent other trace gases, mainly argon. We need the partial pressure of oxygen to be in the range 0.16 to 1.60 bars. The most dangerous gases are carbon monoxide (which is poisonous even at low concentrations) and carbon dioxide. The latter causes some people to feel drowsy at concentrations up to 1 percent, and concentrations of 7–10 percent may cause suffocation (even in the presence of sufficient oxygen), displaying symptoms of dizziness, headache, visual and hearing dysfunction, and unconsciousness all within several minutes to an hour.

So the life support system must ensure that the air:

- Contains sufficient oxygen to support life, consciousness, and work rate.
- Does not contain harmful gases such as carbon monoxide and carbon dioxide or other poisonous or dangerous gases which may be outgassed by various types of materials or produced by chemical processes.
- Although not strictly regarding health, it is important that the environment is free from unpleasant odors derived from chemicals released by human bodies, produced during various activities, or issued by parasites or rotting biological material.

Figure 8.7 A Mars outpost built by sintering regolith using a 3-D printing machine: artist's impression, elevation and plan. (Drawings by Alberto Canarecci from his Master's thesis in Civil Engineering at the Politecnico di Torino, supervised by the author)

Table 8.1 Composition of the air aboard the ISS (nominal values) and of the air exhaled by astronauts.

	Space station air	Astronaut exhalation
Nitrogen	78.0%	74.2%
Oxygen	21.0%	15.3%
Carbon Dioxide	0.0%	3.6%
Water Vapor	1.0%	6.2%
Trace Elements	0.0%	0.8%

The nominal composition of the air aboard the ISS and the composition of the air exhaled by astronauts is reported in Table 8.1.

The requirements in terms of oxygen and inert gases are not particularly severe. A human being needs about 1 kg/day of oxygen and 3 kg/day of inert gas (say nitrogen) if a ratio 1:3 is maintained. This means a crew of six requires 4.82 t of oxygen and 14.45 t of nitrogen for a mission of 2 years, including a 10 percent contingency. It is possible to reduce the amount of inert gas and at the same time cut the total pressure, so that the partial pressure of oxygen is compatible with the Earth norm. This allows a reduction in the total mass of air, while at the same time reducing the weight of the habitat because the pressure differential that it must be able to withstand is reduced.

The air must be continuously enriched with oxygen, whilst carbon dioxide and water vapor must be continuously removed. Water vapor control is perhaps more important for the apparatus than for human health, as too much water vapor risks condensation that can cause electrical short circuits, and too little vapor can increase the possible accumulation of static electricity and the resulting risk of sparks.

Controlling trace elements is important, because the human body can produce methane, ammonia, urea, and other gases that pollute the air in a confined environment.

The composition of the air inside the ISS is continually monitored. Water vapor, carbon dioxide and trace gases are removed, and nitrogen and oxygen are released from tanks by the ECLSS.

On a mission to Mars, just removing carbon dioxide and adding oxygen would lead to a consumption of gases which may be too large, even if the figures mentioned above show that actually it is just a few tons (so long as the nitrogen is recycled). To decrease the amount of consumables that must be carried to Mars, regenerative ECLSS are planned. If oxygen is re-obtained from the carbon dioxide, then there is virtually no need to carry gases from Earth to maintain the atmospheres in the spacecraft and in the outpost. That said, the degree to which the system can be closed it is not yet clear, and how effectively such gases can be recycled is a matter for debate.

At any rate, some oxygen and nitrogen must be carried from Earth at least to make up for losses and small accidents. If an ISRU system is available, oxygen can be obtained from the carbon dioxide in the atmosphere of Mars. The situation for nitrogen is different, because it comprises less than 2 percent of the local atmosphere. Data recently obtained by the Sample Analysis at Mars (SAM) package on the Curiosity rover, detected nitrogen on the surface of Mars. It was detected in the form of nitric oxide, which may be released when heat breaks down the nitrates present in sediments. Nitrogen from nitrates can be used not only as a gas for breathing, but also to produce fertilizers, a thing which may be quite useful when food production is attempted on Mars by cultivating plants.

When operating airlocks, the air should be recovered, not only for sustainability reasons but also to avoid forward contamination. This may be more difficult when performing an EVA from a rover than from the habitat.

8.2.2   Water

The water required for a mission lasting many months is considerably larger than the amount of breathing gases. While each person requires about 2 liters of drinking water per day, the total water requirements, including personal hygiene, laundry, etc., is about 28 liters per person per day. Allowing a 10 percent margin, this amounts to 135 t for six people on a 2 year mission.

It would be quite costly to send to Mars all the water needed during the whole stay. This can be obviated if either a closed loop life support system is used, or if water is obtained from the ice which is at or near the surface of Mars. Advanced ECLSS technologies may reduce the required mass of consumables by recycling water from urine and habitat moisture and also by recycling oxygen from metabolic carbon dioxide. The appropriate degree of ECLSS closure depends upon the number of crew, mission duration, power consumption and mass budget optimization, plus the feasibility and cost.

Permafrost is now believed to be present at moderate depth across most of the planet, and in particular at high latitudes. However, extracting water from the underground regolith layers implies digging a large quantity of regolith, transporting it to the ISRU plant, and then disposing of the spent regolith either by dispersing it across the ground or by using it as the raw material for building a masonry habitat or for covering a metal or inflatable habitat as a radiation shield.

Although it remains to be established that this can be done using autonomous systems without human supervision, it seems that the second alternative is easier than the first. They may be complementary, in the sense that the more regenerative is the ECLSS, the less water must be extracted from the regolith, and there is also the possibility that the best solution may be to use water extraction with non-regenerative ECLSS, although this may cause forward contamination problems.

8.2.3   Nutrition

Humans need about 1.5 kg of food per day per person. Hence a crew of six needs about 7.2 t of food for a 2 year mission (with the usual 10 percent margin). This is not a large mass, but here too, loop closure can be increased by producing food in greenhouses and recycling solid waste. Producing food on Mars is important not only for saving some mass, but above all for improving the morale of the astronauts. Seeing plants grow and having fresh and varied food at the table is quite important from this viewpoint. It is important that the food tastes good and ideally provides some fresh ingredients. The habitat will therefore require a kitchen equipped with a freezer, an oven, a sink, and appropriate cooking and eating apparatus [14]. As yet no space station has included such facilities. Proving the technology will probably have to be undertaken on the Moon. There is also another point.

Producing food on Mars is essential in the perspective of future colonization and any attempt in this direction is an investment for the future. A greenhouse (see Section 8.7) will not only be a place to produce food as a means of decreasing the mass which must be carried to Mars, but a laboratory to experiment with new agricultural technologies to facilitate colonization.

To summarize, the total mass of consumables required for a 2 year mission of six people is 161 t, comprising 4.82 t of oxygen, 14.45 t of nitrogen, 135 t of water, and 7.2 t of food. To carry all this from Earth would add greatly to the cost and the complexity of the mission. The use of a regenerative ECLSS and/or extracting resources from the Martian environment will therefore be an important step towards making a human Mars mission affordable.

8.3   HEALTH

8.3.1   Physical health

As there is no possibility of returning a crew member to Earth for medical treatment in the event of an emergency, the Martian outpost must have a sick bay that is capable of dealing not only with small surgical procedures but also with long term care.

One of the oldest discussion points concerning deep space missions is whether the crew must include a medical doctor. Nowadays, the prevailing idea is that one person must have a high degree of training in medicine and, if not a doctor, must be at least the equivalent of a nurse. In spite of the communication delay, the assistance of a medical team at mission control is considered sufficient to deal with essentially all medical emergencies, so long as one crew member is trained to perform small surgery and dental care. But when Mars is in conjunction with the Sun, radio communications will be interrupted for quite a long period unless there is a relay satellite at one of the Lagrange points (either L4 or L5) in the Mars-Sun system. In fact, the requirement to be able to advise the astronauts during a medical emergency is one of the principal reasons for including such a satellite in the mission plan.

Owing to the mass and space limitations, and the variety of potential problems that may occur, the decision about which medical apparatus, and above all which drugs, to include on a mission will be delicate. The length of the mission and the impossibility of sending supplies make it impractical to rely only on the pre-mission screening and the initial good health of the astronauts, but at any rate their genetic screening may supply indications of the likelihood of specific problems. A minimum list of the medical emergencies that absolutely must be able to be dealt with, includes traumas of all kinds, dental care, infective diseases that can arise from the immunodeficiency that may be induced by weightlessness, poisoning due to faults in the food conservation or life support systems, and of course acute diseases caused by radiation exposure, particularly if nuclear devices are present in space or on Mars.

The inclusion of additive manufacturing machines in the workshop may enable specific parts to be made to deal with dental care, as well as traumatological and surgical emergencies. Surgical robots are a possibility for enlarging the medical autonomy of the crew, although much research remains to be done, both on Earth and in space. It has been estimated that 6.5 m³ must be provided in an outpost to store medical equipment, but this is only a rough guess.

At present, surgical robots are teleoperated by a surgeon in real time. Although this might be possible on the Moon, it will be impossible on Mars. There have been many studies to build whole-body mathematical models of astronauts. With such work, it is possible to use additive manufacturing to build a complete physical model which has an identical geometry and similar material properties. If surgery is necessary, surgeons on Earth can undertake the procedure on the physical model with all their movements being recorded for transmission to Mars and uploading into a surgical robot that will repeat them while operating the astronaut. At present, an accuracy of 1 mm can be achieved in this manner, and it is considered sufficient for most surgery. This technique has even been tested with brain surgery. But there remains a problem. The astronaut's body may have undergone changes since the mathematical model was made, mainly due to the lengthy period spent in microgravity. It may therefore be necessary to take a 3D scanner to Mars in order to make the mathematical model immediately before the surgery. However, the time taken in scanning the body on Mars, transmitting the mathematical model to Earth, and producing the physical model for the surgeon could make it impossible to offer immediate treatment in a life threatening medical problem.

Prior to undertaking a mission to Mars, we should spend sufficient time on the Moon to obtain information about the types of problem that statistically are more frequent in a reduced gravity operation in an artificial environment with an ECLSS of a type fairly similar to that for use on Mars. It will also be useful to learn how to cope with medical emergencies in an outpost whose proximity to Earth makes such things less dangerous.

8.3.2   Psychosocial aspects

Each crew member will typically require to be available for work about ten hours per day, but will be allowed a day of rest every seven to ten days. Fourteen hours are reserved for sleeping, eating, personal hygiene, recreation, exercise, and private communications. The efficiency of a crew improves when the work is varied. For example, errors and accidents are limited if one day the activity is physical and the next day it is cerebral. To avoid fatigue, extravehicular activities should not be undertaken two days in a row. Physical and psychological limitations result from being confined in a small area. As already pointed out, open spaces and individual spaces must be provided. It has been observed that over time, people tend to prefer more complex and larger spaces. Adding windows and integrating loop-like designs can create the impression of larger spaces.[3] A space also appears larger when it is not overly partitioned. Curved walls enhance the impression of a space as being larger than its actual size because the borderlines of horizontal planes and vertical walls dissolve. Windows reduce the sense of confinement by extending the visual horizon, and they foster the impression of watching rather than of being watched.

Buildings that are made using 3D printing permit larger spaces and a more varied design than ones which are delivered to Mars ready-made, so using this technology is

---

[3] B. Imhof, "The Socio-Psychological Impact of Architectural Spaces in Long-Duration Missions," SAE Technical Paper 2003-01-2537, 2003, doi:10.4271/2003-01-2537.

perhaps no less important from this perspective than in terms of the mass, and therefore cost savings. If a 3D printer is used to make the habitat, then it is possible to add other buildings during the stay on Mars to enhance the living conditions for the crew. Leaving the astronauts to decide among themselves which improvements to make in what they regard as their home may be important from the psychological viewpoint.

Humans tend to install their personal spaces in regular patterns, with the greatest possible separation between the individual spaces. On the Mir space station, for example, astronauts usually abandoned the official adjacent private quarters and instead installed their sleeping quarters and private workstations in a convenient spot elsewhere. Not only must the private space be isolated in terms of sound, vibrations, light and viewpoint, but also spatially. Close physical proximity of the crew quarters should therefore be avoided.

The structure of the base must reflect the structure of society in order to be supportive of it. The resources must be shared according to a strict distribution system. Every inhabitant will play a crucial role in managing the establishment, looking after the crew, resources and vital systems, as well as participating fully in the scientific research. Every crew member should be able to make a contribution in every field.

8.4   EQUIPMENT

Humans have only a certain amount of physical force and attention span, especially over a long period in partial gravity, therefore all systems in a functional Mars outpost or habitat must be considered at least complementary to human capabilities, if not actually capable of operating entirely autonomously.

The designers of Mars rovers must remember that low-level teleoperation from Earth is difficult when the communication delay is at its minimum value, worse when at its maximum, and impossible when communications are interrupted for a lengthy period. This holds also for the habitat in general. Setting up the habitat must thus be done autonomously, particularly if the material to build the base is sent in advance and the astronauts are to find the habitat already deployed when they land.

The equipment should be designed in such a manner that it is both intuitive and efficient. Machines that are connected or share a common task must be positioned in order to minimize transitional paths. Critical controls must be accessible from more than one location and there should be easy access for maintenance and repairs. It must also be possible to use portable computers to access systems and control them. This has the advantage of eliminating some of the dedicated workstations, displays, and controls. However, remote controlled activities and other activities with special requirements will require workstations that include the necessary controls, materials, and tools; for example, if remotely controlling a rover or a laboratory.

For extravehicular activities, one or more rovers capable of carrying two persons should be included as part of the base's infrastructure. Its assembly can rely more on human work, as it is not critical to survival, but there must also be a room to control these activities and even to command the rover from a distance.

In designing an outpost for Mars, it is important to keep in mind the different pathways used. Activities occurring simultaneously should be located far enough apart that they will not interfere with each other or create the hazard of crossing pathways. Similarly, emergency exits must be as direct as possible from all points in the base, although it must be remembered that exit is possible only when wearing a space suit. All systems should be designed to follow fail-safe criteria, and should be redundant. Safety considerations suggest the presence of a fully equipped workshop for maintenance and repair, full redundancy for critical and irreplaceable materials, and the presence of a 3D printer to replace broken parts.

It is extremely important that the base does not transmit noise and vibrations. If it is not possible to obtain silence in some areas, and in particular the sleep area, the effect on the crew may vary from anxiety and poor performance to hearing impairment. Since the toilets and the exercise room are likely to be the loudest facilities, they should be located as far as possible from the sleeping compartments. In these terms, a masonry habitat is preferable to either a 'tin can' or an inflatable. Thick walls will cut down on noise, as well as reduce the radiation dose to the crew.

Standards now require all equipment such as seats and tables to be adaptable to people ranging in height from 150–190 cm.[4] They should also be adjustable for individual comfort. For a mission exceeding 6 months, a habitable volume of at least 25 m^3 per person is strongly recommended.[5] Overcrowding and a lack of free volume are a significant contributing factor in human error.

In partial gravity, such as on Moon or on Mars, the whole volume is no longer equally accessible, as is the case when in microgravity, hence horizontal surface areas gain the same significance as they do on Earth. Also, some working area is lost because beds and surfaces must be horizontal. This must be taken into account when calculating the volume and area needed per crew member.

8.5   POWER SYSTEM

Electrical power must be available to run heating, life support, communications, computers, scientific instrumentation, etc. The main energy consumers are the heating, ISRU, ISPP and battery charging systems (e.g. for rovers). A wide variety of lesser consumers will add up to a substantial requirement. The energy for heating will depend on how effectively the habitat is insulated, but the low atmospheric density and the requirements for radiation shielding may result in low energy requirements for heating. If greenhouses run on artificial lighting, they too can be major energy users.

Solar energy may be enough for the habitat itself, but Mars has an average solar radiation flux of 589 W/m^2. Of course, heating is mostly required at night, so a storage system will be required if solar energy is the primary source of energy. With a 24.6 hour sol, the

---

[4] C. M.Adams, "Four Legs in the Morning: Issues in Crew-Quarter Design for Long-Duration Space Facilities," (SAE 981794). 28th International Conference on Environmental Systems (ICES), Danvers, Massachusetts, USA, 13–16 July 1998. Warrendale, Pennsylvania, USA: Society of Automotive Engineers

[5] LIQUIFER Systems Group et.al., "SHEE Self-Deployable Habitat for Extreme Environments: Design Requirements", 2013–2014.

situation on Mars will be much less severe than that imposed by the longer and much colder nights on the lunar surface.

Because it would be expedient for a single power station to drive all the functions of the outpost, it is widely presumed that solar power will be insufficient. Nuclear power might be supplied by RTGs. These units have been used for several space missions, and provide a life span of at least 20 years depending on the isotopes (^{238}Pu is the most commonly utilized and delivers power for fairly a long time). But RTGs have a high mass-to-power ratio and ^{238}Pu is extremely expensive and in very short supply; the true alternative is thus between solar panels and a nuclear reactor.

A comparison of the different surface power strategies has recently been carried out.[6] A solar power system must include non-tracking, thin-film, roll-out arrays and either batteries or regenerative fuel cells for energy storage.

Reasons for using very large solar arrays are:

- The distance of Mars from the Sun.
- Owing to the difficulty of keeping the solar panels oriented sunward, thin-film arrays should be laid on the ground or on the roof of the habitat.
- The need to provide energy continuously, day and night, taking due account of the efficiency of the associated energy storage devices.
- The requirement to provide energy during dust storms.
- The deterioration of the solar panels over time.

After a dust storm it will probably be necessary to remove dust from the solar panels, either robotically or manually. A solar power system seems suitable to outpost configurations where the energy requirements are limited. As ISRU, ISPP, greenhouse and other apparatus including digging and construction devices and robotic rovers are added, nuclear power will ensure continuous power. This is particularly true if the surface crew exceeds three people. A nuclear power plant is also more tolerant of dust than solar arrays. In particular, it will be possible to operate the above mentioned apparatus and charge batteries for vehicles (if these are electric) during the night, when photovoltaic cells are starved and a nuclear plant is still supplying power. Solar arrays should therefore be somewhat larger than strictly necessary, so that they can supply much more power during the day to compensate for the difference.

Current designs for surface nuclear power generators are monolithic in order to simplify their transportation. They deliver a power output of 0.01–1 MW. In addition to the reactor, a monolithic system should include the thermal electrical conversion system, the cold source (i.e. a radiator), the power conditioning and distribution systems, and some means of storage such as batteries and/or fuel tanks for fuel cells. On Mars, energy storage is essential, because a failure would have catastrophic consequences. There must therefore be apparatus to power at least the life support system in the event of a power outage. It may be safer to locate the storage device near to the ECLSS, so that the latter remains powered even if something goes wrong with the transmission line.

---

[6]C. Cooper, W. Hofstetter, J.A. Hoffman, E.F. Crawley, "Assessment of architectural options for surface power generation and energy storage on human Mars missions," *Acta Astronautica* vol. 7–8, pp. 1106–112, 2010.

An alternative to batteries may be a fairly standard emergency generator powered by an internal combustion engine (ICE) such as those commonly used in hospitals, but in this case fed with methane and oxygen produced on Mars. An alternative to an ICE may be a steam engine. A boiler that uses methane and oxygen could operate a steam power generator and also produce hot water to heat the habitat. An ICE may be a good solution if the rovers are powered in the same manner, giving the outpost a general purpose store of fuel and oxygen produced by the ISPP devices. If the rovers have an electric transmission, an interruption in the main power supply of the base might be temporarily dealt with by hooking the electrical system to the generators in the rovers and running their engines.

If more than one mission lands in the same place, the first mission, perhaps a short stay one, might have no electrical backup system. The next mission in the project would carry a second nuclear generator to give the base redundancy. If the first mission suffered a severe power emergency, its crew would probably have to abandon the planet and try to survive in Mars orbit. As soon as a second generator is on site, such a failure will be less critical.

A nuclear reactor is best buried, to ensure radiological protection and to save shielding mass. Conventional fuel and architectures limit development cost and risks. In the design of the power distribution system an important factor is the distance from the power plant to the habitat. If the reactor is buried in regolith, radiological practice indicates 100 m as being a safe distance, but this may rule out low voltage DC transmission. Consequently, the choice between AC or DC operation (or a combination of the two), the frequency of the AC, and the output voltage will depend on the conversion technology and the applications. With a typical aircraft voltage of 28V it will be possible to use off-the-shelf components such as alternators and motors, but it may be necessary to use a higher voltage to limit transmission (ohmic) losses. The primary power distribution of the ISS is 137–173 Vdc and the secondary is at 124.5 Vdc. However, for heritage reasons the Russian segment works at 28 Vdc.

Several small nuclear fission units are currently under development for use in space or on celestial bodies. For instance, NASA is working with DOE on a power unit that is based on a fission reactor and a power-conversion unit consisting of two Stirling engines. Pumped liquid metal is used to transfer heat from the reactor to the engines, which in turn power the generator for a design power output of 40 kW.

The advantage of using small, compact units is that it is possible to have two or more of them and thereby enable an outpost to survive a failure, even if with a reduced power. This concept may also allow a reduction in the number of backup batteries that must be available for use in an emergency, thereby saving mass.

Finally, it must be noted that a nuclear power system on Mars is less complicated than one designed for space, because in space the biggest problem is to get rid of waste heat. This can only be done by radiating heat to vacuum. It requires the waste heat be dissipated at a higher temperature. By reducing the efficiency, this causes the system to generate a higher thermal power to supply the desired electric power. The dispose of waste heat on the surface of Mars is much simpler, even though the low density of the atmosphere limits the scope for convective cooling of the radiator. In fact, the waste heat can be used to heat water, which in turn can heat the habitat, thereby realizing something similar to tele-heating on Earth.

Heat may be conveyed underground, to melt permafrost and produce water which can be pumped to the surface. This will be a much simpler way of extracting water from the regolith than by digging it up.

8.6   IN SITU RESOURCES UTILIZATION (ISRU) PLANT

Exploiting in situ resources on Mars will be much easier than on the Moon, because the main resources on the Red Planet are the atmospheric carbon dioxide and the water which has been demonstrated to be present in large quantities as ice, either on the surface at the poles or at a shallow depth as a permafrost almost everywhere else except in the equatorial zone. Neither of these resources will be particularly difficult to extract.

In contrast, there is no atmosphere on the Moon and the only water supply is intermixed with the regolith on the dark floors of polar craters. The main lunar resource is the regolith, which contains large quantities of oxygen in the form of oxides and other compounds in the minerals. Its extraction implies heating the regolith to high temperatures and then collecting and purifying the oxygen. It will also be possible to extract a variety of metals. And there is the much discussed helium-3 that may one day run fusion power plants. ISRU will involve collecting, processing and then disposing of large volumes of regolith; precisely how much depends on the rarity of the resource being sought. Certainly such processing will require a large amount of energy. One advantage of setting up mining operations at one of the poles will be near continuous sunlight for solar power.

The Moon has been proposed as an important source of propellant for a mission to Mars, but it has been shown (see, for instance [22]) that this has an intrinsically low efficiency. We must therefore ask ourselves if it makes sense to extract propellant from the Moon to power chemical rockets for Mars, when we can use nuclear rockets which need much less propellant, and no oxygen. This is not to say we oughtn't to produce propellant and other resources on the Moon, just that they should not be used, with a low overall efficiency, for Mars exploration. The rationale goes as follows: If the concentration of ice grains in the regolith is small, this suggests that the water that is present in any given locality is so precious that we shouldn't mine it simply to produce LOX and LH2 as propellants, we should leave it for a future when large lunar settlements need the water for life support.

The logic that we should learn to exploit lunar resources in order to be better able to do so on Mars is questionable, because the resources and the extraction processes are so different. On the Moon we can (and perhaps must) do many things which will assist with human Mars exploration, but producing propellant from ice in regolith is probably the least useful activity.

ISRU research for Mars is focused primarily on providing rocket propellant for a return flight to Earth (perhaps initially by a sample return mission and later by a human crew) or for use as fuel on Mars itself. Lifting off from Mars will require many tons of propellant, with the amount increasing as the altitude of the orbit increases, and even more will be required for the escape maneuver to head home. However, there are trade-offs and producing propellant on Mars will substantially reduce the quantity that must otherwise be carried out from Earth. Chapter 7 discussed In Situ Propellant Production (ISPP) in some detail.

The composition of Mars' atmosphere as raw material for ISRU is quite well known and, since it can be easily simulated on Earth, the relevant processes and machinery are relatively simple to test on Earth.[7]

The first point to consider is compressing the atmospheric gases. The pressure and density are so low that the compression ratio to attain a pressure of 1 bar is about 166, and even a much lower pressure will require a multistage compressor. The energy needed to compress the raw material, and the mass and bulk of the associated machinery are by no means trivial.

If the fuel, either methane or hydrogen, is brought from Earth, the easiest solution is to use the Reverse Water Gas Shift (RWGS) reaction

$$CO_2 + H_2 \rightarrow CO + H_2O$$

to produce water, from which oxygen can be extracted by electrolysis. Because hydrogen is also obtained by electrolyzing water, the reaction uses only a small amount of hydrogen from Earth. The net result is oxygen which, among other applications, can be used as the oxidizer for rocket engines.

Another proposal for the production of oxygen is the electrolysis of the atmospheric carbon dioxide

$$2CO_2 \rightarrow 2CO + O_2.$$

The mass saving for a methane/oxygen rocket varies with the fuel-to-oxidizer ratio of the rocket. The stoichiometric ratio between the mass of oxygen and that of methane is about 4, so the reduction of the mass to be carried in order to fuel the Mars Ascent Vehicle should be 80 percent. However, the maximum specific impulse is obtained with a ratio of 2.77, for which the reduction is only of 73 percent. (If the MAV uses LOX and LH2 the ratio is about 1:5 and the gain increases to 83 percent. If water is extracted from the permafrost, it would be possible to directly produce LOX and LH2. This combination delivers a much higher specific impulse than methane/oxygen, but poses the issues of long term storage of hydrogen, particularly its boil-off.)

The reaction to produce methane takes place rapidly in the presence of an iron-chrome catalyst at 400°C, and has been verified by NASA in a test on Earth. If the propellants to be produced on Mars are methane and oxygen, then only hydrogen need be carried to Mars. A typical proposal for ISRU is to use the Sabatier reaction

$$CO_2 + 4H_2 \rightarrow CH_4 + 2H_2O$$

and then electrolyzing the water to obtain oxygen and hydrogen. The latter is recycled into the Sabatier reactor. Since only hydrogen is brought from Earth, the mass advantage is even better.

---

[7] K.R. Sridhar, J.E. Finn, and M.H. Kliss, "In-Situ Resource Utilization Technologies for Mars Life Support Systems," *Advances in Space Research*, vol. 25, 2, pp. 249–255, 2000.

If all the resulting water is recycled, then for 1 kg of hydrogen and 22 kg of atmospheric carbon dioxide, it is possible to produce 16 kg of oxygen and 8 kg of methane. About 22 kg of oxygen is required to burn 8 kg of methane. Therefore the reaction produces less oxygen than is required in order to burn the methane that it produces. To increase the efficiency of the ISPP system, hydrogen can be extracted from water in order to produce oxygen and methane in the correct stoichiometric ratio – which means little, since the best specific impulse is obtained by non-stoichiometric burning.

Mining water on Mars is not as simple as extracting carbon dioxide from the atmosphere. If the water management system of the outpost is not based on a regenerative ECLSS, then it is possible to consider using waste water to produce hydrogen and oxygen as propellants, but this has at least two drawbacks. First, the fuel for the MAV cannot be produced in advance, as would be advisable for safety reason, but only after some water has been used. And second, by neither regenerating water nor obtaining it on Mars would require carrying large amounts of it from Earth.

It will be possible to produce plastics on Mars from water and carbon dioxide. The basic reactions provide the building blocks for more complex reaction series which are able to make plastics. Ethylene is used to make plastics such as polyethylene and polypropylene and can be made from a reaction similar to that involved in the production of methane

$$2CO + 4H_2 \rightarrow C_2H_4 + 2H_2O.$$

The regolith can be used more or less directly to build habitats using 3D printers, and it can also be used as a reinforcement for plastics made on site. Another approach is to produce concrete from regolith and then apply conventional building techniques. It has been suggested that buildings on Mars could be made from basalt, which has good insulating properties. An underground structure of this type would provide ample protection against space radiation.

The main components of the ISRU plant are therefore a Reverse Water Gas Shift (RWGS) reactor that reduces CO_2 by means of hydrogen; a water separator to remove water from CO, and an electrolyzer to dissociate the water into O_2 and H_2. If methane is required, the main components of the In Situ Propellant Production Plant (ISPP) are a Sabatier reactor to reduce the CO_2 to produce methane by means of hydrogen; a water separator to remove water from methane; and an electrolyzer to dissociate the water into O_2 and H_2.

ISRU development should start with an automated testbed demonstration and lead to a pilot propellant plant. The ISRU plants must first be tested on Earth, and then on Mars by an automated mission.

8.7   WORKSHOPS, GREENHOUSES AND AUXILIARY EQUIPMENT

Depending on the length of stay, astronauts must be able to repair their vital devices such as the ECLSS and ISRU plants, and rovers. They will need a purposely designed pressurized portion of the habitat into which faulty equipment can be taken. The ability to produce their own spares would both enhance overall safety and reduce the need for a large stock of spare parts. Additive manufacturing technologies are an important step towards this goal. It is thus likely that a workshop for such manufacturing will be needed on Mars (and perhaps also on the interplanetary spacecraft). If no artificial gravity is used, the two additive manufacturing machines may require to be different. Such machines able to operate in weightlessness are at present under experimentation on the ISS and it is clear that by the time that a human Mars mission is attempted, much experience will have been gained. All tools for maintenance and repair must be present in the habitat and there should be redundancy for critical materials.

It is essential that all devices carried to Mars are designed with this strategy in mind, so that a single machine, of a given size, is able to maintain all the critical equipment (including itself, of course). It will be possible to transmit revised software from Earth to improve the range of components. Although this may sound fanciful, additive manufacturing techniques are undergoing a very rapid development.

The design of all machinery to be used at the outpost must be coordinated to save mass, and the possibility of starting from materials locally available, or at least by recycling waste material, must be specifically addressed. These studies must focus on the differences between additive manufacturing on Earth and in space, taking into account both the technologies and the operating conditions. On Earth, for instance, the mass of raw materials (e.g. powders) is of secondary concern and what matters is the saving to be made on the process and perhaps on the cost of the raw materials. On Mars, however, the most significant cost is carrying things to the planet, which is essentially a case of the mass involved (see Section 12.1).

A greenhouse in which to conduct experiments and grow food would improve the crew diet, with the resulting benefits to their physiological and psychological health. It would also simplify logistical issues related to carrying supplies. Technologies that will achieve food independence are essential to establishing a permanent Mars base and, in the longer term, the terraforming of the planet.

The plants on Earth are optimized to grow in terrestrial conditions and with light that has a spectral range produced by the Sun. On Mars, it would be possible to expose plants directly to sunlight by using a greenhouse with a transparent covering, but this would expose the plants to cosmic radiation. There is no requirement for the greenhouse to have an atmosphere which is compatible with humans, hence the plants might enjoy an atmosphere that is too rich in carbon dioxide for human survival. The fact that the atmospheric pressure on Mars is too low means the local air will need to be compressed. Therefore the greenhouse must be pressurized, even if not as much as the human habitat.

To minimize the degree to which plants are exposed to radiation, it will be necessary to use a non-transparent greenhouse and use mirrors to illuminate the interior, directing sunlight through paths that cannot be followed by either cosmic rays or solar ionizing radiation.

Until the development of LED lighting, the idea of using artificial light to grow plants on an important scale had always been discarded due to the low efficiency. This is no longer the case. The efficiency of artificial lighting is now adequate to grow plants in a greenhouse where a roof protects the plants and the people who tend them from cosmic radiation. Once the concerns about forward contamination of underground sites have been dismissed, the ideal greenhouses may be lava tubes because they are impervious to cosmic radiation, the pressure can be high, the temperature can be easily controlled, and an atmosphere that is rich in carbon dioxide can be maintained. It would even be possible to grow trees as a local source of building materials.

The high efficiency of LEDs offers the benefit of producing light with little heat. A large number of tiny light sources can be located very close to the leaves of the plants, so that the overall efficiency in terms of electric power to food is high. As a result, the production of food requires only power, carbon dioxide from the atmosphere and water, plus fertilizers. From this viewpoint, a plant is just a machine that converts electric power into edible biomass. If there is a nuclear reactor producing sufficient power, then large yields are possible.

Several experiments are being conducted aboard the ISS with good success (Figure 8.8). A number of plants have been chosen for their good yield and possible adaptation to a soil created from the Martian regolith. Potatoes are one possible choice (as in the movie *The Martian*) and asparagus is well suited to soils rich in iron.[8]

Genetically engineered plants could live in an atmosphere more like Mars' atmosphere, and this would be important in the long term. The lower the pressure of the greenhouse and the higher its carbon dioxide content, the cheaper will be the growing of plants and the lower will be the damage in the event of a system failure. Furthermore, plants that are engineered to grow well in a cold climate will be an asset on Mars, especially if they can be made relatively insensitive to radiation.

If all of the missions of the Mars campaign land in the same place, the greenhouse can be automated in order that the plants will continue to grow when the outpost is uninhabited. Or, even better, the greenhouse may be established before the first astronauts land, so that they will find food already growing and, likely, ready to be harvested.

One intriguing option related to growing plants on Mars is the possible use of nitrogen. Since the regolith at the surface has probably never hosted biological processes (or even inert organic materials) some fertilizers will be required, and producing them in situ is potentially interesting.

Figure 8.8 An astronaut checks lettuce plants in a vegetable production system called 'Veggie' aboard the ISS. The purple glow comes from LED lamps which provide the plants with a sense of direction. (NASA image)

---

[8] A. Sommariva, G. Bignami, "Oro dagli asteroidi e asparagi da Marte," Mondadori, Milano, 2015.

The generic heading of 'auxiliary equipment' includes all of those devices required by specific aspects of the mission that aren't listed in previous headings. These include digging and construction equipment, fuel tanks, navigation beacons, telecommunication devices, and many others. They can be determined only in the planning for each specific mission. Since such equipment will accumulate at the base after each mission, a large variety of types and functions can be considered. For instance, the first mission can bring just a small workshop; the second a greenhouse and provisions to enlarge the workshop; the third some machinery to build roads at least to the locations of greatest interest, and so on. Pretty soon the outpost will have considerable capabilities.

8.8   SPACE SUITS

As noted in Chapter 5, even in the lowest-lying terrain on Mars the atmospheric pressure is insufficient for humans to survive without specific protection. A Crew Altitude Protection Suit that can be worn beneath a space suit is effective only at pressures higher than 20 mbar, which is twice the maximum pressure found on Mars. As the pressure varies throughout the Martian year, much lower values can be experienced. This is even more true at elevated sites. Space suits are thus required for external activity on the planet, as they are in space.

The spacecraft must be provided with standard space suits, both for possible EVA while in space and for emergencies. Perhaps the lander can be provided only with space suits designed for working on the planet's surface in order to save mass by not carrying both types.

In principle, a space suit for operating on Mars is not much different from one designed for the Moon. The Apollo space suits (Figure 8.9) worked well, and could be donned and doffed inside the very cramped cabin of the Lunar Module. One of the main problems in designing the Apollo suits, usually referred to as Extravehicular Mobility Units (EMU) was thermal control and insulation. Since the thermal range on the Martian surface is less severe, designing a suit should be less challenging. Because Martian gravity is stronger than lunar gravity, a suit for Mars must be lighter, and particularly the Portable Life Support System (PLSS) worn on the back.

Various designs and technologies are under evaluation for a Mars space suit. NASA has been working on the so-called Z-series as its next-generation space suit. Each iteration of the Z-series will advance the state of the art for astronauts who must work on the planet's surface. The newest Z-2 prototype employs a hard composite upper torso that provides the long term durability required for sustained exterior work. After extensive evaluation of the Z-1 to study different ways of optimizing mobility of the complex shoulder and hip joints, a new style was designed for the Z-2. It is also equipped for use with a suitport. The boots are much closer to those of a suit intended for use in space, and the materials are compatible with a full vacuum. The final design for a Mars suit will be thoroughly tested to assess mobility, comfort, and performance in a very harsh thermal and radiation environment. Whilst it will be configured for the Martian environment, it should be possible to test the suit on the Moon.

Several suits were developed during the Apollo program, some optimized for internal use and others for EVA, both in space and on Moon, with those for use on the lunar surface

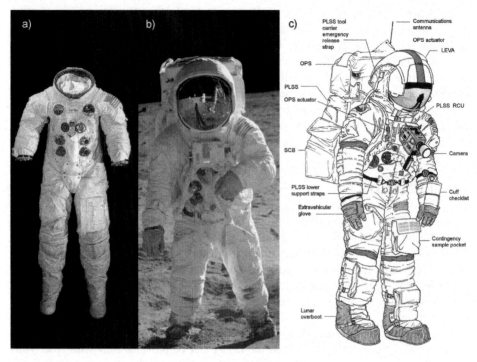

Figure 8.9　Apollo Extravehicular Mobility Unit (EMU). (a) Neil Armstrong's space suit. (b) Buzz Aldrin in his space suit on the Moon. (c) Sketch of the EMU with LEVA: Lunar Extra-vehicular Visor Assembly; OPS: Oxygen Purge System; PLSS: Portable Life Support System; RCU: Remote Control Unit; SCB: Sample Collection Bag. (NASA images)

becoming increasingly flexible to enable the astronauts to bend at the waist and kneel down to pick up rocks. Suits for Mars will need to provide at least a similar freedom of movement. Alternatively, the possibility of designing motorized space suits (something between a space suit and an exoskeleton) may help to reduce fatigue both in space and on planetary surfaces.

The experience gained on the Moon showed that lunar (and hence also Martian) dust is dangerous to space suits. Over time, some joints of the Apollo suits tended to become difficult to move. This was annoying, but not life threatening. Those suits were intended to be used for a short time. For a Mars mission, in particular in a long stay mission, the space suits will have to operate for many months (and probably considerably longer), and so the effect of dust must be well understood and dealt with. One point must be considered, however. Mars dust is likely less dangerous than lunar dust. The minuscule fragments of rock that make up lunar dust possess very sharp edges and tend to lock against each other, making them cohesive. But in ancient times the Martian dust grains were subjected to water and ever since then have been eroded by winds, therefore they are more rounded. And being somewhat smaller than lunar dust they will even more readily penetrate machinery and biological systems, albeit with less dangerous consequences.

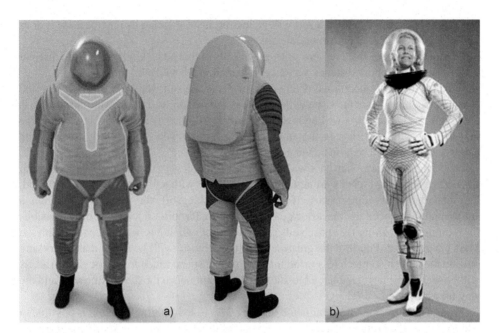

Figure 8.10 (a) The NASA Z-2 space suit. (b) A 'biosuit' for EVA on Mars being worn by the scientist who developed it, Dava Newman from MIT.

Some suits for Mars have been proposed that promise to be very light and comfortable. They are essentially a very thin elastic skin which exerts sufficient pressure on the human body to replicate the experience of Earth-normal pressure. Figure 8.10b shows the 'biosuit' designed at MIT with the cooperation of the Italian Dainese company. This design radically changes the way that a space suit maintains the desired pressure around the body. A standard space suit is essentially only an airtight container into which air is pumped at the pressure required to create an enclosed portable environment in which it is possible for a human to live. This has many advantages, but the suit inflates like a balloon, making all movements difficult, or at least tiring. The alternative is to put a 'second skin' around the body which shrinks and exerts the required pressure mechanically. It works, to an extent, like the 'elastic stockings' or 'socks' which exert a pressure on the legs to improve venous circulation and comfort when standing for lengthy periods of time. Designing a full body suit in this style was a challenge because elastic garments are difficult to don, until MIT developed a self-shrinking material that incorporated Shape Memory Alloy (SMA) actuators with 3D printed structures and passive fabric materials.

The suit is put on whilst in its slack state, then an electric current is applied to heat the material, causing it to shrink tight against the body and exert the desired pressure. Not only does this suit not impede movement (as would a standard space suit) it directly supports the body. The latter will be important during the first days on Mars, if the vehicle was weightless during the interplanetary journey. To doff the suit, a moderate force is sufficient to loosen the garment. This development offers great promise and further research may enable such a suit to support a broken bone or stop a hemorrhage, possibly in an automatic way. It is also likely that such a suit will be less prone to damage by dust because it does not have all the joints of conventional suits.

Obviously, like normal pressure suits, space suits of this novel kind will require thermal control and life support systems, very likely located in a backpack.

On the Moon the operational challenges for the space suit were related to keeping the astronaut cool. On Mars the issue will be to keep the person warm. The Martian surface is extremely cold. The boots will be highly insulated. All other parts of the body that come into contact with materials at the ambient temperature, even momentarily, such as the hands and knees, will require adequate protection.

New kinds of space suit have a deep effect on the design of any device linked with EVAs. For instance, biosuits are not easily compatible with suitports. On the other hand, they make suitports unnecessary. If wearing, and taking off, the space suit becomes a simple and fast procedure, the airlock can be just a small room, so long as it has provisions for thoroughly cleaning the space suits and any object which has been in contact with the Mars environment. Possibly the greatest benefit of a biosuit will be that it makes working outside the habitat so much less tiring that people can spend much longer outside. And, as will be seen in the next chapter, a biosuit will deeply influence the design of rovers and the way they are used.

Protecting astronauts from radiation during EVAs is an important topic. It is possible to add some specific protections, for instance pockets filled with water, in specific parts of the space suit in order to protect some organs that are more sensitive to radiation damage. In this respect, it must be remembered that the most dangerous radiations on Mars are coming from directly above, because this is the shortest distance to space; even though the air is so thin, lines of passage at oblique angles offer increased protection. This selective radiation protection is being exploited by ESA in its PERSEO project. Another factor is that owing to their differences in sensitivity to radiation, the protection for space suits intended for women may be very different from those for men.

8.9 PLANETARY PROTECTION

Protection from contamination, both forward (protecting Mars from Earth) and backward (protecting the astronauts and Earth from Mars) must be implemented in the design of the outpost on Mars. This must be a key aspect of the mission design. In particular, biological materials (including human beings) must be kept separated from the planetary environment.

Forward planetary protection becomes ever more difficult as the variety of biological materials there increases: the presence of a greenhouse, for instance, makes this issue very sensitive. To prevent backward contamination, from possible biological Martian material and from the Martian dust and fines, materials exposed to the Martian environment must be cleaned before being introduced into the habitat or pressurized rovers.

The dust can be controlled using electrostatic precipitators such as are currently being investigated for future lunar activities, and suit-locks are preferred to standard airlocks.

Ever since we started sending probes into deep space, we have worked to understand the potential for biological contamination. The Apollo astronauts returning from the Moon were quarantined until it was certain that there was no threat of back contamination.

Recently, as targets of exploration progressively become more distant, consideration is also being given to the protection of planets in the outer solar system, their moons, and various small bodies, particularly those that may possess subsurface oceans in which life may have developed.

NASA's Planetary Protection policy calls for the imposition of contamination controls for certain combinations of mission type and target body. There are five categories for target body/mission type combinations but only Categories IV and V are described here, following [32].

Category IV includes certain types of mission (typically an entry probe, lander, or rover) to a target body of chemical evolution or origin-of-life interest, or for which scientific opinion says that the mission would present a significant chance of contamination that could jeopardize future biological exploration.

The requirements include:

- Very detailed documentation.
- Bioassays to enumerate the burden.
- A probability-of-contamination analysis.
- An inventory of the bulk constituent organics.
- An increased number of implementation procedures.

The latter may include:

- Trajectory biasing to ensure that spent propulsive stages will pose no threat.
- The use of clean rooms (Class 100,000 or better) during the assembly and testing of the spacecraft.
- Bioload reduction.
- Possible partial sterilization of the hardware that will come into direct contact with the target body, and in rare cases a bio-shield for that hardware.
- Full sterilization of the entire spacecraft.

Subdivisions of Category IV (designated IVa, IVb, and IVc) apply to lander and rover missions to Mars (regardless of whether they have experiments designed to detect life) and missions that land on or access regions of the planet of particularly high biological interest. Category IVc refers to Special Regions. A Special Region is "a region within which terrestrial organisms are likely to propagate or a region which is interpreted to have a high potential for the existence of extant Martian life forms."

In current understanding, this applies to regions where liquid water is present or may occur. Committee (SR-SAG) of the Mars Exploration Program Analysis Group (MEPAG) has been tasked with investigating both the limits to microbial life and the potential for biologically available liquid water on Mars. It first determined that, in order to proceed with identifying Special Regions, the definition of the relevant words needed clarification. In this connection, "propagate" was taken to mean reproduction (not just growth or dispersal), while "likely" was taken to mean the probability of specific geological conditions occurring during a certain time period (rather than the probability of growth of terrestrial organisms). The SR-SAG analysis indicated that the definition of a Special Region is determined by a lower temperature limit for propagation ($-20°C$, including margin) and a lower limit for water activity (a threshold of 0.5, with margin). Furthermore, some remotely sensed features on Mars were categorized as being Uncertain. A later COSPAR Colloquium confirmed 0.5 for the water activity threshold and revised the lower temperature limit of for propagation to −25°C.

More recently, in the light of a new body of information drawn from multiple disciplines, a further analysis of Special Regions on Mars was carried out by a second MEPAG/SR-SAG2 group. This included a review and reconsideration of the parameters used to define a Special Region, and it updated maps and descriptions of those environments that were recommended as being either Special or Uncertain. These environments include natural features, as well as those resulting from future spacecraft landings.

In the latter context, the committee considered the impact of Special Regions on the potential for human exploration of Mars, having regard both to the locations of potential local resources and to places which ought not to be inadvertently contaminated by human activity. Overall, significant advances in our understanding of terrestrial organisms plus the capability now achieved to identify possibly habitable Martian environments, have created a new perception as to where Mars Special Regions may be located and how they should be provided with protection from contamination.

Since it is almost impossible to be certain that the zones of the planet that are visited by humans will not become contaminated, the logical strategy is to send humans only after it has been established that the planet does not host any form of life, and no harm will come from forward contamination. In reality, this would mean postponing indefinitely any human Mars presence on the planet, because it is very difficult to reach any such certainty without resorting to human exploration.

A step-by-step procedure therefore seems more reasonable. A zone in which indigenous life seems unlikely to exist (determined by analyzing the specimens carried back to Earth by a sample return mission) can be selected and everywhere else classified as Special. Humans then land in this permitted location, and send sterile teleoperated rovers into the surrounding areas to bring samples back to the outpost, where they can be studied to look for life. These areas can then be removed from the Special Region when are shown to be lifeless. In this way, it is possible that a fairly large portion of the planet will be opened to human exploration and that Special Regions will be restricted to some particular places, like the bottom of deep canyons, caves, hydrothermal vents (if they exist), etc., where there is a non-negligible probability of finding life. With this strategy, it is likely that either life is found or, in a reasonable time, it is possible to conclude with certainty that Mars is lifeless.

Category V applies to all missions for which the spacecraft, or a spacecraft component, will return to Earth. The issue is to protect Earth from backward contamination caused by the return of extra-terrestrial samples (usually regolith and small rocks). The subcategory called Unrestricted Earth Return is defined for solar system bodies deemed by scientific opinion to have no indigenous life. Missions in this subcategory impose requirements on the outbound (Earth to target) phase. For all other Category V missions, in the subcategory of Restricted Earth Return, the highest degree of concern is articulated by requiring: the absolute prohibition of destructive impact upon return; the need for containment throughout the return phase of all returning hardware which directly contacted the target body, or non-sterilized material from that body; and the need for containment of any non-sterilized samples being returned to Earth. Post-mission, there is a need to conduct timely analyses of the returned non-sterilized samples under strict containment and using the most sensitive techniques. If any sign of the presence of a non-terrestrial replicating organism is found, the returned sample must remain in containment unless treated by an effective sterilization procedure.

These Category V concerns are reflected in requirements which encompass those of Category IV with added continuous monitoring of mission activities, advanced studies, and ongoing research in sterilization procedures and containment techniques. NASA has recently appointed a dedicated panel to consider the possibility of organic contamination involving their proposed Mars 2020 mission in which a rover will look for signs of past life, collect samples for possible future return to Earth, and demonstrate technologies for future human exploration. The science from this rover's instruments is expected to yield essential context to inform decisions about whether samples should be returned to Earth. A report issued by this panel also considered the possibility of organic contamination of Martian samples.

The requirements for the provision of witness plates, archive facilities, and blanks and standards are now under consideration by NASA.

An important aspect of planning human exploration of Mars will therefore be developing procedures to prevent contamination of the outpost by automatic or crewed vehicles which return from Special Regions.

9

Mobility on Mars

On the surface of Mars, astronauts must have at their disposal the means for exploring a suitable area of the planet. A ground vehicle was tested for the first time outside Earth during the Apollo program, but the longer stay and the wider extent of the exploration will make similar vehicles designed for Mars larger, faster, and more complex. In later missions, transportation may also require aerial vehicles. Ultimately there will be an entire transportation infrastructure. Robotic rovers will be required to assist the astronauts in their exploration tasks.

9.1 GENERAL CONSIDERATIONS

In designing transportation devices for astronauts on Mars, the first task is to decide the goals and the extent of the exploration mission, as these will determine the mobility requirements.

The first requirement is that the landing takes place on a site where landing is relatively easy and safe. Since these sites are, generally speaking, not ones which scientists consider to be of the greatest interest, it will be necessary for explorers to travel substantial distances from the landing site. For instance, Valles Marineris and similar deep canyons are potentially very interesting for science, but it will be very difficult to land close to them and, moreover, they will also be difficult to reach by traveling over ground. If the surface of Mars is subdivided into a zone where humans may enter and a zone which can be explored only by robots, then the robots must have the required mobility.

Clearly the type of mission influences the type of land mobility required. A short stay mission cannot include long range exploration travel because there will not be sufficient time. But a long stay mission will have ample time to range widely across a large area of the planet.

Despite the fact that ground vehicles may be supported and propelled by wheels, tracks, legs, snakelike or apodal devices, or other means of locomotion often simply referred to as unconventional, at present the exploration of Mars is performed using robotic rovers which rely on wheels (Fig. 9.1a).

© Springer International Publishing Switzerland 2017
G. Genta, *Next Stop Mars*, Springer Praxis Books, DOI 10.1007/978-3-319-44311-9_9

Figure 9.1 (**a**) MSL-Curiosity is the largest rover ever sent to Mars. It is about 3 m long by 2.8 m wide and 2.1 m tall. It has a mass of about 900 kg and its wheels are 508 mm in diameter. (**b**) The first rover ever sent to Mars was very small with a mass of only 4.5 kg, and was carried by the Mars 2 mission launched by the Soviet Union in 1971. Although the probe entered the planet's atmosphere, no further signals were received and it is believed to have crashed.

Other types of land locomotion devices have been proposed over the years, but never assigned to actual missions, with the exception of the earliest rovers that were carried by the Soviet Union's Mars 2 and Mars 3 missions in 1971. In each case, a lander was to release a rover that would use a pair of ski-like mechanisms to move about 15 m from the lander, to which it would remain connected by an umbilical (Fig. 9.1b). Mars 2 evidently crashed. Mars 3 landed but its transmission ceased after just 20 s and therefore the rover was unable to be tested.

Airborne vehicles for Mars may be supported by aerostatic or aerodynamic forces, but there are other possibilities such as jet sustentation.

Finally, hoppers that take off under rocket propulsion, make a parabolic flight and then land using either the same rocket or parachutes can be used on Mars. Hoppers propelled by springs or electromagnetic actuators have also been proposed, but they are likely to be better suited to bodies with much lower gravity than Mars (e.g. the Martian moon Phobos). Other options such as electromagnetic levitation and propulsion have been proposed but seldom studied in detail.

9.2 ROVERS

Ground vehicles are considered essential to human exploration of Mars, and the outpost needs to be supplied with one or more of the following:

- A small robotic or teleoperated rover or astronaut assistant.
- A small human-rated, unpressurized ground vehicle.
- A human-rated, unpressurized or semi-pressurized ground vehicle.
- A pressurized ground vehicle.
- A large pressurized mobile habitat.

9.2.1   Small robotic rovers

One or more small robotic rovers could accompany astronauts on EVAs to act as assistants. They could also scout ahead of human-carrying rovers, and enter areas that are either too dangerous or off-limits to humans. The degree of autonomy of such rovers can vary from fully autonomous to simply teleoperated. Robotic swarms can also be made autonomous. Due to the proximity of humans, the fully teleoperated option is not only practicable, it allows faster travel. The rationale is that if the most precious resource on Mars is the time available to the astronauts, then it is questionable whether this should be spent driving rovers.

Increasing the autonomy of rovers therefore remains an important goal. In particular, if the landing site is to be the same for all human missions, it would be conceivable (although costly) to land unmanned rovers at interesting points on the planet and then teleoperate them from the outpost. This requires Mars orbiting telecommunication satellites. Similar to robotic rovers are 'assistants' that accompany astronauts during EVAs and support their work. Robots could be built to perform both functions. A centauroid astronaut assistant built by NASA is shown in Fig. 9.2.

Specialized rovers may assemble and repair machinery, assist astronauts with assembly and repair tasks, particularly of the nuclear reactor. Again the degree of autonomy required may range from fully autonomous robots to teleoperated machines The size of these robotic rovers will depend upon a variety of factors. If they are explorers, then the miniaturization of instruments might permit very small rovers, but the small size of the running gear limits the capability to manage obstacles. In the case of a rover which has different tasks, these will define its minimum size.

Figure 9.2 The wheeled version of the Robonaut built by NASA. It is a centauroid astronaut assistant. (NASA image)

9.2.2   Small unpressurized rovers: the LRV

Small unpressurized rovers such as the Lunar Roving Vehicle (LRV) of the Apollo era, may carry astronauts during EVAs on Mars. Because the LRV is the only vehicle ever to be used by humans on another celestial body, it is worthwhile reviewing it in some detail. It was an electric vehicle with four-wheel drive and steering (4WDS) that could carry two people in space suits and a variety of tools and samples at a speed of 18 km/h for a maximum distance of 120 km (Fig. 9.3).

The main characteristics of the LRV were[1]:

- Mass: 210 kg.
- Payload: 490 kg.
- Length (overall): 3099 mm.
- Wheelbase: 2286 mm.
- Track: 1829 mm.
- Maximum speed: 18 km/h.
- Maximum manageable slope: 25°.
- Maximum manageable obstacle: 300 mm height.
- Maximum manageable crevasse: 700 mm length.
- Range: 120 km accumulated over four traverses.
- Operating life: 78 h.

Figure 9.3 The Lunar Roving Vehicle (LRV) of the final three Apollo missions, seen here on the lunar surface. (NASA image)

---

[1] A. Ellery, *An Introduction to Space Robotics*, Springer Praxis, Chichester, 2000; Boeing, *Lunar Roving Vehicle Operations Handbook*, http://www.hq.nasa.gov/alsj/lrvhand.html

On the final Apollo mission in December 1972, the LRV traveled a total of almost 36 km and reached a maximum 7.6 km distance from the landing site.

At the time it was designed, the LRV was a concentration of high automotive technology adapted to the peculiar operating conditions on the lunar surface. It was the first time that a vehicle was built with four steering and driving wheels, each wheel having its own motor in the hub.

The LRV development was deeply conditioned by the mass and size constraints imposed by the requirement that it be taken to the Moon in a side compartment of the Lunar Module (LM) that carried the astronauts. Apart from the need to minimize the rover's structural mass, these constraints forced its designers to employ a foldable architecture. It is likely that these considerations led to the adoption of by-wire steering, which in those days was a completely immature technology.

The ensuing sections provide a short analysis of the various subsystems of the LRV and some considerations of what is still viable today and could therefore be used in designing a small human-carrying rover for Mars, and also those things which were rendered obsolete by technological advances.

Wheels and tires

Pneumatic or solid rubber tires were rejected primarily to reduce the mass of the vehicle. In their place, tires made from an open steel wire mesh were created with a number of titanium alloy plates acting as treads in the ground contact zone. Inside the tire, a smaller, more rigid frame served as a stop to avoid excessive deformation under high impact loads. The outer diameter of the tire was 818 mm (Fig. 9.9a).

Although the possibility of using standard pneumatic tires on a celestial body without an atmosphere (the Moon), or possessing a very thin atmosphere (Mars) cannot be entirely ruled out, the lack of air is not the point (see Sect. 9.2.6). The predicted short duration of use (a single mission and a total driving time of a few hours) allowed such an innovative design to be chosen without the need to perform long duration tests and analyze fatigue factors. (Future rovers operating on Mars will require to be able to travel for tens of thousands kilometers, particularly if several, if not all, missions employ a single landing site. In this case, more traditional solutions are likely.) The training version of the LRV used by the astronauts on Earth had standard pneumatic tires and a sturdier construction. Indeed, if anyone had been permitted to sit on the real vehicle in normal gravity, it would have collapsed under their weight.

Drive and brake system

The LRV had four independent, series wound, DC brush electric motors mounted in the wheel hubs, along with harmonic drive reduction gears.[2] Each motor was rated at 180 W, giving a maximum speed of 17,000 rpm. The gear ratio of the harmonic drives was 80:1.

---

[2] Harmonic drives are reduction gears based on a compliant gear wheel that meshes inside an internal gear that has one tooth more than the former wheel. An eccentric (wave generator) deforms the first wheel so that, for each revolution of the eccentric, the flexible wheel moves by one pitch. Very high gear ratios can be obtained while maintaining good efficiency and little backlash, but the high cost of such devices allows their use only in selected applications, primarily in robotics.

The nominal input voltage was 36 V, controlled by Pulse Width Modulation (PWM) from the Electronic Control Unit. The drive unit was sealed to maintain an internal pressure of about 0.5 bar for proper lubrication and brush operation. The energy for locomotion was supplied by two silver-zinc primary batteries at a nominal voltage of 36 V and a capacity of 115 Ah (4.14 kWh) each. The use of primary batteries derived from the fact that each vehicle required only to be used by a single mission, and for a limited time.

The LRV had four cable actuated drum brakes directly mounted in the wheels. As will be seen in Sect. D, braking torques are low in vehicles operating in low gravity conditions, and drum brakes are a reasonable solution. They are also preferable in terms of dust, as dust can produce problems for disc brakes because they are more exposed to external contamination. Even though the braking power is much smaller than that typical of vehicles used on Earth, the lack of air may cause overheating problems.

The driver interface for longitudinal control was the same 'T' handle which actuated the steering. Advancing the handle actuated the motors forward. Pulling it back actuated reverse, but only with the reverse switch engaged. Actuating the brakes required the handle to be pivoted backward about the brake pivot point. The wheels could be disengaged from the drive-brake system to adopt a free-wheeling condition.

Suspensions

The LRV had fairly standard double wishbone suspensions, with the upper and lower arms almost parallel. No anti-dive or anti-squat provisions seem to have been included. Springs were torsion bars applied to the two arms, and a conventional shock absorber was located on the diagonal of the quadrilaterals. The ground clearance varied between 356 and 432 mm for the fully loaded and unloaded states; a range of 76 mm. These values yield a vertical stiffness of the suspension-tire assembly of 2.40 kN/m, which is a very low value. Assuming that the tire was much harder than the suspension, the natural frequency in bounce was just 0.6 Hz for the fully loaded vehicle, or 1.1 Hz when unloaded.

Steering

The steering controlled all four wheels and was electrically actuated (steering by-wire). The geometry was designed with kinematic steering in mind: Ackermann steering on each axle and opposite steering of the rear axle with equal angles at front and rear wheels. It provided a kinematic wall-to-wall steering radius of 3.1 m. Each steering mechanism was actuated by an electric motor through a reduction gear and a spur gear sector. If one of the two steering devices malfunctioned, the relevant steering could be centered and blocked so that the vehicle could be driven by steering one axle only.

The same handle that controlled the motors and brakes also carried out steering by lateral displacement. A feedback loop ensured that the wheels were steered by an angle proportional to the lateral displacement of the handle, but there was no force feedback except a restoring force which increased linearly up to a 9° angle on the handle, then increased with a step, and finally increased again linearly with greater stiffness.

Steering control is perhaps the most outdated part of the LRV, although in a way it was a forerunner of the 4WS and steer by-wire systems currently available. The 4WS logic was based strictly on kinematic steering. This could be justified by the low maximum speed, but only to a point. In the low lunar gravity the sideslip angles are much larger than on Earth for a given trajectory and a given speed. Hence the very concept of kinematic steering is applicable only at speeds much lower than on Earth. Because centrifugal acceleration is proportional to the square of the speed, the top speed V_{max} of 18 km/h is equivalent (from this viewpoint) to the speed at which dynamic effects start to appear. No modern 4WS vehicle has such prominent rear axle steering. Today, a much more sophisticated rear steering strategy is present in even the simplest vehicle employing all-wheel steering.

A second difference is that today steer by-wire systems are reversible and the driver interface is haptic, meaning that there is an actuator supplying feedback to allow the driver to feel the wheel reaction, just as in conventional mechanical steering systems. The fact that it is possible to drive without a force feedback is proven by the vehicles in videogames and radio controlled model cars (the control of the LRV has surprising similarities with that of R/C model cars) but it is considered unsafe and difficult to operate a full size car in this manner.

Because astronauts needed much training to operate the LRV, and it was impossible to operate it on Earth, a special trainer was created that simulated aspects of its performance.

9.2.3   Small unpressurized rovers for Mars

Tests have been performed in Mars analogue sites using small single-seat vehicles similar to quads (Fig. 9.4). These may be the best choice for short range mobility, because of their simple construction, ruggedness, and mobility. If a single-seat quad can carry two people in an emergency, this would allow two people on a pair of quads to venture out from the outpost to a distance greater than that from which they could safely walk home in the event of one of the vehicles breaking down. A small rover may be carried aboard a pressurized rover in order to enable astronauts to access difficult places. This could prove a life saver if the pressurized rover malfunctioned or was disabled by an accident.

The likelihood of using quads on Mars depends on the type of space suits worn. If light and comfortable suits are used, it will be possible to use small unpressurized vehicles even for medium distance exploration. Indeed, perhaps motor bikes may be appropriate. Two-wheeled motor vehicles provide excellent off-road capability and are light, rugged, long lasting and easily fixed; all of which are characteristics that will be valuable on Mars.

It is fairly likely that a quad, or as they are often called (particularly by the military) an All-Terrain Vehicle (ATV), will be able to be transformed for Mars use. For instance, the one shown in Fig. 9.5a already possesses non-pneumatic elastic wheels that will be suitable on Mars. The challenge is to modify the engine. Either the internal combustion engine needs to be adapted to work with locally produced methane and oxygen, or it will have to be replaced by one or more electric motors and the associated power system.

The principal characteristics of small quads for civilian use are listed in Table 9.1. They are lightweight, despite their fairly large pneumatic tires. Although some mass savings may be achieved in adapting the structures to the weaker Martian gravity, such savings are likely to be diminished when transforming the engines to run off methane/LOX.

Figure 9.4 A picture taken during a test using quads on a Mars analogue terrain in Austria. The driver is wearing a sort of space suit.

Figure 9.5 (**a**) A military ATV with non-pneumatic tires. (**b**) A dune buggy.

Table 9.1 Main characteristics of three small quads for civilian use

	ICE		Electric
Engine type	2 strokes	4 strokes	DC, 48 V
Capacity/Power	50 cm^3	250 cm^3	1 kW
Size (L×W×H)	1250×660×650 mm	1780×1080×1120 mm	1350×700×650 mm
Mass (approx.)	55 kg	115 kg	70 kg

So-called dune buggies are similar to quads but larger, usually being able to carry two people (Fig. 9.5b). They are used widely for traveling on sand and they should be at their best on the Martian regolith. An ATV often includes a roll-bar as a safety feature. If one is incorporated into a small Mars rover, it will probably support a radiation shield to fend off radiation coming from directly above (as noted earlier, this is the most dangerous angle of incidence because the radiation passes through the shortest path through the atmosphere).

If a quad is provided with sufficient autonomy or it can be teleoperated, then it can be used as an astronaut assistant, with the bonus that the astronaut can ride it during a lengthy traverse or in an emergency.

9.2.4   Medium-size unpressurized and semi-pressurized rovers

A lot of work has been devoted to designing rovers for Mars. In most cases, the result was a vehicle that was quite different from terrestrial vehicles but performed basically similar tasks. But this isn't necessarily the case, and it seems that, in these design exercises, the experience gained in more than a century of designing and building off-road vehicles has been essentially underestimated.

The mobility requirements may include:

- Crossing the Martian landscape at a nominal speed of 25 km/h, with the capability to reach higher speed where the ground permits.
- Climbing 20° slopes and their crests (if necessary at a reduced speed).
- Avoiding or negotiating obstacles with a height of 50 cm.
- Providing autonomous or remote control and handling.

Figure 9.6 shows a typical medium sized Mars rover, as depicted in the science fiction movie *The Martian*. It expresses the essence of what is usually imagined for a rover operating on Mars. Like most designs for Mars, it has six wheels. One may wonder why. Actually, six wheels are useful to reduce the pressure on the ground, which is surely not important on Mars where the low gravity and the seemingly total absence of humidity and products of biological origin in the ground make four wheels sufficient.

Figure 9.6 A fictional Mars rover in the movie *The Martian*, with either pneumatic or solid rubber wheels.

Figure 9.7 A Humvee managing a moderately difficult off-road passage.

The experience of off-road vehicles on Earth shows that four wheels are a good solution except for very large and heavy trucks; and also in the case of six wheels, the three axles are best not equi-spaced, but located in the conventional way of a single axle at the front plus two close together at the rear, with only the front steering. Steering on several axles may be useful for high speed sport driving (although it is not used in races), but it represents an undesirable complexity in off-road driving.

A feature which may differentiate vehicles designed for Mars (or the Moon) from those that we use on Earth, will be their width. On Earth the width of vehicles is limited by existing infrastructures, while on other worlds it may be chosen more freely. Wider vehicles may have several advantages in terms of dynamic and static stability and the possibility of having large habitable compartments.

A medium sized rover for use on Mars may ideally employ the architecture of military reconnaissance vehicles, like the light trucks usually referred to as High Mobility Multipurpose Wheeled Vehicles (HMMWV) and commonly called the Humvee (Fig. 9.7). The one in the figure has a mass of about 2500 kg, a size of 4.57 m (length; wheelbase 3.30 m)×2.16 m (width)×1.83 m (height). It is powered by a V8 diesel engine of 6.2 l capacity yielding 142 kW through a 3 or 4 speed automatic transmission. A vehicle of this kind could be turned into a Mars rover just by using non-pneumatic elastic wheels instead of the pneumatic tires and by using either an electric power-train or an internal combustion engine running on oxygen and methane produced on Mars. In the latter case, owing to the lower gravity and the lower maximum speed, it is likely that a 3 l V6 engine would be sufficient. A possible alternative is also the use of a steam engine of the type proposed in recent years to reduce the fuel consumption and the air pollution of terrestrial road vehicles. Owing to the lower gravity and the removal of the unnecessary features, a significant mass saving could likely be obtained even if the roof, and partially the sides, may be reinforced with radiation shielding, possibly in the form of a tank filled with water or perhaps a sheet of hydrogen-rich plastic.

Whether a medium-sized rover ought to be non-pressurized, semi-pressurized, or fully pressurized depends on the type of space suits worn and on the type of work the astronauts have to undertake during an EVA. If traditional space suits are used, it may be important that the astronauts be able to travel in a shirt-sleeve environment, and this argues for a pressurized rover. On the other hand, if the space suits are comfortable like that of Fig. 8.9b then even long journeys can be made in a non-pressurized vehicle. However, it is not only a matter of comfort. If the task is to travel for short distance, get off to collect samples, get on again to continue the sampling trip, then a pressurized rover would require them to don and doff the space suit very often; something which is best avoided. Safety would dictate that astronauts travel with their suits on, since in a crash a pressurized rover might lose its atmosphere more rapidly than its occupants could don their suits. However, wearing suits inside a pressurized compartment may contaminate the vehicle with dust or, even worse, by Martian biological matter (if any exists).

If journeys exceeding one day are planned, the vehicle must be provided with a pressurized compartment in which the astronauts can spend the very cold night and have their meals. In this case, the rover may need to have two compartments: a pressurized one and a non-pressurized compartment. A medium sized rover based on the vehicle in Fig. 9.7 could accommodate two or perhaps even three suited astronauts on the front seat while traveling, with a pressurized compartment in the rear for halts. The airlock may be located between the two compartments or, perhaps even better, it will be an inflatable structure located on the back wall. Some military vehicles that offer protection against chemical, bacteriological or nuclear contamination, have their cabin entirely sealed off from the environment and (apart from the fact that the pressure difference between the inside and the outside is smaller) they could be considered as a viable model for a pressurized or semi-pressurized Martian rover.

9.2.5 Large pressurized rovers

A large pressurized rover is often regarded as key to sustaining exploration by enabling large areas to be traveled in a shirt-sleeve environment. It must have auxiliary systems, such as an airlock or a suitport to enable astronauts to work outside and a variety of robotic tools which can perform exploration tasks in an efficient manner. A pressurized rover is a very complex system because it must meet requirements similar to those for the habitat, but be mobile and have long range capability.

The pressurized rover habitability requirements may include:

- Providing a pressurized shirt-sleeve environment.
- Storing all crew resources, accommodation, and facilities for periods of 10–14 days.
- Accommodating scientific instruments.

What is here referred to as a large pressurized rover may be as large as a city bus, if not even larger, with a mass of at least 8 t. An example of a reduced scale demonstrator of an 8 t rover is shown in Fig. 9.8. It has a cylindrical pressurized compartment and an inflatable airlock (the white part in the figure) at the back. In fact, a very large pressurized rover would morph into a mobile habitat. Extending the exploration range by relocating the

Figure 9.8 A reduced scale demonstrator of a large pressurized rover built by Thales Alenia Space.

habitat, albeit slowly, is an interesting option. But this would become increasingly difficult with increasing scale of the outpost. If the outpost has ISRU systems, a greenhouse, etc., with the associated power plant, then the difficulties involved in making it mobile rapidly become overwhelming. At any rate, a large pressurized rover may serve as a backup habitat.

The issue of the allowable exploration range for rovers is contentious. On the Moon the distance to which the LRV was permitted to drive was restricted to that from which it would be possible for the astronauts to walk back to the LM after a rover breakdown. With the limited mobility of their space suits and the brief duration of their portable life support systems, it was not very far. Indeed, their radius of action diminished as the EVA progressed and they drew upon their consumables. In planning EVAs, it made sense to start off by driving as far away from the LM as possible and collect samples as they made their way back, timing their halts so that they did not violate the walk back limit. Although this was satisfactory for the Apollo excursions which never ventured more than a few kilometers, it would not be very applicable on Mars.

The exploration range on Mars could be extended by having two rovers, each capable of hosting both crews in an emergency. An alternative may be a light, unpressurized rover that serves as a 'lifeboat.' Also, it is debatable whether an architecture such as the one shown in Fig. 9.8 is actually required. The best solution may be a single front steerable axle and one or two non-steerable rear axles, depending on the mass. The pressurized compartment may be based on a cylindrical shell in order to withstand the pressure differential. Placing two seats, perhaps including the driver, outside the pressurized compartment would allow astronauts to make short trips without donning and doffing their space suits.

9.2.6   Running gear

There are basically three options for the running gear: wheels, tracks, and legs. For Mars, it is possible to exclude legs, and many reasons recommend using wheels rather than tracks [25].

Wheels may have pneumatic tires, wire mesh tires like those of the Apollo LRV (Fig. 9.9a), elastic wheels like those presently developed for robotic rovers (Fig. 9.9b), or non-pneumatic wheels similar to those developed for normal civilian (the tweel by Michelin) or military use (Figs. 9.5a and 9.10). The latter wheels have a rubber tread, which must be either rebuilt using a suitable rubber formulation or replaced by another material (leather seems to be a good solution for operation in a vacuum).

Figure 9.9   (**a**) A wheel of the Apollo LRV. (**b**) An elastic wheel for a small rover designed by the author.

Figure 9.10   A non-pneumatic elastic wheel used on a military reconnaissance vehicle.

It should be possible to design pneumatic tires for use on Mars, even though they may require specific rubber formulations in order to cope with the low temperatures, ultraviolet and ionizing radiation, and operation at very low pressure. However, they may have limited working life due to wear and tear. Perhaps it is possible to design them so that they can use parts, in particular those that are more likely to wear, which can be built on site by additive manufacturing.

If pneumatic tires are used, several spares will be needed; something which may not be simple. Various different tires may be needed, because a large pressurized rover may need different tires from those of a lighter vehicle. If the tire industry can devise tires which are specifically designed for Mars, taking into account the fact that the reduced gravity allows tires to be lighter, perhaps this option might become feasible. In order not to have to extract nitrogen from the Martian atmosphere, it would also be interesting to experiment with tires which inflate with carbon dioxide.

Wire mesh tires of the LRV proved to be light and they performed well in the harsh lunar daytime environment, but they are unlikely to be suitable for long term applications, at least in their present configurations. Despite the low mileage the LRV tires had to withstand, on one Apollo mission a tire failed. This was not critical, because the tires were designed to continue working on an internal emergency rim, but it would have severely limited the performance of the vehicle on a long mission.

Elastic wheels such as those currently used on small robotic rovers, may be not entirely suitable for vehicles carrying humans. But this should be thoroughly tested on Earth by using load relievers to simulate reduced gravity, and perhaps also out on the lunar surface.

9.2.7   Power-train

Rovers may be powered by rechargeable batteries or by hydrogen/oxygen fuel cells, using hydrogen produced by an ISRU plant from Mars water. If the outpost is powered by a nuclear reactor, then the rover's batteries can be recharged, or fuel may be produced, by night or at a time when the power produced exceeds that required by the habitat. In the case of a solar powered habitat, the batteries of idle rovers may be plugged into the power system of the outpost to increase the overall storage capacity.

Robotic rovers intended to operate far from the outpost and from astronauts, may be powered by RTGs, as in the case of the Curiosity rover.

If the rovers are driven by electric motors, then these can be located in the wheels, as was done for the LRV and for most of the robotic rovers. However, this solution causes an increase of the unsprung masses of the vehicle, which for a vehicle carrying humans may have a bad effect on comfort and in general on performance when managing rough terrain. Motors in the wheels also have the drawback of making it difficult to use a variable transmission ratio to the wheels, which may be necessary to give the vehicle the desired trafficability and hill climbing performance, but has the advantage of eliminating the need for a mechanical differential; this may be replaced by an 'electronic differential' function in the motor control units.

Unlike the LRV, which had brush motors, it will be possible to use modern brushless motors, which will increase the power-to-mass ratio and allow electric motors to operate in vacuum; or better, in the Martian atmosphere.

An alternative to electric power is to use a conventional internal combustion engine that is fueled by methane and oxygen, or perhaps hydrogen and oxygen, produced by the ISRU plant. This solution has the advantages of high energy density and relies on our century-old experience of very reliable and low cost internal combustion engines. In this case, it will be possible to use a conventional automotive driveline that is light, reliable, and easily fixed, in particular because the complexities of modern internal combustion engines are mostly linked with pollution regulations that will have no validity on Mars.

Running internal combustion engines on methane (or hydrogen) and terrestrial air is a straightforward process, but on Mars the engine cannot use air as oxidizer; it needs oxygen. This will cause the engine to run too hot, which must be avoided. Apparently the simplest thing will be to mix the oxygen with atmospheric carbon dioxide, but this has its drawbacks because the atmospheric pressure is so low that the carbon dioxide will have to be compressed, which will require heavy compressors and a lot of power. Carrying a supply of compressed carbon dioxide is not a practical solution because to burn 1 kg of methane at a manageable temperature requires about 4.7 kg of oxygen and 11.8 kg of carbon dioxide. The simplest solution seems to be to take the exhaust gases, separate the water for recovery as a valuable resource, and recirculate part of the carbon dioxide. Exhaust Gas Recirculation (EGR) is a common automotive practice that can easily be adapted for use on Mars.

At any rate, low gravity and the absence of relevant aerodynamic forces, at least at the speeds attainable by ground vehicles, makes the power needed for locomotion much lower than is the case on Earth, thereby permitting the use of smaller engines and reduced energy consumption.

9.2.8   Performance and comfort of vehicles in low gravity

The primary difference in operating a ground vehicle on Mars and on Earth is the reduced gravity. In vehicles which use the forces exchanged between the wheels and the ground for propulsion, braking, and trajectory control, the performance will depend essentially on the vertical forces exerted on the ground and hence (in the absence of aerodynamic forces) on weight.

Inertia forces are proportional to the mass of the vehicle, but the forces exerted on the ground are proportional (at least using a linear model) to its weight. The performance of the vehicle in terms of acceleration, braking deceleration, and centrifugal acceleration when it is moving along a curved trajectory decreases (more or less linearly) with decreasing gravitational acceleration. Taking into account only the ability of the wheels to exert longitudinal and lateral forces on the ground, a vehicle operating on Mars accelerates with an acceleration that is about 1/3 of that attained by the same vehicle on Earth and, what is worse, the braking distance is about three times longer. The ability to manage a curved trajectory is similarly decreased for the vehicle on Mars.

Devices like anti-lock systems (ABS) and Vehicle Dynamics Control (VDC) which were recently added to terrestrial road vehicles for increased safety, will be even more important on Mars because they can improve performance in difficult conditions.

On the other hand, because both the forces exerted by the wheels and the component of the weight along the slope decrease in the same way, the ability to manage rough terrain will not be affected by low gravity. Mars' regolith is generally better than similar terrain on

Earth, since what constrains off-road performance is the presence of water in the ground, from wet sand to mud, and above all materials of organic origin. None of these are present on Mars. Very fine dry sand may cause traction problems, but this can be managed.

Comfort is little influenced by the gravitational acceleration of the planet on which the vehicle travels. The criteria for bounce and pitch motions, suspension damping, and other suspension characteristics developed for vehicles on Earth will also apply on Mars. There is even an advantage. One limitation to spring softness in vehicle suspensions derives from the need to limit suspension travel with changing load. In low gravity, if the springs are designed with dynamic considerations in mind, the static deflection under load is small and presents no constraints on suspension softness. The only limit in this area is the need to avoid bounce and pitch frequencies that are too low, because they may induce motion sickness.

The human side is another matter. We know little about how a body which has adapted to being weightless during a long voyage through space will react to vibration under low gravity conditions. The usual guidelines may not apply as they do on Earth. The LRV, for example, had a bounce frequency in its fully loaded condition that was too low for comfort, but there are no reports of any astronaut suffering from motion sickness while driving on the Moon. Further studies are required, but they must be conducted on site because low gravity cannot be properly simulated on Earth.

However, low gravity does cause an unwanted effect on bounce and pitch motions. The wheels tend to lift off the ground, as can be seen in the footage taken on the Apollo missions. It is obvious that at reduced gravitational acceleration, inertia forces become more important with respect to weight and increase the difficulty in maintaining a good wheel-ground contact when traveling on uneven ground.

9.3 BALLOONS AND AIRSHIPS

Aerostats on Mars will have to be large because the aerostatic forces are quite low as a result of the low density of the atmosphere. However, both balloons and airships have been suggested, in particular for robotic exploration.

Balloons have no propulsion and are simply carried by the winds, whereas airships have a means of propulsion and can be steered along a given course. Airships are of three types: rigid, semi-rigid, and non-rigid (also known as blimps) depending on whether their shape is kept by a rigid structure or simply by the gas within them.

The force that a fluid exerts upon a body which is immersed in it is called the Archimedes force (see Sect. D). To supply the required buoyancy, on Earth aerostats are filled with either helium or hydrogen. As shown in Sect. D.3, the difference between these two is marginal. However, in a non-oxidizing atmosphere like that on Mars there is no danger in using hydrogen aerostats, so this gas, which is both cheaper and easily produced in situ, can be utilized with a slight improvement in performance. If the gas is heated to cause it to expand, this will create lift. Similarly, if the gas is cooled, the balloon will sink. An interesting concept is a balloon that heats strongly and expands by day, thereby rising in the atmosphere, then lands when it cools down at night having traveled with the prevailing winds.

Figure 9.11   An artist's impression of a balloon for automatic Mars exploration. (NASA image)

As shown in the above mentioned section, to carry the same mass in equal conditions a balloon on Mars must have a volume about 81 times that of a similar balloon carrying the same load on Earth. An aerostat on Mars will be some 4.3 times larger than its counterpart on Earth. As a consequence, its structure must be made of lighter materials and it must be designed very accurately.

It is unlikely that hot air balloons will be used on Mars. A hot air balloon uses the same gas as the ambient atmosphere, simply heating to lower its density. On Earth this is easy, since air is heated by combustion (early balloons used straw as a fuel, modern ones use propane), but this is impossible in the non-oxidizing atmosphere of Mars. Nevertheless, hot air balloons in which the atmospheric gas is heated by sunlight have been suggested by NASA, together with regular balloons, as potential robotic explorers (Fig. 9.11).

Non-rigid airships have been suggested a number of times for Mars exploration, both as robotic devices and to carry people. Perhaps the principal objection against the use of airships is the very strong winds. To travel against such winds, the airship would need to be able to hold a relative speed greater than that of the wind itself, which is difficult. Furthermore, docking an airship to a pylon, or landing it on the ground, may be impossible in winds such as can occur on Mars. Another challenge would be to build the hangars in which to store large airships.

At any rate, powered airships in which the electricity is produced by solar cells covering a large portion of the envelope have been suggested both as robotic devices or as transportation for astronauts. It may be even more expedient to use internal combustion engines powered by methane/oxygen because the higher specific power may be sufficient to overcome the winds.

Both balloons and airships have a long history of technological developments, and there may be valuable lessons to be learned from the balloons that travel in the stratosphere, where the air density is even lower than it is at the surface of Mars.

9.4   AIRCRAFT

Fixed wing drones—both unpowered (gliders) and powered—have been suggested for Mars exploration. NASA considered launching a glider to Mars in 2003 to celebrate the hundredth anniversary of the first powered flight by the Wright brothers. The drone was to initiate its mission at high altitude as soon as it entered the Martian atmosphere and take measurements and pictures until it landed in an uncontrolled manner (Fig. 9.12a). However, this concept was not pursued. Powered drones have also been designed (Fig. 9.12b). If they use electric motors and are powered by a battery, then in the case of a manned mission the battery can be recharged at the outpost.

As shown in Sect. D.4.2, an aircraft must fly much faster in the thin Mars air than in Earth's atmosphere. The ratio between the speeds is about 5.7, and that between the power required to fly on Mars and on Earth is 2.2. These figures have been obtained for equal mass and wing areas. To reduce the speed and the power, it is possible to increase the wing area. If the wing area is increased by a factor of 4.7, then the power is the same on both planets and the ratio of their speeds reduces to 2.6. These values hold if the mass of the aircraft is assumed to be the same. This can be achieved by taking into account the fact that the reduction in weight on Mars reduces the structural stresses.

Consider an Unmanned Aerial Vehicle (UAV) designed for flying at low altitude in the Martian atmosphere. Assume the pressure is 600 Pa and the temperature is $-50\,°C$ (223 K). The aircraft has a wingspan of 10 m, a mean wing chord of 1.5 m, and a ready-to-fly mass of 150 kg. Assuming a lift coefficient at take-off (with the flaps extended) $C_L = 1.4$ and a

Figure 9.12 Mars fixed wing drones. (**a**) A glider. (**b**) A powered aircraft. (NASA images)

drag coefficient $C_D = 0.15$, then the minimum take-off speed and the power to achieve take-off are respectively $V_{min} = 61.6$ m/s $= 221$ km/h and $P = 3.73$ kW. The latter value doesn't include the power needed to accelerate, and greater power will be required to fly at a safely higher speed. However, an aircraft on Mars can fly faster than one on Earth at equal power.

It is possible to imagine astronauts on Mars using something like an ultralight aircraft to make quick reconnaissance flights around their outpost, or to travel from one place to another. However, owing to the low air density, take-off and landing speeds are high and it is debatable whether it would be safe to employ unpaved runways. As the outpost grows, it will become feasible to prepare a runway to enable aircraft to make reconnaissance flights of surrounding areas. An interesting use of aircraft on Mars would be to move from one outpost to another if each mission lands in a different place. Aircraft would be attractive in this role, because this mode of transport is much faster than on the ground. To achieve the necessary range, rather than using battery powered electric motors the plane can be powered by internal combustion engines and the ISPP plants at the various outposts may maintain stocks of propellant. This would facilitate a network of outposts spanning virtually the entire planet.

9.5   HELICOPTERS AND MULTICOPTERS

Rotary wing aircraft (helicopters or autogyros) are rarely studied for planetary exploration, undoubtedly owing to the difficulty of flying such machines in the low density atmosphere of Mars. However, the possibility of taking off and landing using a very short run (autogyros) or even vertically (helicopters or multicopters) is an important benefit; as indeed is the ability to hover.

A configuration for an unmanned rotorcraft that has become quite popular is the so-called quadrotor or quadcopter, generally comprising a cruciform structure with a rotor at the end of each of the four arms. Quadcopter UAVs or drones usually have fixed pitch rotors which in the smallest models reduce to four propellers rotating about vertical axes. Their control is much simpler than that of either a single or a twin rotor helicopter, which require a variable pitch with both collective and cyclic pitch control. A miniature quadcopter is shown in Fig. 9.13a (it is just a few centimeters across) and a larger machine (the Parrot AR. Drone) is shown in Fig. 9.13b.

The control of a quadcopter is achieved by varying the relative speed, and then the thrust, of each rotor. This is easy if electric motors are used. Referring to Fig. 9.13a, rotors 1 and 4 rotate in one direction and rotors 2 and 3 rotate in the opposite direction to balance the reaction torques and hence eliminate the need for a tail rotor (as is required by a single rotor helicopter).

- Reducing the speed of rotors 1 and 3, and increasing that of rotors 2 and 4, produces a roll rotation (about the x axis) to the left, whilst continuing to balance the torques.
- Reducing the speed of rotors 1 and 2, and increasing that of rotors 3 and 4, produces a pitch rotation (about the y axis) to dive, whilst continuing to balance the torques.
- Reducing the speed of rotors 1 and 4, and increasing that of rotors 2 and 3, produces a yaw rotation (about the z axis), with the direction of the yaw rotation depending upon the direction of the rotation of the rotors.

Figure 9.13 (**a**) A miniature quadcopter UAV. (**b**) The Parrot AR. Drone, which is a commercial quadcopter UAV.

A quadcopter can be easily controlled by using a simple control electronics and sensors (generally rate gyros). It will display a good maneuverability, fly in any direction, and turn on the spot. The design is scalable all the way up to large machines.

It should be possible to develop a small quadcopter capable of flying in the low density atmosphere on Mars. It could be carried by a wheeled rover, take off vertically to hover a few meters above its parent, powered and controlled through an umbilical. The batteries and the controller need not be on board the quadcopter. A camera on the aircraft would enable the rover to see much farther than it otherwise could. And if the aircraft possessed an antenna, this would enable the rover to remain in contact with its human supervisor at longer range.

As shown in Sect. D.4.4, for a multicopter that has a given mass and a given rotor area, the ratios between the rotor speed and the power needed in order to fly on Mars and on Earth are respectively 5.7 and 2.2, but the power can be decreased by using larger propellers. For instance, if the rotor diameter is increased by a factor of 2.17 the power remains the same and the propeller speed is increased by a factor of 2.6. Multicopters therefore seem the ideal form of drone for Mars.

In the same way, small human carrying helicopters, similar to the ultralight helicopters used on Earth, may be the ideal flying machines for Mars, at least until runways are built.

Quadrotor or twin rotor machines for carrying humans, such as the hoverbike shown in Fig. 9.14, might prove the best devices for improving mobility on Mars in the future.

9.6 HOPPERS

Hoppers powered by rocket engines have been suggested several times as a possible means of transportation for Mars, both in crewed and autonomous variants.

A robotic hopper rover powered by a radioisotope thermal rocket engine was proposed in 2015 by a research group led by Leicester University and the Astrium company. This would be about 2.5 m across, have a mass of 1000 kg, and carry at least 20 kg of science instruments. Carbon dioxide would be extracted from the atmosphere and compressed in

Figure 9.14 The hoverbike is a multirotor device for carrying humans. In the future such devices might greatly improve mobility on Mars.

order to liquefy it. This would be pumped into a chamber that was heated by a radioactive source so that it would expand explosively through a nozzle, supplying a thrust sufficient to cause the robot to leap a distance of up to 900 m. In this manner, the robot could manage terrain that was far rougher than could be navigated by more conventional locomotion systems. Locations in deep craters and canyons could be reached and samples could be obtained from places which could not be reached otherwise.

Hoppers powered by chemical or nuclear rockets could be large enough to carry humans, and could greatly enhance mobility on Mars. As related in Sect. 7.5.2, designs employing NIMF (Nuclear rocket using Indigenous Martian Fuel) have been proposed.

10

The ground segment

The ground segment is particularly important and politically critical in complex international space missions. Each participating country will request ground facilities be present on its own territory, and even non-spacefaring nations may already have some. Ground facilities generate jobs and local authorities will try to host them.

10.1 LAUNCH ASSETS

An initial mass in low Earth orbit (IMLEO) of the order of many hundreds of tons will require superheavy launchers, both for direct launch and on-orbit assembly. The NASA Space Launch System (SLS) vehicle will be quite suitable for a human Mars mission, since its most powerful configuration (Block II) is expected to have a capability of placing 130 t in LEO and sending 45 t to Mars. Other launchers in this class, existing or in planning, are the Long March 9 by China and the Energia by Russia. In the USA, the SpaceX company is privately developing its Falcon Heavy launcher which, although somewhat smaller, is nevertheless intended to be able to facilitate a Mars mission.

Appropriate vehicle assembly buildings, crawlers and launch pads will be designed and built by these countries. The issue of how to organize the many launches that will be necessary for an international Mars mission is political, and is not dealt with here. This is also the case for subcontracting work. Maintaining a steady industrial workflow and workforce will be critical, because mission scheduling depends on the time constraints of the launch windows, and long intervals may occur between periods of intense activity.

If there are long gaps between orders for rockets, their makers might find it difficult to keep their personnel and corporate know-how together. In addition, in most architectures for Mars missions the launch windows may require several rockets to be launched in a single year. As it usually takes many months to create and assemble such large rockets and the working teams are busy for long periods, it might be necessary to duplicate the facilities and train up additional teams, and that might significantly impact the complexity and costs of the mission.

© Springer International Publishing Switzerland 2017
G. Genta, *Next Stop Mars*, Springer Praxis Books, DOI 10.1007/978-3-319-44311-9_10

Clearly, things will be easier if the rockets used for Mars missions have other uses. For instance, if they are also used to maintain a lunar outpost or commercial enterprises either on the Moon or on the asteroids. The field of launch services will also benefit from the synergy between Mars and lunar operations.

10.2   COMMUNICATION CENTERS

Deep space communications are very different from those on Earth or in LEO. Owing to the distance between a spacecraft en route to Mars or an outpost on its surface, large, high gain antennas are required in order to increase the signal-to-noise ratio. NASA has its own Deep Space Network (DSN). Conceived in 1958 to serve the early lunar and planetary missions, this is operated by the Interplanetary Network Directorate of the Jet Propulsion Laboratory (JPL), which is part of the California Institute of Technology. It operates independently of the single missions and of the various NASA centers that develop deep space missions. The DSN also has the responsibility of maintaining its antennas and receivers and of developing the hardware and the software needed to fulfill its tasks.

The DSN is a worldwide network of large antennas and communication facilities located at three main sites that are separated in longitude by about 120° so that at least one has a line of sight to any direction in space at any moment. They are:

- The Goldstone Deep Space Communications Complex located in the Mojave desert, in California.
- The Madrid Deep Space Communication Complex, 60 km west of Madrid in Spain.
- The Canberra Deep Space Communication Complex, 40 km southwest of Canberra in Australia.

Each Deep Space Communications Complex features a 70 m dish antenna. Figure 10.1 shows the Goldstone site. All communication centers are linked to the JPL control facility in Pasadena, California.

The main functions of any DSN site during a human Mars mission will be:

- Telemetry and tracking.
- Command and control of the space vehicles and satellites involved in the mission.
- Communication with the crew.
- Data storage.

Together, the DSN facilities provide the two-way communications needed to control missions in deep space.

The limited coverage of the DSN in the southern hemisphere led to the loss of some data during the Voyager missions. This may not be an issue for Mars missions because the planet remains close to the ecliptic plane. However, the need to maintain a continuous contact with the astronauts en route to Mars or on its surface may require the duplication of some of the antennas and other apparatus if the DSN is required to service other missions at the same time.

In 1959 the Soviet Union began to construct a network to communicate with its planned missions to Venus and to Mars. The early stations were built using a set of eight 16 m

Figure 10.1 The Goldstone Deep Space Communications Complex. (NASA image)

dishes arranged together using makeshift structural elements made from submarine hulls and bridge trusses and bearings from the gun turrets of battleships. The network had three antennas, two for receiving and one for transmitting. In 1978 the system gained a pair of 70 m dishes, one located at Yevpatoria in the Crimea and the other at Ussuriisk in Primorsky Krai. A 70 m dish at Suffa in Uzbekistan was never completed, but there is a 64 m antenna at Bear Lakes, near Moscow. There was never any intention to cover the entire sky, so communication with deep space missions is possible only for limited periods.

ESTRACK is the European Space Tracking network. It is operated by ESOC (European Space Operations Centre) for the European Space Agency (ESA). Ten stations are owned by ESA:

- New Norcia Station (Australia).
- Perth Station (Australia).
- Redu Station (Belgium).
- Kourou Station (French Guayana).
- Cebreros Station (Spain).
- Maspalomas Station (Gran Canaria, Spain).
- Villafranca Station (Spain).
- Kiruna Station (Sweden).
- Santa Maria (Azores, Portugal).
- Malargüe Station, Argentina.

It can also call upon antennas operated by other organizations:

- Malindi (Kenya).
- Santiago (Chile).
- Svalbard Satellite Station (Norway).

These sites provide three 35 m dishes, seven 15 m dishes, and various smaller antennas, all of which are remotely operated by the ESTRACK Control Centre (ECC) sited at ESOC in Darmstadt, Germany.

Japan, China and India, which are starting to send missions into deep space, are creating their own tracking stations.

Since a human Mars mission will be an international enterprise, several communications centers from different countries could participate to form a deep space network with multiple redundancy. At any rate, a single coordination center is necessary and its designation may be problematic for mission planning.

Communications satellites around Mars are essential to maintain a continuous contact with the astronauts. These must be deployed in advance. However, to deploy them from Mars orbit is not very costly and they represent a small fraction of the total cost of the mission. Another means of improving communications is to station a relay satellite in one of the Lagrange points that lead or trail Mars in its orbit around the Sun (or better, a satellite in each point) to prevent a communication blackout when Mars is in conjunction, on the far side of the Sun from Earth.

10.3   MISSION CONTROL AND ASTRONAUT TRAINING CENTERS

The general public is used to seeing rooms full of operators at consoles, controlling robotic and human missions (Fig. 10.2).

But the control room is only a small portion of a mission control center (MCC). This is a complex facility that hosts a staff of flight controllers and support personnel who monitor all aspects of a mission, receiving downlinked telemetry and sending commands to the ground stations for uplinking to the vehicle. Often a mission is controlled, in its various phases, by several control facilities. For instance, many NASA missions are initially controlled by the Launch Control Center (LCC) at NASA's Kennedy Space Center on Merritt Island in Florida. For Space Shuttle flights, once the vehicle was airborne responsibility was transferred to the Mission Control Center (MCC-H) at the Lyndon B. Johnson Space Center in Houston. For a crew heading for the ISS in a Soyuz spacecraft, the primary facility is the Russian Federal Space Agency (RKA) Mission Control Center in Korolev, near the RKK Energia facility. Some Russian missions are controlled by the Titov Main Test and Space Systems Control Centre in Krasnoznamensk, Russia.

The activities of European astronauts on board the ISS are controlled from the European Space Operations Centre (ESOC) in Darmstadt, Germany. But Automated Transfer Vehicles making resupply flights to the ISS were controlled by the ATV Control Centre (ATV-CC) at the Toulouse Space Centre (CST) in Toulouse, France. The Columbus Control Center (Col-CC) which supervises operations in the Columbus research laboratory of the ISS is located at the German Aerospace Center (DLR) in Oberpfaffenhofen, Germany.

Figure 10.2 The ISS Russian Mission Control Center at Korolev, in the Moscow Oblast. The picture was taken on October 25, 2012, as Expedition 33 arrived on the space station. (NASA image)

Chinese missions (robotic and human) are controlled from the Aerospace Command and Control Center located in a suburb northwest of Beijing. Indian missions are controlled from the AsiaISRO SHAR Mission Control Centre located at the Satish Dhawan Space Centre in Sriharikota, India. Japanese ISS missions are controlled from the JEM Control Center and the HTV Control Center at the Tsukuba Space Center (TKSC) in Tsukuba, Japan.

In the USA, there are several other control facilities. Some of these, like the Goddard Space Flight Center in Greenbelt, Maryland, which controls the Hubble Space Telescope, are directly managed by NASA. In contrast, the JPL is operated by a university, and is federally funded to manage NASA's robotic deep space missions. Other control centers are privately managed by aerospace industries, most particularly the Boeing Satellite Development Center (SDC) Mission Control Center in El Segundo, California; the Lockheed Martin A2100 Space Operations Center (ASOC) in Newtown, Pennsylvania; and the Space Systems/Loral Mission Control Center in Palo Alto, California.

An international venture such as the ISS has a complex network of control centers (Fig. 10.3). Note that in this figure there are no control centers in China or in India, because they are not participants. (American legislation prevents China from playing a role.) If the number of countries participating in a human mission to Mars is larger, then the number of control centers is likely to be much larger.

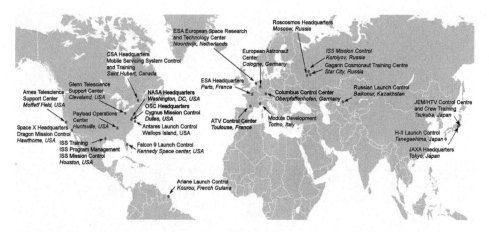

Figure 10.3 The various control centers for ISS missions.

Specific buildings have to be built for the various teams who will follow and control the mission. The organization of the teams has yet to be determined, but the list of expertise that they must provide is a long one: astronautics and flight control, power systems, mechanics, electronics, programming of embedded systems, life support systems, communications, crew support (physiology and psychology), science and exploration, data storage and processing, mission planning, command, etc. And there must be all the control equipment and facilities, including the testing and simulation of specific procedures before these are carried out by the astronauts. And, importantly, the various teams should be trained to work together towards a common set of goals.

There must also be an infrastructure to support the selection and training of astronauts. In the beginning, most astronauts were selected from the ranks of military pilots and their skills were centered on being able to fly the spacecraft, even though there was discussion about how autonomous the vehicle should be and to what extent the astronaut should be a 'passenger' or a 'pilot.' Then the focus shifted toward astronauts who possessed higher academic qualifications in engineering, the physical and life sciences, and mathematics, leading to the introduction of the terms 'mission specialist' and 'payload specialist.'

At present, most training facilities in the countries that participate in the ISS project are focused on preparing astronauts to play a part in flying the Russian Soyuz spacecraft, living in the space station and operating experiments. The process starts with basic training, which lasts about 2 years. The candidates learn about technical subjects, such as the life support systems, orbital mechanics, payload deployment, Earth studies, space physiology and medicine, and so on, as well as soft skills that include a thorough knowledge of languages. In particular, non-Russian astronauts must learn Russian, particularly if they are to play an active role in flying the Soyuz spacecraft.

Next is a further 2 years of advanced training that focuses specifically on the Soyuz and the ISS, performing EVAs, robotics, operations using the remote manipulators, experiment activities, and maintenance tasks. This requires facilities in which the relevant environments are reproduced, such as neutral buoyancy tanks to simulate, as far as is possible, weightless conditions. This second phase of the training process qualifies the candidates as astronauts.

Once an astronaut is assigned to a crew, they start the specific training for that mission. The primary training facilities for ISS astronauts are the European Astronaut Centre (EAC), headquartered in Cologne, Germany; the NASA Lyndon B. Johnson Space Center, Houston USA; the Gagarin Cosmonaut Training Center in Star City, in Russia; and the Tsukuba Space Center in Japan.

The only nation that currently launches astronauts into space without participating in the ISS is China. Their astronauts are all air force officers with substantial experience as pilots. After a general training program lasting about 4 years, there is mission specific training. The Astronaut Center of China (ACC) is in Beijing.

India does not yet have a fully approved human space program, but it is likely that the Indian Space Research Organization (ISRO) will soon start training astronauts, either on its own or in cooperation with NASA and Roscosmos.

With the advent of space tourism, private companies such as Virgin Galactic and Xcor Aerospace are developing non-governmental astronaut training programs. They are starting with suborbital flight, but may advance to orbital flight and perhaps even to flights around the Moon. The more adventurous companies may seek to offer lunar landings. However, the selection and training of space tourists is different from that of professional astronauts.

All the training since the Apollo era has presumed that in the event of an emergency the astronauts can get back to Earth in just a few hours. This will not be possible when we resume flying to the Moon and advance to Mars.

For a mission to Mars, astronauts will need to be trained to live and work in the unique conditions that prevail on that planet, on missions lasting several years, and so far from Earth that instant communications and a quick return in case of emergency are impossible. There, astronauts must deal with an emergency on their own, perhaps with advice from colleagues on Earth but without receiving any material assistance. Hence astronauts on Mars will need to be given training as physicians, dentists, scientists, engineers, technicians, pilots, geologists, and biologists. As there is no possibility of including all of the required specialists in the crew, it is likely that each astronaut will be cross-trained in at least two specializations, therefore the time to train the crew will be much longer than for ISS crews. The general training can be done in the manner described above, using the existing facilities, but there will have to be specialized facilities specific to Mars missions. If in the meantime we resume operations on the Moon, it may be possible to share specialized partial-gravity simulation facilities.

Actually, at present some analogue environments, like the NASA Extreme Environment Mission Operations (NOAA NEEMO), the NASA Desert Research and Technology Studies (Desert RATS), the NASA Flight Analog Research Unit in Houston, and the Haughton-Mars Project (HMP) are being used with the long term objective of human deep space missions. For instance, a total of fifteen people (who for some reason are known as aquanauts) have trained for possible future missions to asteroids. These simulation facilities are described in Sect. 10.5.

The ISS will be a very useful training environment, but the best simulations for many aspects of the Mars environment will be achieved on the Moon, and this is one of the most important reasons for establishing an outpost on our nearest neighbor as a stepping stone to Mars.

Apart from physical simulation, virtual reality simulators will be used, in part due to their cost effectiveness, but mainly for the impossibility of simulating on Earth some peculiarity of space or the Martian environment.

10.4   TESTING KEY MARS MISSION HARDWARE

10.4.1   Interplanetary propulsion systems

Whatever the choice of propulsion, whether it be chemical, solar electric, nuclear thermal, or nuclear electric, it will be necessary to build new systems and test them on the ground before space testing. Facilities to test chemical engines and electric thrusters are available now, but for specific needs it might be necessary either to make important adaptations or to build new facilities. Also for electric propulsion, test rigs for ion or plasma thrusters are already built. At present, vacuum chambers for plasma thrusters rated up to 200 kW are in use. There may be a need to build vacuum test facilities for large plasma thrusters, if instead of using an array of small plasma engines it is decided to use a small number of thrusters with powers in the MW range, but this is a straightforward construction task.

NTP may pose more serious problems, because test facilities on Earth are complex and costly. It is often claimed that testing nuclear thermal rockets cannot be done on the ground because it would violate laws and treaties which prohibit the release of radioactive particles into the atmosphere, or at the very least would make it extremely expensive to do safely. But this is simply not correct. There are several sites in the USA and in countries which were once part of the Soviet Union, in which there are underground test rigs for nuclear weapons and these comply with the treaties and laws. These are much more complex than the rigs that would be needed for nuclear rocket engines, because there is no nuclear detonation and the release of radioactive particles will be very limited. As was shown by an above ground test where a NERVA class rocket was deliberately exploded, the contamination is limited to the immediate neighborhood of the explosion point. And, of course, the underground facility is designed to contain a release of radiation.

In one possible layout,[1] the installation is based on a borehole of the same type as those used for nuclear weapon testing (Fig. 10.4). The rocket is secured at the top of the pit and discharges its plume vertically downwards. Cooled by a spray of water, the exhaust is left to seep through the pores of the rock in which the hole is drilled. The geological character of the Nevada Test Site (NTS) is well understood. It performed underground nuclear weapons tests from 1952 to 1992. There are a number of unused holes which can be converted to nuclear rocket testing. This concept has been called Subsurface Active Filtration of Exhaust (SAFE) and, following the above mentioned paper, promises to allow the testing of nuclear thrusters for a capital cost of about $50 million and an operational cost of about $1–2 million per test.

---

[1] S.K. Bhattacharyya, "A Rational Strategy for Nuclear Thermal Rocket Development," 47th AIAA/ ASME/SAE/ASEE Joint Propulsion Conference & Exhibit, July–August 2011, San Diego, California, AIAA 2011-5945

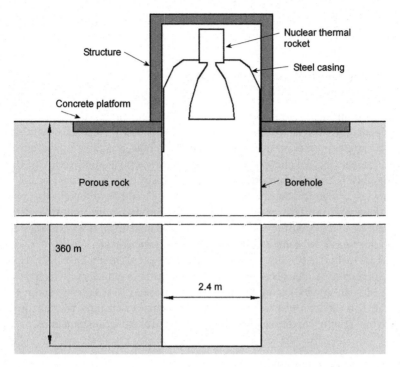

Figure 10.4 A sketch of a simplified testing facility for nuclear thermal rockets.

The cost of ground testing nuclear thrusters is therefore not prohibitively high and the development of NTP with a specific impulse of about 900 s is feasible, and a higher figure could be achieved if the project takes advantage of progress in the field of materials since the 1970s.

10.4.2   Power systems

If NEP is selected, the reactor will have to be specifically designed and tested on Earth, but this won't require the construction of specific facilities, except perhaps for the radiator that must work in conditions very different from those typical of power systems on Earth.

Similar considerations apply for the nuclear reactor that powers the outpost, but in this case its automatic deployment must be accurately tested in a facility that simulates Mars. If the outpost is to be powered by solar panels, their deployment must be tested in a simulated environment. Solar panels on Mars may have problems in terms of long term performance. Dust storms will mask the panels, so devices designed to clean them must be tested. In this regard, it is encouraging that the solar panels of the automated landers and rovers of the last 20 years have performed extremely well, particularly because as dust slowly masked them, they were blown clean by passing dust devils. Nevertheless, for an outpost it will be a serious challenge to maintain solar panels operating at high efficiency.

Several facilities with simulated solar lighting are in operation, so it will be possible to update one of these to include a simulation of the Martian atmosphere and dust deposition.

10.4.3   In situ propellant production systems

In most mission architectures, the propellant for the ascent from the Martian surface is to be produced from local resources, such as by extracting carbon dioxide from the atmosphere or perhaps hydrogen from water in the permafrost. Local resources can also be used to obtain a variety of other products. Various tools (robots, compressors, heaters, coolers) and chemical reactors (Sabatier reactors, water electrolyzers) therefore need to be designed and thoroughly tested as appropriate to their criticality for the success of the mission. Before any test is made on the surface of Mars, there should be ground testing on Earth in order to determine the best experimental conditions, production rates, storage, power requirements, and the robustness of such systems. The chemical reactions are well understood and there has been some testing but not at the desired scale.

In addition to that, depending on the strategy, the systems might have to be deployed by robots. Experiments will be necessary to test the robustness of the deployment procedures. Although Martian gravity cannot be simulated on Earth, its atmosphere and regolith can be replicated in order to test full scale ISRU systems. Load relievers cannot precisely reproduce low gravity conditions, but they can be useful in certain situations.

10.4.4   Life support systems

The first life support systems for human spaceflight were developed right at the start of the Space Age, and have been progressively improved over the years. However, no mission has been conducted without frequent resupply and exchanges of crew members. Life support systems used on space stations are therefore not entirely appropriate for a mission to Mars because they are not optimized for spending long periods in deep space and on that planet.

The key issue is to optimize the recycling rates of bioregenerative and closed loop life support systems. Most life support systems are based on chemistry and physics. Biological devices can use micro-organisms to purify water. Closed loop environmental control life support systems have been built at an experimental level. One of these is the ESA Melissa project, but several facilities based on this concept have been built and used all around the world. Although very promising, this approach poses problems because living organisms depend on many parameters that cannot easily be controlled.

An extensive test campaign will have to be carried out to assess the robustness of such systems. ENVIHAB is a DLR laboratory in Cologne, Germany, that specializes in medicine, space physiology, and psychology to prepare future human spaceflight. Russia and China have their own life support systems. The NASA Johnson Space Center has expertise in life support systems (air revitalization, water recovery, waste processing) and the design and testing of space suits for different situations. Hence all spacefaring nations will be able to pitch their experience to a collaboration on the Mars mission.

The preliminary tests will be performed on Earth, but the existence of a lunar outpost will be very helpful.

The medical issues include preventing physiological weakness due to spending long times in microgravity, mitigating radiation effects, and using health monitoring systems. These issues are still being investigated, and solutions remain to be developed. Simulation facilities like the one in Sect. 10.5 will be essential, and new ones will likely be needed.

10.4.5   Surface habitat

The habitat on the surface of Mars might, or might not be very similar to the crew module used in deep space. They ought to be designed with as many commonalities as possible in order to minimize cost and to reduce the astronauts' need for cognitive adaptation in the transition phase. Several simple surface habitats have already been built on Earth. However, the habitat module will need to be optimized to minimize mass, to sustain the crew, and to cope with the low atmospheric pressure, the substantial temperature excursions and of course the dust.

Recently, this aspect of the mission has been entirely revolutionized by the advent of additive manufacturing. Testing this approach on Earth is not only possible, but also fairly easy and inexpensive. Moreover, the use of additive manufacturing for low cost and readily assembled buildings is currently being investigated for general civil engineering applications, so for Mars it is more a matter of developing new architectures for this specific role than of developing new technologies. We simply have to confirm that the process can use the local materials. In this regard, there is the possibility of producing large quantities of Mars regolith simulant on Earth and defining the appropriate binding material (if any is required). With that done, a whole outpost can be built which, after demonstrating the construction technologies, will be available for training the astronauts.

The synergy with a lunar habitat is complete, because the same (or at least very similar) technologies will be developed and the same outpost on Earth can be used for training the astronauts for both Moon and Mars. Then the Mars astronauts will train on the Moon in an even harsher environment than that of their intended destination.

10.4.6   Testing of other devices and equipment

The previous sections dealt with systems that require extensive testing on Earth before being tested in space (or better, on the Moon) prior to carrying out a full test on Mars during robotic missions. The extensive ground testing will require either converting facilities currently used for other purposes or, more likely, the construction of entirely new test facilities.

There are other important elements which will require testing. The Mars Ascent Vehicle, for instance, will need ground testing in much the same way as was done for the LM of the Apollo program. Even though Earth gravity is much stronger than that on Mars, tests using partial loads can be performed if ascent flight tests are deemed necessary. In addition, much testing of such a vehicle can be performed in LEO.

A new deep space habitat module must be designed, since the NASA Orion spacecraft is able to host astronauts for only three weeks, at most. A specific facility might have to be built for the construction, integration, and testing of that module (see Sect. 10.4.4).

The entry, descent and landing (EDL) system is a critical one. The TRL of such systems for heavy vehicles entering a planetary atmosphere is very low. Wind tunnel testing is feasible in some facilities, as is already done for robotic probes. A vehicle of the mass and size needed to carry people has never been tried before. Scaling by applying similarity theory could help, but the final qualification will require actual testing in the Martian atmosphere.

As already stated, the issue of testing EDL systems may not be critical if nuclear (NEP or NTP) or solar (SEP) propulsion is employed, but if the interplanetary transfer is performed with chemical propulsion then the EDL system needs to be thoroughly tested, first in wind tunnels, then probably in the upper atmosphere of Earth, and finally on Mars. This is one of the few systems for which the Moon, which lacks an atmosphere, cannot assist with testing.

Specific space suits may require to be designed for the Mars mission, mainly to enable astronauts to operate out on the surface. Existing space suits will be tested in order to verify their appropriateness for various phases of the mission. Dust might be an issue. The Apollo astronauts found lunar dust to be an unexpected challenge. It was abrasive, it was cohesive, it infiltrated the outer gloves and stuck to everything, it made movements more difficult and posed a risk to zippers and ring seals. The facility that tests the surface habitat might also be used to test the space suits. It will also test vehicles. Whatever the final choice (unpressurized, pressurized, small or large), the vehicles will have to be tested in Mars-like environments on Earth and candidate astronauts will train to use them efficiently and safely.

Numerous innovative devices, tools, and scientific equipment will have to be designed, built, and tested in simulated Mars conditions. These range from the greenhouse to additive manufacturing machines, and from communication systems to specific computer peripherals and software. Perhaps the most challenging thing, particularly for a multinational program, will be to minimize the number of different components and to unify the standards applied. Standardization will be of paramount importance in such a complex enterprise.

10.5 SIMULATIONS

Accurately simulating the Martian environment is an essential prerequisite for testing the efficiency of systems (life support, recycling, power, space suits, rovers, robots, scientific tools), their robustness in a harsh environment, and the extent to which they can be reliably operated by people who are constrained by limited mobility (pressurized suits, rigid gloves, etc.).

The procedures for surface deployment, EVA preparation, exploring distant areas, using tools, communicating with the habitat and ground mission control, avoiding contamination, removing dust, maintenance operations, rescuing, preparing for Mars ascent, and others must also be tested and rehearsed in simulated Mars conditions.

Perhaps the field in which environment simulations will be most essential is in dealing with human factors:

- Physiological issues: nutrition, health monitoring, medical emergency.
- Psychological issues: confinement, isolation, stress, facing dangerous situations.
- Social issues: collaboration, conflicts, task sharing, leadership.
- Cognitive issues: selection of crew members based on background skills, training, skill development, maintaining competencies.

A number of Mars analogue simulations have already been carried out on Earth. These include:

- The pioneering Biosphere 2 was built between 1987 and 1991 and so-numbered because the first biosphere is that of Earth itself. It is a large structure with an area of 12,700 m² and contains a complete and self-sufficient ecosystem. Its explicit goal was to study isolated biospheres with the long term goal of the colonization of other planets or the construction of space settlements. It is in Oracle, Arizona, and contains a coral reef, a desert, a savanna, a forest, cultivated fields, and a habitat that includes living quarters and laboratories (Fig. 10.5). Its first mission was undertaken between 1991 and 1994; the second was initiated in 1994 and interrupted after 3 months as a result of severe psychological problems. Columbia University used it between 1995 and 2003 for scientific purposes, and later it was transformed into a tourist attraction. It ceased to be pressurized in 2006. Despite the failure of the project, it demonstrated it was possible for a time to live in a self-sufficient small Earth-like ecosystem.
- The Mars Society's cylindrical habitats are located in different desert areas for several years to test rovers, exploration procedures, human/robot cooperation, and to evaluate human factors (Fig. 10.6). Many reports have been published.
- NASA Desert RATS (Research and Technology Studies) placed stations in different desert sites within about 120 km of Flagstaff, Arizona, to allow engineers & scientists from NASA centers and partner organizations to undertake dress rehearsals for future missions to other planets. Various dedicated rovers and robots have been field tested. Other studies promoted by NASA are being performed in Antarctica (the Antarctica Analog Studies).
- Human Exploration Research Analog (HERA). This habitat at NASA Johnson Space Center in Houston, Texas is an artificial analogue of a planetary base.

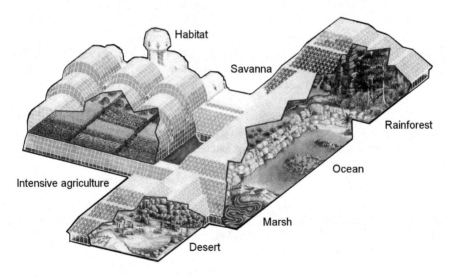

Figure 10.5 A schematic cutaway of Biosphere 2.

Figure 10.6 The Mars Desert Research Station in Utah, USA. (The Mars Society image)

- Ground Experimental Complex (NEK). It is an artificial analogue of an interplanetary vehicle located in Moscow, Russia. It has investigated human factors during a 500 day simulation in a confined environment.
- Hawaii Space Exploration Analog and Simulation (Hi-seas). Set up to simulate food and culinary routines during long duration spaceflight, this is intended to be used for more general purpose studies in the coming years.
- Pools and undersea assets. Several swimming pools around the world are dedicated to training astronauts for working in microgravity conditions. Underwater experiments have been undertaken in Marseille Bay by COMEX (Compagnie Maritime d'Expertises) to simulate EVA activities.
- The Self-Deployable Habitat for Extreme Environment (SHEE) is a European project to develop a planetary habitat testbed for terrestrial analogue simulations.
- The Austrian Space Forum (OeWF or ÖWF) simulations were conducted in analogue field tests in different European locations. For instance, the Mars 2013 campaign was a month-long Mars analogue field test in Morocco held in February 2013. A number of experiments supervised by a Mission Support Center in Innsbruck, Austria were carried out by international teams under simulated Martian surface exploration conditions.
- The Collaborative REsearch And Training Experience (CREATE) project of the Natural Sciences and Engineering Research Council (NSERC) of Canada funded a number of experiments that were conducted in various Canadian locations in order to test robots and EVA.
- The Mars Yard is an artificial analogue of the Martian Surface located in Stevenage in England.

Finally, several aerospace industries (like ALTEC in Torino, Italy) and universities have small and medium sized areas dedicated to Mars simulations.

11

Timeframe and roadmap

Human missions to Mars are still a goal for a more or less distant future. If we want to transform them into reality, then we must state detailed roadmaps which address all the still unknown aspects and focus our efforts on the many details remaining to be settled. The roadmap must not only address technological and scientific aspects, but must also concentrate on feasibility, affordability, and risk control if we are to transform a dream into an exploration campaign. It is likely that the road to Mars will involve the exploration and development of the Moon and that the experience gained on the Moon will be essential to later going on to Mars in a safe, affordable, and consistent way. As a means of assessing how feasible each project for human Mars exploration is at any moment in time, an index, the Human Mars Mission Feasibility Index (HMMFI), has been proposed.

11.1 PREPARATORY MISSIONS AND ROADMAP

A human mission to Mars needs to acquire new knowledge, to develop new technologies, and to test so many systems that have never operated in the desired conditions, that there must be a number of preparatory steps.

The International Space Exploration Coordination Group (ISECG) published a Global Exploration Roadmap (GER) leading to the first human mission to Mars; its last version is shown in Figure 1.13 [26]. A step-by-step approach is described to develop key exploration technologies and capabilities. Important stepping stones include a mission to a near-Earth object and a mission to the Moon. Similar proposals can be found in the literature. However, there are still ambiguities about the precise configuration of the first human Mars mission that create uncertainties in the roadmap.

For instance, if it is decided to use nuclear propulsion, either thermal or electric, related milestones need to be inserted into the roadmap. On the other hand, for architectures that use chemical propulsion, missions that develop aerocapture and EDL systems must be inserted into the roadmap.

© Springer International Publishing Switzerland 2017
G. Genta, *Next Stop Mars*, Springer Praxis Books, DOI 10.1007/978-3-319-44311-9_11

11.1.1   The role of the ISS

The ISS plays an important role in the GER. At present, the use of the space station has been extended to 2024, and may well continue in operation until the end of the decade, so it will be available, as stated in the GER, to perform "general research and exploration preparatory activities."

Apart from general studies into the effects of a prolonged exposure to microgravity on humans, these activities cannot be defined in detail until after the selection of the architecture for the Mars missions. At present, there is no agreement between the participants in the ISS program about such matters. Indeed, there is not even a consensus on a strategy to reach an agreed mission design. So the research must cover a wide spectrum. This has the advantage of clarifying certain scientific and technological issues. For instance, testing a 200 kW plasma thruster to reboost the ISS, as planned, will be essential in obtaining data required in choosing the propulsion system most suited for going to Mars.

Other examples are research in artificial gravity that can be performed using a centrifuge or a rotating laboratory that is added to the ISS, inflatable habitats, and additive manufacturing in reduced gravity.

Another point is the continuation of the experience of international cooperation on which the ISS is based, because a human mission to Mars is sure to be an international enterprise.

In addition to the ISS, the GER refers also to "commercial or government LEO platforms and missions." This includes the future Chinese space station, since the Chinese government has several times stated that their station will be open to international collaborations. Therefore it will probably host experiments that will be useful to the definition phase of missions to Mars. Having started late, the Chinese space station will probably outlast the ISS.

Among other government LEO platforms, there may be Indian satellites and perhaps their crewed spacecraft.

Commercial platforms may include crewed orbiting spacecraft and, eventually, orbiting hotels. The experiences of such companies could help to define private contributions to Mars exploration.

11.1.2   Robotic missions to discover and prepare

Telescopic studies before the Space Age told us very little about conditions on Mars. It was not until we began to send robotic probes that we made progress. The data from flyby probes, from orbiters and from landers and rovers on the surface progressively disproved many of the wrong ideas. With each new mission we are still gathering knowledge about the planet's surface, its atmosphere, and the conditions that explorers will find there.

To a modern reader it may seem strange that 70 years ago, Mars was thought to be a cold and dry planet with an atmosphere of nitrogen and some carbon dioxide at a surface pressure of 10 millibars, and there appeared to be vast areas that were rich in plant life which thrived and diminished with the seasons. Many people even believed that the planet was inhabited by intelligent beings and that the canals, although there existence and their nature was disputed, might really be present. This was essentially the planet described by the great astronomers of the late nineteenth century.

Now we know better, but we still need to gather further information before we will be able to safely land there and establish an outpost.

Some space agencies (NASA, ESA, Roscosmos, ISRO) are planning robotic missions designed to clarify many points, in particular whether life simply cannot exist on the surface. Answering this question will be very important in designing a human mission, because it will clarify the dangers of contamination, backward and forward. Other issues to be resolved are the actual quantity of radiation at the surface, the exact chemical composition of the regolith, the feasibility of extracting water from the permafrost, and the locations of caves and lava tubes that might prove useful.

However, in the context of "robotic missions to discover and prepare," the GER lists not only robotic missions to Mars but also probes to the Moon and to the asteroids. This part of the roadmap culminates in one or more Mars Sample Return (MSR) missions, because it is considered essential to make a thorough analysis of the materials on the Martian surface prior to sending humans there, and such an analysis can only be performed in the best laboratories on Earth. However, although plans for this very difficult mission have been made several times it has always postponed.

Recent work suggests that what is widely called a "Heavy Mars Sample Return Mission," meaning a robotic mission that is carried out by using the same hardware designed for the human mission, could play an important role in the preparatory phase for the qualification of interplanetary vehicles, ISRU systems, and lifting off from Mars. Because the return vehicle can carry a good quantity of samples from the planet, this test flight could thus double as the MSR mission with substantial programmatic savings.

11.1.3 Human missions beyond LEO

On this theme, the GER includes three subheadings. In order, they are:

- Explore near-Earth asteroids.
- Extended duration crew missions.
- Humans to the lunar surface.

The sequence raises questions. In particular, it is difficult to understand why the missions to the near-Earth asteroids and the extended duration missions (in all likelihood these would be to the Lagrange points of the Earth-Moon system, where a space station might be located) should precede missions to the lunar surface.

As already noted, before spending several years in space and on Mars, it is necessary to perform long duration missions on the ISS. But this is insufficient. The environment in LEO differs from that in deep space, primarily due to the radiation, hence missions beyond LEO are required.

We know a lot about the effects of LEO conditions on human organisms, and in particular about microgravity, but we know little about the effect of reduced gravity, such as is available on the Moon or, better for preparing for a mission to Mars, in a rotating habitat for which the strength of the artificial gravity can be varied as part of the research effort.

Lunar missions are essential for learning how to live on another world. The brief periods spent on the Moon by the Apollo astronauts taught us a lot, but that experience is insufficient for planning a human mission to Mars. Section 11.2 discusses the importance of missions to the Moon.

The GER follows the above mentioned three points with "missions to deep space and Mars system" and then "sustainable human missions to Mars surface." But it is not clear what the former are. They are not clearly aimed at the near-Earth asteroids that were mentioned before. They could be Mars flybys or missions into orbit around the planet, possibly with visits to its satellites (as discussed in Chapter 7).

11.1.4   Missions to test new technologies

Missions of this kind are not explicitly stated in the GER. However, they are very important because, depending on how the Mars mission is designed, it may rely on technologies which have a low TRL. Generally speaking, it is unlikely that we could go to Mars using entirely off-the-shelf technologies. For instance, if the interplanetary vehicle uses nuclear propulsion then the nuclear rocket or the power generator (depending on whether NTP or NEP is used) will have to be tested in space, but if chemical propulsion is used, the mission architecture would very likely make use of aerocapture or aerobraking and, particularly if large vehicles are used, that technique will have to be tested in space.

Life support systems, ISRU, and many other essential systems will require to be tested, either in space or on Mars, and all of this activity requires specific missions.

In the case of Apollo, there were four crewed missions prior to the historic landing on the Moon. One (Apollo 8) orbited the Moon ten times, and another (Apollo 10) was a dress rehearsal right to the point at which the powered descent would be initiated at an altitude of 10.6 km, thereby enabling their successors to concentrate their training on that final phase. However, it is unlikely that a similar approach will be followed in the case of Mars, since it would involve a journey of 17 months. It is likely that a final rehearsal mission will be done without astronauts on board, and will test all the automatic systems. In the case of an Apollo mission, commands had to be input directly by a human on board the spacecraft but the ships for a human mission to Mars will almost certainly be able to operate in an automatic manner and be able to be tested without requiring a crew to be present.

As noted above, a fully automatic test may take the form of a Heavy Mars Sample Return Mission. All the important systems are taken to Mars. Once on the surface, rovers will collect samples and load them into the ascent vehicle. The transfer of the samples to the Earth Return Vehicle and its landing on Earth (or docking at a space station) will closely follow what will be done later with the crew. Such a mission would thoroughly test all the essential systems in conditions very close to those of a human mission, while also returning samples for detailed analysis. What is more, the habitat, the rovers, and other hardware will remain on Mars and be available for a future crew. At the very least, they would constitute a precious mine of spare parts.

All in all, for any specific Mars mission architecture, the roadmap highly depends on the complexity, the cost, and the duration of the qualification phase, and the sustainability of the program.

In planning a roadmap, psychological issues must also be taken into account. Even if it is difficult to justify a Mars flyby only from the technological and programmatic points of view, it might be interesting to undertake such a mission simply to prove to the public, and the skeptical decision makers, that it is feasible (e.g. the Inspiration Mars Foundation's proposal by Dennis Tito described in Section 1.7).

11.1.5   Timeframe

It is very difficult to estimate a timeframe for the first human Mars mission. The variables that could affect this are so numerous and so unpredictable that any suggested date is just a guess. It seems clear that it is unlikely that a single space agency will take the initiative for a mission of this kind. If the mission is to be undertaken by a public organization, then it will probably be an international endeavor.

Dates for human Mars missions have been offered many times. All these predictions have proven overly optimistic. So too, in all likelihood, will be the recent suggestion of a timeframe for a program starting in the early 2030s and running to the end of that decade. The opposition of Mars in 2037 will be a very favorable one, but it is unlikely that this opportunity will be taken. A more likely date would be in the late 2040s, but even this may be too optimistic.

A recent study by The Planetary Society [31] suggests a more optimistic approach. It was presented at the Humans Orbiting Mars workshop that was held in Washington DC on March 31, 2015. It basically advocates an approach that is essentially based on NASA plans, plus a loosely specified collaboration with other space agencies and private ventures. The project relies upon the SLS launcher and the Orion spacecraft, which are planned to be tested in a number of launches between 2016–2027. The solar electric propulsion that is to be used to deliver cargo to Mars orbit, will be tested in 2020. The EDL system will have been tested by 2026. The first human expedition is scheduled for the 2033 launch opportunity. This will achieve orbit around Mars and perhaps visit Phobos, but not land on the planet. A test of the Mars lander on the Moon is scheduled for 2035. The first human landing on Mars in 2039 will be a short stay mission. The first long stay mission is to follow in 2043. This timeline is shown in Figure 11.1.

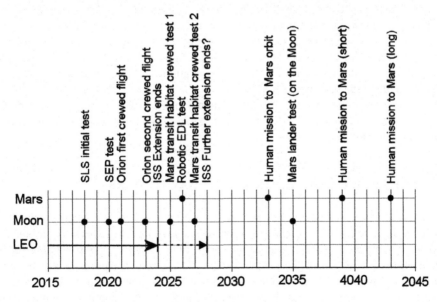

Figure 11.1 The timeline leading to the first human Mars mission proposed in a recent study by The Planetary Society.

Figure 11.2 An artist's impression of the Mars Colonial Transporter approaching Mars. (SpaceX image)

Perhaps even more optimistic are the plans announced by the SpaceX company in April 2016 to land on Mars a robotic Red Dragon spacecraft (an upgrade of the Dragon spacecraft which is currently being used to resupply the ISS, designed for Mars). The schedule calls for launch in 2018 using the company's Falcon Heavy rocket. The Mars Colonial Transporter (Figure 11.2) is planned for the mid-2020s, but it is unlikely that this date will be kept.

Other predictions are more pessimistic, with dates between 2040 and 2050 (or perhaps 2060). Most of these proposals envisage the first mission being a landing on the planet. But a wide variety of factors could move the predictions in either direction: for example, unpredicted technology advances, a change in the political climate, and an improvement or a worsening of the economic situation.

The true drivers that pave the way to human exploration of Mars may be a significant reduction in the cost of access to LEO and a start to asteroid mining or, better, the mining of the Martian satellites.

11.2 THE ROLE OF THE MOON ON THE WAY TO MARS

The role of the Moon in human Mars exploration is a controversial one. Many people think that it will be practically impossible to colonize Mars without the experience of building an outpost on the Moon. On the other hand, the fate of the Space Exploration Initiative that was announced in 1989 has led people to argue that a complex program that includes exploration of the Moon will inevitably result in an unaffordable dream which will collapse even before starting. Even if the result is not a total cancellation of human Mars exploration, it is claimed that a significant investment in the Moon can only delay by many years the primary goal of a human journey to Mars.

It must be acknowledged, however, that often the reasons for why the Moon should be a stepping stone to Mars are stated in an unclear manner, and, worse, there are statements in the literature which are readily contradicted, such as using the Moon to produce propellants for a Mars mission or as a spaceport to launch the spacecraft to Mars.

In the case of simplified Mars mission architectures, a mission to the surface of the Moon may be seen as a complication to be avoided. However, returning to the Moon before going to Mars may be useful, both for psychological reasons and to increase the global experience of human missions beyond LEO. It is possible to say that this will probably not cause an actual delay, since other reasons could easily prevent launching a Mars mission in a short time.

11.2.1   The Moon Village approach

It is often stated that when the commitment to the ISS ends, a human Mars mission will be a good candidate for the next international effort in space, hopefully with a larger number of participants. Even countries that are currently not spacefaring, may join such a venture with the objective of contributing in a manner that matches their ambitions and areas of expertise. However, the funding for the ISS program is currently guaranteed only to 2024 (although it may continue for several years beyond that) and this does not really allow sufficient time to start a fully developed program aimed at a human mission to Mars.

In fact, when funding for the ISS is terminated, the partners in that venture plus some other countries may well initiate a joint program to build a lunar outpost. This was recently described by Johann-Dietrich Wörner, the ESA Director General, as a "Moon Village." He said this would see "different actors joining together in the same place, be it different states, individuals or private companies, to establish an infrastructure on the Moon that has the ability to do first-class fundamental research." Hence the idea of returning to the Moon in a coordinated manner is basically a political project to establish an international venture on at least the scale of the ISS, if not considerably larger. The concept of a village suggests that each participant can pursue their own goals in their own way, benefitting from the common infrastructure and exploiting the synergies created by having everything occur in the same place. Among the goals of the Village is science: first "Moon science, but also cosmology, with a telescope on the far side" of the Moon, possibly a radio telescope, human biology and medicine, and much more.

Technology is deeply involved too, both as an enabling factor and as a subject of study. Actually, the very concept of the Moon Village is made possible by the existence of additive manufacturing machines that not only enable dwellings to be built of a size that would have otherwise been impractical but also to shield them from cosmic radiation. In this manner, it will be possible to spend long periods on the Moon without incurring costs that are likely to render the entire project unaffordable. The Moon Village, having been made possible by this recent technological development, will be a source of further advancements, most of which will be essential for traveling to Mars and staying on its surface in an affordable way.

However, the Moon Village is not only a place to experiment with new forms of political cooperation in space, and nor is it merely a scientific and technological laboratory. By being open to the involvement by private ventures it will also be a place to develop new forms of space business. Perhaps the simplest business model is space tourism, hence one of the most popular facilities may be a hotel.

Figure 11.3 The Moon Village. (a) How a building obtained using additive manufacturing might look. (b) A cutaway of the same building showing the thick regolith walls that protect its occupants from cosmic radiation.

Figure 11.3 presents two images showing the Moon Village, or better, just one of many buildings of the village.

Other economic activities are associated with lunar resource utilization. This depends on where the Village is built. If it is near one of the lunar poles, then the extraction of water will be one of the most important businesses simply because "water is the currency of space" [30]. The amount of water, its locations, and the difficulty involved in extracting it have yet to be determined, so one of the first activities for the villagers will be to prospect for water. If it proves to be a scarce resource, then perhaps it would be unwise to mine it indiscriminately to make propellants that can be obtained in other ways, and it should instead be used to sustain and expand the Village itself. If it turns out that extracting water from the lunar regolith is impractical or uneconomic, then it may be possible to draw water from a near-Earth comet and transport it to the Moon. If, however, large quantities of water are found and are readily extracted, then yes, one of the basic commercial activities of the Village will be the production of propellant to be used by other activities on the Moon and also to support ventures such as the mining of near-Earth asteroids.

Another business opportunity will be mining the lunar regolith for minerals. As a result of analyzing the specimens brought to Earth by Apollo astronauts and by the three automated Russian probes that returned samples between 1970–1976, we know a lot about the chemical composition and physical properties of the regolith. Field studies were carried out by the astronauts and particularly by the only professional geologist to reach the Moon, Harrison Schmitt on Apollo 17 (Figure 11.4).

Obviously the composition varies from one site to another, but it is possible to state average compositions for the maria and for the highlands (Table 11.1). Most of the regolith consists of oxides like SiO_2, Al_2O_3, TiO_2, CrO_3, FeO, MnO, etc. Hence the most abundant element is oxygen. The most common metals are aluminum, titanium, iron, etc. In terms of resource utilization, large amounts of energy will be required to extract oxygen and metals from the regolith.

The solar wind implants some volatiles (hydrogen, helium, nitrogen) into the regolith. Their amounts are reported in the table, expressed in parts per million (ppm). The content is highly variable from one specimen to another, so these values are really indicative averages. In particular, the helium-3 is very variable. Of course, these values are based on specimens that were collected at low lunar latitudes. It is possible that the abundances of volatiles are greater in the colder regolith in the polar regions.

The most abundant elements are hydrogen, nitrogen and carbon, but the one that has prompted the greatest excitement amongst proponents of exploiting lunar resources is helium-3. Mining helium-3 may become an important industry in the future, but only if we develop power plants that use nuclear fusion, and in particular the type that can be fueled by helium-3.

Figure 11.4 A geologist on the Moon. Harrison Schmitt is shaking soil out of a rake after making a swath through the regolith to collect pebbles. He was the only professional geologist ever to reach the lunar surface. The picture was taken by Gene Cernan at Taurus-Littrow on December 11, 1972 during the Apollo 17 mission (NASA image)

Table 11.1 Average elemental compositions of the lunar regolith in the maria and the highlands (in percentage), and the volatiles implanted in the regolith by the solar wind (in ppm).

Elements	Maria	Highlands
Oxygen	45%	45%
Silicon	21%	21%
Iron	15%	6%
Calcium	8%	10%
Magnesium	5%	5%
Alluminium	5%	13%
Titanium	1%	—

Volatiles	ppm
Hydrogen	50
Nitrogen	50
Carbon	50
Helium (total)	20
Helium 3	0.004
Argon	2
Krypton	1
Neon	1
Others	3

It is possible that rare earth elements will be found in useful concentrations in certain small areas of the Moon, so prospecting for these elements may be important too. And then there is an abundance of helium-4 that could also be useful.

Perhaps the materials from metallic asteroids that struck the Moon in recent times (by which geologists mean the last billion years) may be very important. They are surely present, although they may be hidden by the regolith that covers almost everything on the Moon. They are potentially rich in platinum-group elements and will also contain iron, cobalt, and other metals. Meteoric iron is much easier to obtain than iron from oxides in the regolith. When a mining industry is established on the Moon, it may be easier to mine these metals from the asteroidal deposits on the lunar surface than to get them directly from asteroids in space. The possibilities of establishing a mining industry on the Moon have been discussed at length by many people, with generally positive recommendations.[1] The revenue that lunar commercial activities create will promote a space economy in general, and make Mars exploration more affordable.

11.2.2 Synergies between the Moon Village and Mars exploration

As noted above, it is often said that a lunar outpost will produce the propellant required to go to Mars, and that the Moon will serve as a spaceport to launch missions farther into space. But these statements must be assessed with care and skepticism, in particular the

---

[1] http://www.space.com/28189-moon-mining-economic-feasibility.html#sthash.oNr2q8yk.dpuf

first one. We are sure that oxygen can be produced from the regolith but the cost of doing this will be quite high because large amounts of energy are required. Perhaps oxygen from the Moon will be used as propellant only if its production is powered by nuclear reactors. But an industry on this scale is unlikely to be established in the time remaining before we would like to initiate the human exploration of Mars.

If water is found on the Moon in quantities large enough to decide to use it also to produce propellant for general use, things may be different and both fuel and oxidizer may be produced. But it must be remembered that fuel produced on the Moon must be transported to where it is needed for Mars missions, and this will impose a cost, mostly in terms of the fuel required by the tankers.

As regards using the Moon as a spaceport, it appears that this is convenient only if the spacecraft are built on the Moon. This may be possible, but in a much more distant future. It does not seem expedient to launch vehicles from the Earth's surface, take them to the Moon, and then launch them to Mars.

The synergies between exploring and utilizing the Moon, and human missions to Mars are (in the opinion of the author) of a different kind. The Moon is an ideal training ground for Mars, and it permits the development and testing of relevant technologies in a cheaper and safer manner and in a location from which it is possible to return home in just a matter of days in the case of accidents or failures. All the technologies associated with the habitat, the life support systems, and the transportation systems, can be tested for periods sufficient to qualify them. The same holds for the countermeasures against radiation and the technologies for growing plants and producing food. In some cases, the analogy can be pushed to the details. The lunar dust is dangerous for humans and machines to at least the same extent as Martian dust, even though the two have some different properties. The protection of delicate machinery, and the procedures for avoiding dust contamination are also similar.

The role of robots and other automatic machinery in close contact with people is another aspect of a Mars mission which can be rehearsed on the Moon.

On the Moon, as well as on Mars, prolonged stays and the use of machinery that requires large amounts of energy will mandate the use of nuclear power. In fact, this will be even more important on the Moon, where the nights are very long and very cold. The design, construction, and operation of nuclear plants on the Moon will provide vital experience for planning a mission to Mars.

The case cannot be pushed too far, however. ISRU and ISPP on the Moon will not only be much more difficult than on Mars but also very different, hence little can be learned on this point.

11.3   HUMAN MARS MISSION FEASIBILITY INDEX

The idea of introducing an index which can be computed for any given proposal for a human mission to Mars, in order to assess in a quantitative manner whether is realistically feasible, was proposed by Giuseppe Reibaldi, responsible for human spaceflight of the International Academy of Astronautics. It was then elaborated in [32], from which the following lines are reported.

The Human Mars Mission Feasibility Index (HMMFI) is able to include technical, human, programmatic, political, and sustainability parameters. Each year it can be updated to take into account the evolution of the different parameters.

1. Technical parameters (a maximum of 10):

   • Environment Control and Life Support System (ECLSS).
   • Propulsion.
   • EDL.
   • Mission design.
   • Launcher.
   • Landing technology.
   • Re-entry.

2. Human parameters:

   • Radiation.
   • Low gravity.
   • Psychological.
   • Physiological.

3. Programmatic parameters:

   • Political climate for cooperation to a human Mars mission.
   • Governance of a global cooperation.

4. Sustainability parameters:

   • Commercial market probability of a human Mars mission.
   • Government budget affordability.

This is just a preliminary suggestion, and a detailed study can identify a larger number of relevant parameters. Some parameters may be considered irrelevant. For instance, it might be decided that it is possible to travel to Mars without any form of artificial gravity, therefore the parameter about low gravity can be neglected.

11.3.1 Proposal 1: averaging algorithm

Different types of parameters can be suggested. A first possibility is to use an index going from 1–9 (as in the TRL) or 1–10. Whilst in Technical parameters the same definitions as in TRL may be used, it is likely that different definitions will be used for Human, Programmatic and Sustainability parameters. It must be noted that, even for the TRL, different definitions are used in different industries, or sometimes by different organizations working in a given field.

Once the parameters have been decided and values assigned, then the simplest options are to sum them or to compute the average, although doing so has the drawback of treating all parameters as being equally important. A better solution may be a weighted average, but to do this the weights must be given and that is something that requires a deep study and introduces an arbitrary evaluation of which parameters are of the greatest importance.

11.3.2   Proposal 2: multiplying algorithm

Another suggestion, given here as a possible example of an implementation of the index, is to assume that all systems proposed to implement a human Mars mission are technically feasible and then to consider two parameters:

- The development cost, including tests and qualification.
- The time needed to achieve that development.

These parameters are thus not primarily technical, but essentially financial and political. The feasibility index for a given system (which is already assumed to be technically feasible) can be defined in terms of the probability of its development and qualification being financed. Intuitively, in first approximation, it is inversely proportional to the costs and to the duration of the development and qualification of that system.

The two parameters are numerically evaluated as:

- C: the cost in $billion ($B) required to reach technical feasibility (TRL = 9).
- T: number of years for the development and qualification.

Obviously, these definitions are entirely arbitrary and the evaluation of the value of the parameters is very subjective. The index for the i-th system can be computed from the cost and the development time using an empirical equation like:

$$F_i = \frac{1}{\dfrac{C_i T_i}{100} + 1}.$$

Hence the feasibility index is zero for a very immature technology (infinitely expensive, with a very long qualification phase) and 1.0 for an off-the-shelf technology (cheap and only a short development and qualification phase).

The denominator is a completely arbitrary number which is related to the units used for expressing cost and development time. The figure 100 has been chosen so that a technology requiring $10 billion to be developed in 10 years has $F_i = 0.5$.

The global feasibility index of a mission can be given by the product of all the systems feasibility indices:

$$F_i = \prod_{i=1}^{n} F_i.$$

Combining the various indices by multiplying them is entirely arbitrary, justified mostly by the fact that in order to have a global index of 1 (mission feasible) it is necessary that all indices are 1 (all components are ready). Even a single index of 0 (a single component is not feasible) gives a global index of 0 (mission unfeasible).

Clearly, this approach is much more justified for technical parameter than for parameters of other types, but it can be adjusted to suit all parameters. Maybe instead of considering the product of cost and time, it would be possible to devise a product of other features and modify the figure 100 accordingly.

Another point is that the qualification phase of several systems might be carried out during the same mission. For instance, a Heavy Mars Sample Return Mission could contribute to the qualification of the propulsion stages for the TMI maneuver, aerocapture, EDL systems, and ISRU. It may therefore be possible to identify savings that are not taken into account by this simple formulation.

11.3.3 Example

An example with 13 technological factors is reported in Table 11.2. It is clear that the index is strongly non-linear, so that a case whose feasibility is not very problematic seems to have a very low index. This, however, may prove to be a small problem if the index is used to compare different solutions or to follow the change of the index over time.

The non-linearity of this definition of the index makes it impossible to compare the index obtained in this way with an index obtained using global parameters. For instance, by assuming a cost of \$65 billion and a time of 15 years, the example yields an index of 0.093; much higher than the mentioned value of 0.02.

The mentioned example is merely a first attempt to define the Human Mars Mission Feasibility Index. If this index proves useful, then a discussion involving all the interested parties must be started and a serious study undertaken.

Table 11.2 An example of a Human Mars Mission Feasibility Index.

Main systems	C (\$B)	T (years)	HMMFI
Heavy launcher and ground infrastructures	10	5	0.667
Space assembly and staging operations	1	3	0.971
Interplanetary propulsion stages	5	5	0.8
Deep space habitat	5	5	0.8
Aerocapture	5	5	0.8
Entry, descent and landing systems	15	15	0.308
Surface habitat	2	5	0.909
Surface power and energy management	2	5	0.909
Surface mobility, including robotics	1	3	0.971
ISRU, O_2 for MAV, surface power included	1	5	0.952
Mars ascent vehicle	3	10	0.769
Earth return vehicle	10	10	0.5
Long duration LSS and human factors	5	10	0.667
Global indicators	65	15	0.02

12

A look to a more distant future

As technology advances, human Mars missions become increasingly easy and their cost reduces, and the simultaneous growth of the global economy makes it more affordable. At present the space economy is primarily focused on LEO operations, but the development of a deep space economy will allow an increased number of nations, space agencies and, above all, private enterprises, to participate in long range exploration. Possible technological breakthroughs may facilitate this trend but they are not strictly essential, as technologies that are the logical development of the present ones are all that are required in order to explore Mars. Eventually, large numbers of people will regard Mars (and perhaps the Moon) as their 'home world.'

12.1   TECHNOLOGICAL ADVANCES

Up to now, it has been taken for granted that people from planet Earth will eventually reach Mars, explore it, and construct dwellings there. Hence the questions were why we should do this, how we should do it, and when.

The inevitability of human Mars exploration is based on one simple consideration. With new technological developments that will occur entirely independently of any decisions about space travel and space exploration or colonization of the planets, the challenge of traveling to Mars will become increasingly easy and inexpensive until, at some point, it will be within the scope of small companies or even non-governmental organizations and advocacy groups. These new technologies will be pursued by commercial markets for other goals, and hence will not depend on decisions made by space agencies or other governmental organizations.

One of these technologies is nuclear energy generation, which is in any case required in order to satisfy the world's increasing energy requirements. Public acceptation of NTP or of NEP will certainly mark an enormous step in the right direction, but facilitating human Mars exploration requires not only nuclear propulsion. When will nuclear fusion become a reality? In popular reporting, fusion power is always forecasted as being a generation away! What is sure is that fusion for a thermal rocket will be simpler than for a power station on Earth. In all likelihood, fusion power generators will be much lighter than fission generators. By greatly increasing the specific impulse of NTP, fusion will reduce the travel time to Mars. This can only make things easier. In recent years, Lockheed Martin announced the intention to build a relatively small fusion generator for general use. In reference to NEP, the company gave an optimistic estimate that this generator would enable a spacecraft to reach Mars in a month.[1] And because the pay-off is potentially so significant, other companies are pursuing the same strategic goal. And of course, research continues on more conventional approaches to deep space travel.

The use of antimatter would be another great step forward. Its energy density is much higher than fusion and so it would make space travel even faster and easier. But there is no realistic estimate of when antimatter will be produced in the required amounts to make it a viable fuel for a spaceship. However, schemes to use tiny amounts of antimatter to catalyze nuclear fusion have been proposed and designs for fusion rockets that employ quantities of antimatter consistent with what might be possible in the near future have been reported.[2]

Advances are always occurring in the field of materials science. For example, recently discovered graphene, a form of carbon, will be of benefit to space travel. Radiators made of this material may reduce the specific mass α of a nuclear generator to 1 kg/kW, a fairly low value which would substantially reduce the travel time.

Together with new materials, new construction technologies, possibly derived from the former, will also be important. Bottom up construction brings about a revolution, not only in terms of performance, but also in a reduction of costs. While until a decade ago bottom up construction was pursued only at the nano- or microscale, today additive manufacturing is operating at the macroscale (albeit at its 'small' edge) and looks very promising. Both space agencies and private companies (e.g. NASA and SpaceX respectively) are beginning to use additive manufacturing in the construction of rocket engines and other aerospace hardware, with the prospect of large reductions in mass and cost. The main benefit so far derives from being able to print very complex shapes. For example, it is possible to reduce the number of components of a rocket engine from several thousands to just a few tens. The possibility of making all manner of parts in space or on another world is a very attractive use of additive manufacturing, especially when using locally sourced materials. And of course, as discussed previously, it would even be feasible to construct habitats on the Moon or on Mars.

All of these things are at the experimental stage, but when they become routine they will revolutionize all fields of engineering. This will be particularly true in the aerospace field. For example, in the case of deep space exploration it will allow things which would otherwise be impractical or simply unaffordable.

Nanotechnologies are also full of promise, or perhaps we should say there are as-yet unfulfilled promises. Ever since the mid-1990s people have predicted that nano-engineered materials will allow us to build much lighter and cheaper spacecraft, and even the panacea

---

[1] http://www.lockheedmartin.com/us/products/compact-fusion.html

[2] R.A. Lewis, G.A. Smith, et al., "Antiproton Catalyzed Microfission/Fusion Propulsion Systems for Exploration of the Outer Solar System and Beyond," First IAA Symposium on Realistic Near-Term Advanced Space Missions, Torino, June 1996.

of reusable single-stage to orbit launchers. All these promises may materialize in the future, but for now there are few applications for nanotechnology in the mechanical or aerospace fields.

Carbon nanotubes are now a reality and they are approaching the performance that may allow us to realize one of the dreams of space enthusiasts: the space elevator that will carry things up a cable into space. It is still too early to know whether space elevators will remain a dream. Some studies delivered encouraging results (e.g. [27]), but there remain technological and economic uncertainties about the feasibility of such a vast infrastructure. However, since the advantages of having a space elevator system are so significant, they will be discussed in detail in Section 12.3.

Except perhaps for space elevators, the topics referred to above are potential near-term extensions of what we currently know theoretically and can reasonably expect to implement. Sooner or later, such technologies will become reality. And that will make human missions to Mars and eventually the colonization of the planet easier. But some breakthrough technologies have been predicted which could completely change the way in which we travel through space. While it would be a big mistake to wait for such breakthroughs before starting human Mars exploration, we must always recognize that, if they become a reality, they will not only make easier or more affordable activities that we are already able to undertake using other means, but will also make possible things which are now mere dreams. Some of these breakthrough technologies will be discussed in Section 12.4.

12.2 ECONOMICAL FEASIBILITY

Two parallel trends favor human exploration of Mars. Just as technological advances reduce the cost of such a venture, the general growth of the economy will make it more affordable. Despite the inevitable economic crises that will cause short term and probably localized setbacks, the general trend is toward an ever increasing total volume in the world economy. Not only will it become possible to renew human activities on the Moon at less than Apollo cost in the 1960s, measured in real terms, the fact remains that even if the cost had remained constant, it would represent a smaller fraction of the total national economy of the nation which undertook this new program. Nevertheless, despite serious proposals being advanced over the years, there has been no human return to the Moon. This indicates that the real issue is not the cost, but a lack of political will.

Today, a human mission to Mars is often treated as an international venture in which a number of space agencies join forces to pursue a common goal. This can be seen from two points of view. On the one hand, it could be an exercise in sharing costs in order to realize a mission that no single party would be able to achieve, with the political benefit of improving cooperation and the mutual understanding. From this point of view, it would be a really good idea to award the ISS the Nobel Peace Prize. The ISS is the largest cooperative enterprise ever pursued in times of peace. And a common outpost on the Moon (a Moon Village) and later a joint human mission to Mars would be further steps in that direction. On the other hand, with the passage of time we may well reach the point at which single space agencies, or possibly even single private organizations, can undertake human Mars exploration. This would greatly simplify the enterprise by avoiding the difficulties linked with organizational, technological and political aspects, especially if it were left entirely to private players, but the political benefits of promoting international cooperation and understanding would be lost.

It is therefore possible that the opportunity for a truly international exploration program will reach a peak at some point in time, and if this opportunity is missed then the developing situation will permit single nations, or perhaps single private organizations, to pursue such a challenge.

Up to now, only the size of the economy has been considered, without thinking about its qualitative aspects. Over time, the space economy will grow and an ever-increasing share of the economy will move from Earth first to LEO and then to deep space. This will place in the hands of the players of the space economy the resources and the know-how that is needed to proceed with exploration. Those who hold that deep space exploration in general, and human Mars exploration in particular, are so expensive that they will forever remain beyond what a private player can achieve, miss the point that both the cost of exploration and the economic (and technological) possibilities of private players must be assessed in their dynamic aspects. As soon as the former falls below the latter, private players will seize control of the future of humankind in space.

The definition of new business cases will, without doubt, be a powerful multiplier of the space economy and, at the same time, it will yield economies of scale that drive down costs. A mature LEO business that includes a variety of satellites and launch services is already present. Even space agencies are using private launch services, for example to resupply the ISS on an ongoing basis.

At present, at least two emerging business can be identified:

- Space tourism. Three sequential steps can be identified for space tourism: suborbital operations, orbital tourism, and finally lunar tourism. Clearly, it is too soon to speak realistically about Mars tourism. However, we can say that the other forms of space tourism will have a large effect on Mars exploration.
- Space mining. Resources can be found on near-Earth asteroids (the easiest to reach) and then on the Moon. The utilization of space resources may, at least at the start, have no direct bearing on Mars exploration (the spacecraft for Mars might not, at least in the short term, be built or fueled using materials mined on the Moon) but the existence of this industry in space will inevitably have a strong indirect effect.

The economic utilization of the Moon will therefore strongly influence the creation of a spacefaring society and open the way for private Mars exploration.

12.3   THE SPACE ELEVATOR

A space elevator is essentially a cable with one end affixed to a platform on the surface of the Earth and the other end attached to an orbiting platform. A common misbelief is that in order for the elevator to remain vertical, the platform at the top end must be in geosynchronous orbit above the equator, but this is not the case because the cable pulls downwards on the platform, and this actually requires it to be in a higher orbit so that the platform pulls outwards and thus maintains the cable taut. Geosynchronous orbit (GEO) lies 42,164 km from the Earth's center. Depending on the proposal, the radius of the orbit of the platform may be up to 100,000 km (Figure 12.1). And of course, for the upper platform to remain stationary in the sky, its orbit must be equatorial.

The base platform must be located on the equator, but the elevation of the site is not very important because the length of the cable is such that the savings to be achieved by having the lower platform on high ground are negligible relative to the total costs. Given a lack of suitable locations at high elevation, and also to avoid legal issues, the current feeling is that the base platform should be at sea, on something similar to an oil platform.

As the detailed workings of a space elevator have been described in [27], only its role in making Mars exploration easier will be reported here.

A space elevator is a very effective gateway to the solar system, and has the potential to make both robotic and human exploration faster and cheaper.

The obvious advantage is that of reducing the cost of access to space, allowing objects to be hoisted into orbit without expending propellant, simply the energy to power the motors that pull the payload along the cable. The presence of a space elevator radically changes the approach to spaceflight: reaching GEO becomes easier than reaching LEO. This counterintuitive situation arises from the fact that if a payload is released from the cable at low altitude (a term which in this context includes LEO and even higher orbits) its velocity is too low and it begins to fall to Earth. To accelerate it to orbital velocity, it would have to be released with a rocket motor and this would be only slightly smaller than the rocket that would be needed to launch the payload from the ground. If it is released at GEO altitude (which, as explained above, may well be far short of the upper platform) then it will have the correct orbital velocity. The absolute angular velocity of Earth is 7.2921181 $\times$ 10^{-5} rad/s, and the speed at which an object is released at the geosynchronous radius of 42,164 km from the Earth's center is 3.0743 km/s.

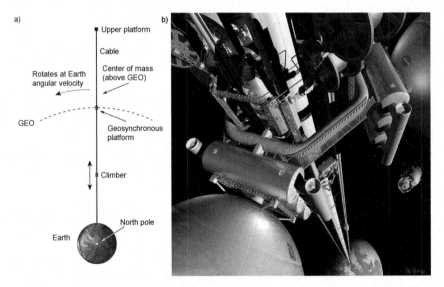

Figure 12.1 A space elevator. (a) A sketch. (b) An artist's view of the climber seen from the upper platform, looking down towards the Earth. (NASA image)

One significant point is that a payload riding a space elevator up to GEO will take much longer to get there than for a conventional launch and a significant fraction of this time will be spent in the Van Allen Belts, which are most intense above the equator, so either the climber will need to be screened or the items it carries must be insensitive to particle radiation.

Above GEO, the velocity of a payload upon release will cause it to enter an elliptical orbit with an apogee above GEO. If release occurs at a radius of 53,127 km, the payload will have a speed equal to escape velocity, and no additional energy will be required in order to leave the Earth's gravitational well. At higher altitudes, the payload departs the gravitational well with a hyperbolic excess speed, which increases with increasing release altitude. For instance, if the radius is 63,378 km, the payload has sufficient energy for a trajectory to Mars, at least if we make the assumption that Earth and Mars are in circular solar orbits. Similarly, any inbound spacecraft needs much less braking to dock with the upper platform than to enter the Earth's sphere of influence in the conventional way.

However, the space elevator is necessarily in the plane of the Earth's equator whilst the trajectory to Mars must be in a plane close to the ecliptic, which is offset by the 23.5° angle of the Earth's spin axis. After the payload is released, it must perform a propulsive maneuver to change its orbital plane. This situation depends on the launch opportunity and the exact date of launch, and such a maneuver may consume a significant quantity of propellant.

Starting the propulsive phase of an interplanetary voyage at a point far away from Earth will be important for a nuclear vehicle, and particularly for NTP. Reducing the cost of putting payloads into high orbits will stimulate the development of innovative propulsion systems like nuclear thermal rockets, thereby reducing the need for ground testing and associated political issues. A space elevator therefore becomes the ideal complement to nuclear propulsion, and the combination of the two would really open the solar system to humankind.

The low thrust interplanetary trajectories for SEP and NEP can be optimized for a space elevator.[3] For instance, consider a fast ship to transport a crew to Mars in 4 months. Without a space elevator, the value of the optimization parameter is $J_{tot} = 72.4$ m^2/s^3 (Figure 6.13), but with a space elevator $J_{tot} = 17.9$ m^2/s^3. The propellant fraction m_p/m_i is therefore practically halved by the existence of a space elevator.

Of course, space elevators can be built on the Moon and on Mars. Owing to Mars having weaker gravity, a space elevator would not require such extreme materials properties as a space elevator on Earth. It would greatly simplify returning from Mars to Earth. If both planets have space elevators then two-way journeys would become quick and easy.

A doubt remains, however. If at some time after the establishment of a space elevator the cost of launching payloads to orbit is greatly reduced, will a space elevator remain attractive, particularly in view of the slow ascent rate? And will there also be a reduction of traffic to orbit as a result of industries moving off planet? Will a space elevator become obsolete in the same way that many railway services did when road transportation became prevalent?

---

[3] G. Genta, P.F. Maffione, "Quick interplanetary trajectories in presence of a space elevator," 9th IAA Symposium on The Future of Space Exploration: Towards New Global Programmes, Torino, July, 2015.

12.4   POSSIBLE BREAKTHROUGH TECHNOLOGIES

It is often said that to become a spacefaring civilization we require breakthrough technologies, particularly in the propulsion field. In 1996, NASA launched the Breakthrough Propulsion Physics Project (BPP) within the context of the Advanced Space Transportation Plan to study various proposals for revolutionary methods of space propulsion which could not be realized without breakthroughs in physics. After spending a relatively small budget of $1.2 million, the project was concluded in 2002 and the results were summarized in a book by its director, Marc Millis.[4]

The main research areas were:

- New propulsion methods which avoid, or drastically reduce, the need for propellant; in other words, to progress beyond present methods of rocket propulsion. This implies finding a new way to move objects, by manipulating inertia, or gravity, or by looking for interactions between matter, fields, and space-time.
- A way to achieve the maximum possible speed and hence cut flight times by orders of magnitude and, possibly, to circumvent the relativistic speed limit. This implies finding a way to move a space vehicle at speeds close to the maximum speed limit for motions through space, or through the motion of space-time itself.
- New ways to generate huge quantities of energy aboard space vehicles. This goal may have a very broad range of other applications.

The aim of the project was "to perform credible progress toward incredible possibilities" such as antigravity and Faster Than Light (FTL) drives.

The BPP was the only effort performed by NASA in this direction, but the space agency is not the only organization pursuing these goals. Research of a similar nature, even though often pursued in a less serious manner, was undertaken by a number of universities, companies, and even individual researchers. From time to time there are claims – more on the internet than in peer reviewed journals – that one or other of these devices has been tested and confirmed to work.

Recently, for instance, a propellant-less propulsion system known as the EmDrive was said to have been tested several times by a British researcher named Roger Shawyer, by a team of Chinese researchers, and finally by an American scientist named Guido Fetta, whose device was tested at the NASA Johnson Space Center in Houston.[5] The tests reportedly produced a thrust but there is no agreement as to the level of amount. At the same time, papers have also been published asserting that this thruster violates fundamental physical laws. And there have been other claims about devices of various types, mainly involving anti-gravity.

Although we should be open to new ideas and proposals, such spectacular claims should be received with skepticism, at least until repeatable proofs are submitted to genuine journals and, above all, verified by independent researchers.

---

[4] M. Millis and E.W. Davis (editors), *Frontiers of Propulsion Science*, American Institute for Aeronautics and Astronautics (AIAA), Reston, 2009.

[5] http://ntrs.nasa.gov/search.jsp?R=20140006052

The point here, is that such breakthroughs are not required in order to achieve the goal of landing humans on Mars and not even for colonization. If such things prove to be feasible, they will make tasks easier, but we should not wait for them. What is required is to develop known technologies, particularly nuclear thermal or electric propulsion. But of course, going to Mars is not simply a matter of refining propulsion technologies. Many other fields of research are just as important, if not actually more so, including radiation protection, life support systems, and artificial gravity. Except in a few cases, technologies advance gradually and therefore waiting for revolutions is pointless.

12.5   TERRAFORMING MARS?

The environment on Mars is quite forbidding. Even if humans can live in an outpost or, later, in a colony, the challenge is to create an artificial environment which is completely separated from the outside. But this task must not be overstated. Even on Earth, most humans live in an artificial environment. A building on the surface of Mars will not be much more artificial and separated from the outside than an air conditioned shopping center or another dwelling in some very cold or very hot place on Earth. The same is true for a Moon Village.

The idea of 'terraforming' Mars has been discussed for several decades. The term was introduced by Isaac Asimov, and has been widely used by science fiction writers to describe a process by which the physical and environmental characteristics of the surface of a planet are transformed to make it more suitable for human life. The idea has such vast implications that it may surprise readers to find that it appears in serious scientific/technological studies. It was only recently realized that terraforming a planet may be easier than had been thought. And so the word terraforming has gained a place in our technical vocabulary, just as have many other words invented by science fiction writers: robot being one of the better known. The following description of a possible terraforming of Mars is derived from [18].

Even before starting the actual terraforming process, we could try to seed the surface of the planet with some life forms that can endure the very harsh Martian conditions. After the first human landing (or even before), the first attempts to grow plants in protected environments on Mars can be made. As discussed in Chapter 8, terrestrial plants might thrive in a pressurized greenhouse. As far as we know, the regolith contains absolutely no organic matter. It must be enriched with fertilizer and with the organic substances that plants require. It is likely there will be oxidizing agents that are harmful to plants, so these will first have to be removed. If water obtained by melting Martian ice is used, that might have to be purified. Genetic engineering might be used to adapt plants to accept a lower pressure in the greenhouse or a soil which has not been so heavily modified. But, sooner or later, an attempt to farm outside of a greenhouse will have to be made. The ethical issues raised by that step will be considered below.

Some very primitive terrestrial bacteria can survive in extreme environmental conditions, but it is unknown whether, if transferred to Mars, they could withstand the very cold night-time temperatures, radiation, and the ultraviolet light flux caused by there being no ozone layer in that planet's atmosphere. Research to identify the best bacteria has begun at the NASA Ames Research Center in California. If none is identified, perhaps it will be possible to genetically engineer the best candidate. There is no doubt that the terraforming process could be greatly accelerated if the conversion of regolith into soil could be initiated prior to the environment changing much.

Bacteria have been found living inside nuclear plants, even on the fuel rods. Micrococcus Radiodurans can survive radiation doses 10,000 times that which can be withstood by a human being. It may be possible to genetically engineer this micro-organism to obtain many products, particularly medicines, that will be important to the first Martian colonists.

Owing to the characteristics of the planet, the work of terraforming can be divided into two phases: first increasing the atmospheric pressure, perhaps by heating the surface; then making it breathable. The first step seems to be easier than the second. There is evidence that the atmosphere of Mars was much thicker in the very remote past. It has even been suggested that the surface pressure was twice the present atmospheric pressure on Earth. As a result the planet might have been a lot warmer. We must remember that at that time the Sun issued much less energy, so the opposite opinion, that Mars was much colder at that time, is also reasonable. The first phase would be a sort of planetary restoration project aimed at giving back to Mars the atmosphere which it possessed some four billion years ago.

A way of doing this is to use the greenhouse effect, which means capturing some of the heat that is otherwise radiated by the planet to space, thereby increasing the temperature at its surface. Some gases (collectively known as greenhouse gases) are very efficient in this regard, particularly carbon dioxide and water vapor, and even more so methane and ammonia. Mars' atmosphere is mostly composed of carbon dioxide, plus traces of water vapor, but its density is insufficient to produce the desired effect. However, heating the surface could liberate some of the carbon dioxide locked away as carbonates, and this would start to increase the atmospheric pressure. In turn, the greenhouse effect would further increase the temperature. Numerous ideas have been put forward, ranging from using large orbiting mirrors to scattering a dark dust on the polar caps, and from exploding thermonuclear bombs underground to sending an asteroid crashing onto the planet, or even digging very deep wells to extract heat from the inner part of the crust of the planet. But these methods are probably insufficient.

By releasing chlorofluorocarbons (CFCs) into the atmosphere it may be possible to raise the temperature at its surface. Since the molecules of these gases aren't entirely inert and are decomposed by sunlight on a time scale of about 200 years, a continuous production of CFCs would be required in order to maintain a temperature higher than the present one. It has been estimated that an annual release of about 100,000 t would be needed in order to maintain the equatorial zone of Mars at a temperature similar to that in a temperate zone on Earth.[6] This is only a fraction of the amount that was being produced annually on Earth in the 1970s and led to the creation of the Antarctic ozone hole that caused so much concern world-wide. To create the required concentration of CFCs within 40 years would need an initial production ten times greater than that plus 5,000 MW of electrical power, equivalent to the output of five medium-sized terrestrial power stations. It therefore appears feasible that we could heat the surface of Mars in this manner or, if a longer time

---

[6] R.M. Zubrin, D.A. Baker, "Mars Direct, A Proposal for the Rapid Exploration and Colonization of the Red Planet," in *Islands in the Sky*, Wiley, New York, 1996.

is acceptable (namely 80 years) this could be done by running three medium-sized power stations and a plant whose output was the same as overall production of CFCs on Earth in the 1970s. This is clearly possible, because the actual energy which heats Mars comes from the Sun, and the role of humankind is that of tampering with the thermal regulation system of the planet.

This effect could be achieved using perfluorocarbons (PFCs), which have the advantage of not dissociating ozone. Hence, as soon as oxygen began to form in the Martian atmosphere, ozone would also start to form, which in turn would begin to attenuate the ionizing ultraviolet light that reaches the surface.

A number of changes would be triggered by warming Mars. The dry ice (i.e. solid carbon dioxide) in the polar caps would sublimate into the atmosphere. In about 40 years the surface of the planet would have a pressure of about one third of that on Earth. This would allow colonial astronauts to dispense with their space suits and move around on the surface with just a mask and a bottle of oxygen, as scuba divers do.

The ice in the permafrost would slowly melt, and water would flow across the surface and down the beds of the ancient rivers to form lakes. Evaporation would start a fresh water cycle, with the formation of rain and snow. In winter, snow would add to the water ice at the poles. However, it is not known how long all this would take, because it depends mainly on how far the permafrost is below the surface. The increase of carbon dioxide and water vapor in the air would both aid the greenhouse effect and hence reduce the rate at which CFCs would need to be produced.

By this point the restoration would be complete, and the planet would have regained its ancient aspect. Although history shows that warm and wet conditions on Mars are unstable, it took a very long time for the planet to lose its atmosphere and cool down. The colonists would be able to maintain stability by periodically boosting the atmosphere with a small quantity of CFCs.

It is not known whether life developed when Mars was a warm planet rich in water, nor indeed whether any hardy life forms are still present in well-protected niches. But when Mars has been restored to its former glory (or possibly before that, as pointed out above), humans will attempt to grow terrestrial plants outside. In an atmosphere containing so much carbon dioxide, vegetation will thrive, even more so if humans select, or genetically modify, certain species to make them more suitable to the Martian environment. It is likely that many plants and trees could be introduced early in the process, when the temperature and the pressure are only slightly above their present values. Agriculture will thrive on the vast open plains and the rougher regions could be forested. But plants have another important function – they produce oxygen. Originally, there was no oxygen in the atmosphere on Earth, it contained only carbon dioxide and nitrogen. Oxygen was introduced by micro-organisms and later greatly increased by vegetation.[7] This could also occur on Mars. It has been computed that the process would require only several centuries. This is nothing in terms of the geological history of a planet, and is not long for civilization. To obtain a Martian atmosphere which humans could breathe freely would be a wonderful audacious achievement.

---

[7] J.I. Lunine, *Earth: Evolution of a Habitable World*, Ch. 17, Cambridge University Press, Cambridge, 1999.

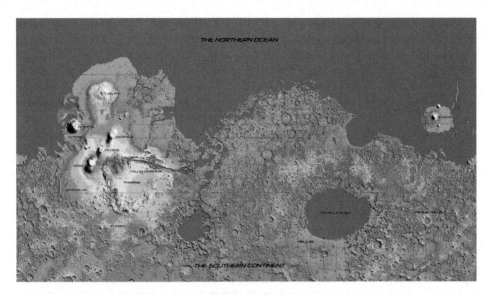

Figure 12.2 A hypothetical map of a fully terraformed Mars, with a large ocean covering the whole of Vastitas Borealis.

This is, however, a controversial issue. In order to render the atmosphere breathable by humans, the concentration of carbon dioxide must be reduced drastically, and that would in turn reduce the greenhouse effect. Some people argue that the choice will be between a warmer planet with an atmosphere that requires humans to carry oxygen bottles and masks, and a very cold atmosphere at a low pressure that has a higher concentration of oxygen.

If these issues can be resolved satisfactorily, then Mars could one day have a breathable atmosphere and a flora and a fauna similar to Earth's, albeit adapted to the local environment. The planet could support a population of several million humans within a time similar to that which separates us from the arrival of Christopher Columbus in America. Mars would then be the first planet to be terraformed by colonizers from planet Earth. An interesting (and epic) fictional account of an endeavor to tailor Mars to human requirements has been told by Kim Stanley Robinson.[8]

A hypothetical map of a fully terraformed Mars, with a large ocean covering the whole of Vastitas Borealis is shown in Figure 12.2.[9] It would probably be impossible to obtain enough water to create such a large ocean from Martian resources, so this situation would be possible only if some comets were made to impact in order to obtain their water.

The scenario described above seems to be feasible. But there are other options that would reduce the power needed to produce the CFCs. First, large mirrors could be put in orbit around Mars to heat the surface using reflected sunlight. This can be done at almost

---

[8] K.S. Robinson, *Red Mars*, *Green Mars* and *Blue Mars*, Bantam Books, New York, 1993, 1994 and 1996 respectively.

[9] http://quanto.deviantart.com/art/Terraformed-Mars-53595798

no cost if solar sails are employed to reach the planet, because when these are no longer needed as propulsion devices they can be left in orbit and oriented in such a way as to melt the polar caps, liberating carbon dioxide into the atmosphere. The greenhouse effect would then amplify that heating. If only a few solar sails were available, then only a small effect would be attainable. It would be necessary to use large mirrors to obtain the desired large scale. The mirrors could be built in space with aluminum mined from either the Moon or an asteroid and then propelled to Mars by the radiation pressure of sunlight.[10]

Alternatively, it might be possible to change the orbits of small asteroids or comets and drop them onto Mars. This could be done by a combination of relatively low thrust electric thrusters and gravitational assists from a giant planet. The kinetic energy of the asteroid or comet would be converted entirely into thermal energy, causing localized heating around the point of impact and releasing the materials from which it was made. A comet nucleus that is rich in ice could significantly increase the amount of water vapor in the planet's atmosphere, and an asteroid rich in ammonia (if one is found) would release this gas into the atmosphere, thereby increasing the greenhouse effect and creating a protective layer against ultraviolet light. It may appear to be a weird idea to crash asteroids and comets on a planet in order to terraform it, but this mimics the natural mechanism of planetary formation. It is thought that during its accretion, the Earth received water and carbon (the latter perhaps already having formed organic compounds) from falling cometary nuclei.

The techniques described here for terraforming Mars are perhaps too primitive to work precisely, but they are the result of just a few years of deliberation of a completely new theme. They are based on the greenhouse effect, and as such they are an application to Mars of what has recently been learned about the effects of human activity on the Earth's environment. When the time comes to progress from speculation to feasibility studies and then to actions, all these subjects will be known in much greater detail. And it is possible that simpler, more effective, and perhaps more economical methods will be found. For example, nanotechnologies offer a tantalizing promise in this field (as they do in many others).[11] Molecular automata, with their incredible ability to replicate themselves and then operate on a very large scale with negligible costs, could change the chemical composition and the characteristics of a planet's atmosphere in times a fraction of those mentioned above. Here the aim is simply to show that it should be possible to terraform Mars, rather than to describe the detailed ways in which the many and varied technical issues could be solved.

To terraform Mars will be a huge undertaking, greater than some engineering feats of the past such as building the Great Wall of China or excavating the Suez Canal. And, apart from giving humanity a new planet on which to settle, it is bound to produce an invaluable body of scientific and technological knowledge. One important result this will be detailed insight into the mechanisms that regulate a planetary environment. In turn, this will assist humankind in cleansing the Earth of the effects of the industrial revolution. Moreover, this exercise will be invaluable in the future when other planets need to be terraformed, possibly under even more difficult circumstances.

[10] R.M. Zubrin, C.P. McKay, "Terraforming Mars," in *Islands in the Sky*, Wiley, New York, 1996.
[11] K.E. Drexler, *Engines of Creation*, Anchor Press, New York, 1986.

One problem of a non-technical nature remains. Assuming that terraforming a planet is a feasible process, is doing so advisable or morally acceptable? The first concern relates to the massive production of greenhouse gases and particularly CFCs. The production of CFCs was banned on the Earth because of the erosion of the ozone layer, but using CFCs elsewhere in the solar system could be beneficial. In a wonderful irony, chemicals that could irretrievably ruin the Earth's atmosphere, turning it into a hell resembling Venus, may be able to restore Mars to its ancient grandeur.

Is terraforming worthwhile? And acceptable? In his book *The Search for Life on Other Planets*, Bruce Jakowsky lists the following points in connection with terraforming Mars and introducing an active biosphere there.[12]

Seven arguments in favor are:

1. A thick atmosphere of carbon dioxide, even though it would be non-breathable, would greatly assist colonists by enabling them to explore the Red Planet equipped only with breathing apparatus rather than with full environmental space suits.
2. Locally generated biomass would be an important source of energy, food, and other useful materials for colonists.
3. Such an activity would provide a long term challenge on which humans could focus, with a goal that is both useful and desirable for humans.
4. Such a project would be an essential prerequisite to any future human colonization of Mars.
5. An active biosphere on Mars would provide a refuge for many forms of life on another planet in the solar system, safe against the event of war or a natural global catastrophe that might destroy life on Earth.
6. Much of the research would be highly relevant to addressing environmental problems at home and to understanding the intricacies of its biosphere.
7. Becoming a spacefaring civilization is less threatening than military developments or an arms race at home, and would provide a worthy outlet for international cooperation and competition and/or technology developments.

Seven arguments against are:

1. The time scale is so long that governmental institutions would not be able to maintain the necessary commitment to such a project.
2. It is not clear that there are significant economic benefits, especially in the short term, that would be commensurate with the cost and effort involved.
3. Scarce human and economic resources would be drawn from other worthwhile projects, such as addressing social and terrestrial environmental problems.
4. Something could go awry during the course of the project that could damage the new Martian biosphere beyond repair, leaving us in a worse situation than if we had never intervened.
5. Humans have made such a bad job of managing the Earth's environment that it would be presumptuous to imagine we are wise enough to succeed on another world.

---

[12] B. Jakowsky, *The Search for Life on Other Planets*, Cambridge University Press, Cambridge, 1998.

6. If terraforming was successful, then Mars might become a tempting target for military and/or economic exploitation, thereby creating more sociopolitical problems than we have at present.
7. The evolution of a Martian biosphere could be inherently unpredictable, and might be detrimental to humans or (by back contamination) to Earth.

An interesting general observation about humankind introducing changes to the Martian environment has been made by McKay and Haynes, "If and only if no potentially viable forms of life are found should we attempt to introduce immigrant species from Earth... What would be the greater good, Mars barren or Mars endowed with life? ... Should the Martian biosphere be tended to ensure at least early development in a manner agreeable to Homo Sapiens? "[13]

The terrestrial (and perhaps a Martian) ecosystem arose naturally by the actions of many factors which play different and contrasting roles. One of these factors is human-kind. The fact that humans are intelligent does not deprive them of their right to play a part in the game, but it places upon our species the burden of behaving wisely by trying to predict the consequences of its action. Yet, as we saw in Section 2.4 with regard to the so-called precautionary principle, this responsibility must not paralyze us.

The first planet of the solar system to be terraformed was Earth, long ago, by the primitive life that changed an atmosphere from one based on carbon dioxide to one rich in oxygen. In a sense, at that time the oxygen was a pollutant. To wonder whether they had a right to do that is clearly meaningless.

If there is no life at present on Mars, all the ethical problems may seem to stop with the question of whether humankind has the right to play some active role in shaping that infinitesimal part of the universe over which it exerts influence. But if micro-organisms are found on Mars, the whole issue must be considered from two points of view. How might the terraforming operations affect them? And what would be the impact of their presence on the entire project? The changes might make them evolve, perhaps enabling them to complete the process that was halted in the distant past. But the changes imposed by humans are likely to be too sudden to allow life to adapt to the new conditions, and therefore the risk that it would not survive is great. Consequently, it is important that any life present on Mars be studied in great detail before we start any work aimed at changing the environment. All the practical measures necessary to protect and preserve what can only be considered as an extreme and special case of biodiversity absolutely must be taken.

A radical point of view is based on the observation that, even if life still exists on Mars, we can be certain that it failed to produce an extensive biosphere. Therefore to supersede it with more successful life-forms that could spread across the planet would be consistent with the basic logic of evolution whereby the fittest forms of life survive and propagate.

These are useful starting points for a debate. And since terraforming Mars is a distant prospect, there is plenty of time to acquire a deeper understanding of all the issues involved.

It has recently been proposed that the Moon might be terraformed. Would terraforming the Moon be an important learning project on the way to terraforming Mars? The answer is likely to be negative for the reason that these two bodies are so different that whatever

---

[13] C.P. McKay and R.H. Haynes, "Should We Implant Life on Mars?," *Scientific American*, p. 108, December 1990.

Table 12.1 Characteristics of the Moon and Mars which are important in terms of terraforming.

	Moon	Mars
Gravity	very low	low
Atmosphere	none	very thin
Temperature	correct	low
Rotation	way too slow	correct
Radiation	high	high
Life	none	possible

we learn from terraforming one will hardly apply to terraforming the other. Table 12.1 summarizes their main characteristics from the point of view of terraforming.

The first difference between them is gravity. That of the Moon is too low to retain an atmosphere for long (at least in cosmic terms). If the Moon is terraformed, the atmosphere will have to be continuously replenished to compensate for the very slow (on a human scale) losses. The use of a heavy buffer gas (xenon, krypton) instead of nitrogen has been suggested because it would be less easily swept away by the solar wind, but it is difficult to imagine us obtaining such large quantities of these gases. And in any case, nitrogen is rare on the Moon.

The Moon does not possess an atmosphere to initiate the terraforming process, and lunar volatiles are present in ppm quantities. Water is present in the regolith in polar craters but we have yet to determine its locations and concentrations by direct sampling. One proposal is to augment the Moon with water and volatiles from comets, but in practice it would be difficult and could conflict with the early settlements that would surely exist when terraforming began.

Heating the surface of the Moon would not be required, in contrast to the situation for Mars. The rotation of the Moon is far too slow, with nights and days lasting two Earth weeks. This may cause large night-day temperature variations, with problems regarding atmospheric circulation. The possibility of crashing small asteroids on the Moon to increase its angular velocity has been proposed, but the proximity of the Earth would cause tidal effects which would act to slow the rotation again. The natural state for the Moon is synchronous rotation. No such problem exists for Mars. Neither body possesses a magnetosphere so, even after the terraforming was complete, the surface would be subjected to cosmic and solar radiation. The presence of oxygen in the enriched atmosphere would create an ozonosphere that could stop ultraviolet radiation, but achieving this would be more difficult for the Moon than for Mars.

The only advantage for the Moon is that we are certain there is no life, so there would be much less concern about ethical restraints than for terraforming Mars, where the situation is still ambiguous.

As a conclusion, terraforming the Moon is more acceptable but much more difficult than terraforming Mars, and the difference between the two bodies means it is unlikely that we could learn from the Moon lessons applicable to Mars. Terraforming Mars is something that may, in the long term, be attempted. Terraforming the Moon may be a project for a much more distant and indeterminate future.

13

Example missions

A few example missions are studied in this section. They include a very simple mission based on chemical propulsion, a more complex mission also based on chemical propulsion, a very similar mission based on NTP, an NEP mission, and an SEP mission. The chapter concludes with an example regarding a very advanced NEP mission, using a light and powerful nuclear generator. They are all long stay missions, but their durations depend on the type of propulsion employed.

13.1 MINIMAL CHEMICAL MISSION

This mission is meant to be a first mission to Mars, and one of its aims is to put a small crew on Mars as soon as is practicable. It envisages two cargo spacecraft being launched during a given launch window, and then a further cargo spacecraft and a crew ship launching in the following window.

This section is based on research by Salotti et al.[1][2] No assembly in LEO is performed, so the proposal qualifies as a Mars direct mission. This limits the mass of each Mars spacecraft to what can be lifted into orbit by a single heavy lift launcher, which means 100–130 t. As a result, the ERV is transported to Mars orbit in two parts and assembled there.

The three (or possibly four) astronauts are assumed to remain on Mars for about 500 days, and therefore it can be considered as a conjunction (long stay) class mission. All vehicles use chemical propulsion with cryogenic propellants (LOX/LH2) and zero-boil-off technologies are assumed to have been developed, including both thermal insulation and active cooling using cryocoolers. All vessels use aerocapture to enter Mars orbit and, in order to simplify aerobraking and descent to the planet all payload is carried in capsules of

---

[1] J.M. Salotti and R. Heidmann, "Roadmap to a Human Mars Mission," *Acta Astronautica* vol. 104, n. 2, pp. 558–564, 2014.

[2] J.M. Salotti, R. Heidmann and E. Suhir, "Crew Size Impact on the Design, Risks and Cost of a Human Mission to Mars," Proceedings of the IEEE Aerospace Conference, Big Sky, Montana (USA), pp. 1–9, March 2014.

© Springer International Publishing Switzerland 2017
G. Genta, *Next Stop Mars*, Springer Praxis Books, DOI 10.1007/978-3-319-44311-9_13

less than 40 t at Mars entry. The deceleration phase is achieved by using either rigid heat shields or small Hypersonic Inflatable Atmospheric Decelerators (HIAD). The mission will use ISRU/ISPP. The oxidizer (oxygen) to return the MAV to orbit will be produced on the surface, and possibly also the fuel (methane). The expected IMLEO is of the order of 500 t.

All computations were performed in terms of the 2035 launch window for the cargo and 2037 for the crew. These are fairly advantageous windows because the opposition of 2035 is perihelic. The outbound leg is a fast free return trajectory that needs a somewhat higher ΔV, but this is deemed to be manageable because it does not have a major impact on the IMLEO. As shown in Table 13.1, the IMLEO of each spacecraft is estimated at about 125 t.

Since chemical propulsion is used, the propulsion system may be regarded as being off-the-shelf. There are, however, elements which will need to be developed and hence should be considered as critical:

- The aerocapture/aerobraking device, because the mass of the vehicles requiring to be slowed down is too large to use off-the-shelf technologies.
- The zero-boil-off system owing to the use of LH2 as fuel.
- The ISRU/ISPP plant to produce either the oxidizer or both fuel and oxidizer for the MAV.

13.2 LARGER CHEMICAL MISSION

This architecture was studied in the NASA DRA5 report [24]. It is meant to be a first mission to Mars, and is based on two cargo spacecraft launched in one launch window and a crew ship in the next window. All computations were performed in terms of the 2035 launch window for the cargo and the 2037 window for the crew; hence they refer to fairly advantageous opportunities. Chemical propulsion with cryogenic propellants (LOX/LH2) was chosen, and zero-boil-off technologies were assumed, including both thermal insulation and active propellant cooling by cryocoolers. The total number of crew carried to Mars is six.

All interplanetary vehicles start from LEO at an altitude of 407 km, and upon arrival enter a highly elliptical orbit of Mars with a 250 km periareion, a 33,970 km apoareion, and a period of about 1 sol. The cargo ships travel a 202 day trajectory fairly similar to a Hohmann transfer. The ΔV of 3,660 m/s for the TMI maneuver is quite close to the figures in Table B.1. The MOI maneuver is performed using aerobraking, and so it requires quite a low ΔV of 1,076 m/s. No return to Earth is considered for the cargo ships. The crewed ship travels a much faster 174 day outbound trajectory that requires a ΔV of 3,950 m/s for the TMI maneuver, again a value quite close to the figures in Table B.1. No aerobraking is planned for MOI, which requires a much higher ΔV of 1,794 m/s. This is much lower than that reported in Table B.1 because the arrival orbit has a much higher energy. The crew stays on Mars for 539 days. The return journey lasts 201 days and the TEI maneuver requires a ΔV of 1,562 m/s.

The mission timing is:

Total mission time	Time spent in space	Time on Mars	Ratio t_{planet}/t_{space}
914 days	375 (174+201) days	539 days	1.44

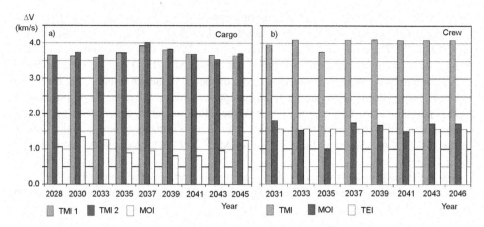

Figure 13.1 The ΔV for the various maneuvers in the launch opportunities from 2031 to 2046. (a) A cargo vehicle: both aerocapture MOI (TMI1) and propulsive MOI (TMI2 and MOI). (b) The crew vehicle (only propulsive MOI). (Redrawn from [24])

These values were computed for the 2035/3037 launch opportunities but (as shown in Figure 13.1) would not be much different if different dates were chosen.

13.2.1   Cargo spacecraft

One of the two cargo ships will carry the habitat into Mars orbit, while the other one carries the Descent and Ascent Vehicle (DAV). Each spacecraft will include a cylindrical aeroshell (a total of 103 t) plus two propulsive stages (103.6 t each) for a total IMLEO of 310.2 t. This does not include the reboost modules which maintain the orbit during vehicle assembly. An aeroshell module initially acts as an aerodynamic shroud to protect the payload during the ascent from Earth into orbit and later as an aerodynamic lifting body for Mars aerocapture, entry, and descent. This is a clever use of a large piece of hardware that would otherwise be jettisoned during launch.

All vehicles need to be assembled in LEO and require a reboost module to perform attitude control and orbit adjustment while orbiting the Earth. These modules are abandoned shortly prior to TMI. In the original mission plan, seven Ares V heavy lift launchers were to place into LEO all the modules to assemble the two cargo vessels, plus two reboost modules. Each launch was expected to lift at most 104 t. This is less than the intended capacity of the SLS Block II and other launchers of that class. The total mass into orbit is 717.3 t, with the first launch set to occur 270 days prior to the scheduled TMI.

Trans-Mars Injection module

Identical TMI modules are used on the cargo and crew vehicles, and each cargo vehicle has two of them. In the NASA study five engines with 110 kN of thrust, a specific impulse of 462 seconds, and a burn time of 700 seconds are used. The TMI maneuver is performed

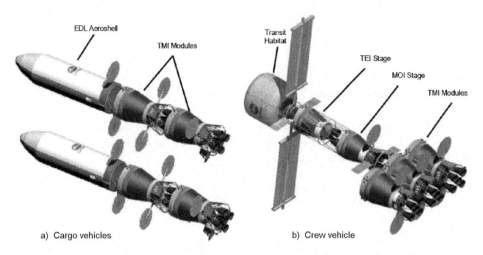

Figure 13.2 The scheme of the vehicles. (a) The two cargo vessels. (b) The crew ship. (NASA image)

using two burns, then the module is jettisoned. Each module is autonomous in terms of electrical power during launch and the assembly process in LEO. The two cargo spacecraft are illustrated in Figure 13.2a.

The mass breakdown, including the Reaction Control System (RCS) which controls the attitude of the spacecraft, is as follows:

Dry mass	Propellant mass	RCS	Total stage mass
15.1 t	86.2 t	2.3 t	103.6 t

13.2.2 Crew spacecraft

The crew spacecraft consists of the Crew Exploration Vehicle (CEV), the transit habitat, three TMI propulsion modules, one MOI propulsion module, and one TEI propulsion module. As with the cargo vehicles the TMI maneuver is performed using two burns, in this case with the two outboard TMI modules performing the first burn and the center module performing the second burn. The transit habitat is provided with a fairly large Photo Voltaic Array (PVA) for the life support system, all functions associated with the crew, and the cryogenic propellant cooling system (the latter is not required by the cargo vehicles, since they will not have any cryogenic propellants on board after the TMI maneuver initiates the interplanetary transfer). The CEV is used to launch the crew to LEO prior to TMI, and then can remain docked to the transit habitat throughout the mission until it is boarded by the crew to perform a direct entry of the Earth's atmosphere. The three TMI modules of the crew ship are identical to those of the cargo vehicle, but carry 91.1 t of propellant.

The plan envisaged five Ares V heavy lift launchers placing into LEO all the modules to assemble the crew vehicles, plus its reboost module. Each launch was expected to carry about 100 to 108.5 t. The first launch occurs 210 days before TMI. The total mass launched into orbit is 534.5 t, including 0.6 t for the crew riding a human rated launcher several days before TMI.

Mars Orbit Insertion module

The MOI module of the crew ship uses two of the same engines as the TMI module. It also supplies Guidance Navigation and Control (GN&C) when docked to the stack.
   The mass breakdown is as follows:

Dry mass	Propellant mass	RCS	Total stage mass
10.3 t	50.2 t	5.3 t	65.8 t

Trans-Earth Injection module

The TEI stage is a scaled-down version of the MOI module, with two engines of the same type as the other stages. It is responsible for orbital maneuvering at Mars, and providing GN&C for the stack after the MOI module is jettisoned. The crew ship is illustrated in Figure 13.2b.
   The mass breakdown is as follows:

Dry mass	Propellant mass	RCS	Total stage mass
11.4 t	24.1 t	7.3 t	42.7 t

Payload

The payload of the vehicle consists of the transit habitat (41.3 t) and the CEV that delivers the astronauts at the start of the mission (10.6 t).

13.2.3   Overall mission characteristics

This is a split mission, with two cargo vessels being launched in a single launch opportunity and the crew ship following at the next opportunity, 2 years later. All spacecraft are propelled by LOX/LH2 chemical rockets. The cargo ships use aerocapture at arrival. Both must enter the same elliptical orbit around Mars. The crew ship uses a propulsive maneuver to enter that same orbit. The cargo vessels bring a total payload of 206 t, but because this includes the aeroshells the net mass on Mars is about 120 t.
   The IMLEO for each cargo vessel and the crew ship (not including the reboost modules) is 310.2 t and 486 t, respectively. However, the total mass requiring to be placed into orbit is 1,251.8 t. So the mission will employ twelve superheavy lift launchers plus one human-rated launcher.

Since chemical propulsion is used, the propulsion system may be considered as off-the-shelf, but there are other components whose development must be considered as critical:

- The aeroshell and in general the aerocapture/aerobraking device, because the vehicle that is to be slowed down in this manner has a mass of about 100 t. This may need a long and costly development program.
- The zero-boil-off system that must preserve the LH2 fuel for a very long period.
- The ISRU/ISPP plant to produce the oxidizer (oxygen) for the MAV.

13.3   NTP MISSION

This is another mission studied in the NASA DRA5 report [24]. Its profile is the same as for the previous mission, hence a direct comparison can be made. The cargo containers are to be aerocaptured on arrival at Mars, while often a propulsive maneuver is considered for an NTP vehicle. The mission is powered by two NERVA-derived nuclear thrusters, each supplying 111 kN of thrust with a specific impulse in the range 875 to 950 seconds. They are fueled by LH2 in zero-boil-off tanks. Again all ships start from an Earth orbit at an altitude of 407 km and upon arrival at Mars they enter into the same elliptical orbit as previously. Consequently the same values of ΔV and of travel time apply.

13.3.1   Cargo spacecraft

Each of the two cargo vessels consists of a cylindrical aeroshell containing the payload (a total of 103 t, distributed in the same way as previously), plus a propulsive stage (96.6 t) and an LH2 tank (46.6 t) for a total IMLEO of 246.2 t. Figure 13.3a illustrates a cargo ship.
The mass breakdown of the propulsive stage and the LH2 tank is as follows:

	Dry mass	Propellant mass	RCS	Total stage mass
Propulsive stage	33.7 t	59.4 t	3.6 t	96.6 t
LH2 tank	10.8 t	34.1 t	1.7 t	46.6 t

As previously, all vehicles are assembled in LEO and require a reboost module. A total of five superheavy lift launchers are needed to place into orbit all the elements to assemble the two cargo vehicles (the reboost module is excluded from the IMLEO stated above for a cargo ship). Each launch should put between 96 and 103 t in LEO. The first launch occurs 180 days before TMI, and the total mass placed into orbit is 492.3 t.

13.3.2   Crew spacecraft

The crew ship consists of two Crew Exploration Vehicles (CEV) with their Service Modules (SM), the transit habitat, a propulsive module, an LH2 tank, and a drop LH2 tank. One of the CEVs will be used to land the crew on Mars and the other will perform a direct entry into the Earth's atmosphere at the end of the mission. The three TMI modules are the same as for the cargo vehicle, but carry 91.1 t of propellant. The dry mass of the LH2 drop tank includes 14 t of the drop tank proper, plus 8.9 t of the 'saddle truss' which is needed to keep the drop tank in-line.

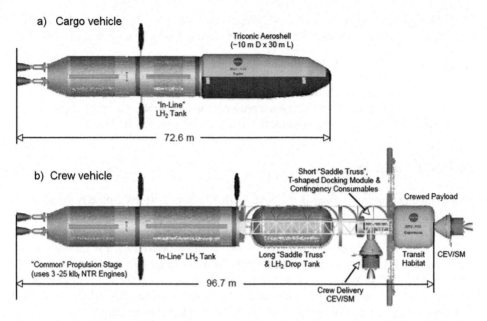

Figure 13.3 The scheme of the vehicles. (a) A cargo vessel. (b) The crew vehicle. (NASA image)

The mass breakdown is as follows:

	Dry mass	Propellant mass	RCS	Total stage mass
Propulsive stage	41.7 t	59.7 t	4.9 t	106.2 t
LH2 tank	21.5 t	69.9 t	–	91.4 t
LH2 drop tank	22.9 t	73.1 t	–	96.0 t

Payload

The payload consists of the transit habitat (32.8 t) and the CEV with the crew (10.6 t), plus other hardware for a total of 62.8 t.

Four heavy lift launchers are needed to place into LEO all the modules to assemble the crew vehicle, plus its reboost module. Each launch should put between 62.2 and 106.2 t in LEO. The first launch will occur 150 days before TMI, and the total mass placed into orbit is 356.4 t, including 0.6 t for the crew riding a human rated launcher a few days before TMI.

13.3.3   Overall mission characteristics

All the data for the timing of the mission, the launch windows, and the crew and payload to Mars are the same as in the chemically powered mission. The IMLEO of the vehicles is 246.2 t and 356.4 t, respectively, for each cargo vessel and the crew vehicle, and the total mass to be placed into orbit is 848.7 t, which requires eight heavy lift launchers plus one human-rated launcher for the crew. However, this figure is uncertain because the altitude of the orbit in which the nuclear rocket can be started remains to be determined. At any rate, owing to the lower number of launches, the waiting time in orbit is greatly reduced and so the study did not predict the need for any reboosting modules. (This alone would allow a reduction in the IMLEO of about 144 t that could go some way toward compensating for the higher starting orbit.) However, because the recommendation has been criticized, we have included them here.

The critical components that need to be developed are:

- Although the nuclear rocket is derived from the old NERVA engine that was tested on the ground, the propulsion system will require a fairly long and costly development.
- The aeroshell and in general the aerocapture/aerobraking device, because the vehicle that is to be slowed down in this manner has a mass of about 100 t. This may need a long and costly development program.
- The zero-boil-off system that must preserve the LH2 fuel for a very long period.
- The ISRU/ISPP plant to produce the oxidizer (oxygen) for the MAV.

13.3.4   Comparison between the NTR and the chemical mission

The two missions are identical in terms of mission profile and payload. Their IMLEO estimates compare as follows:

	Cargo vehicles	Crew vehicle	Total launches	Total IMLEO
Chemical mission	2×310.2 t	486.0 t	12	1,251.8 t
NTP mission	2×246.2 t	356.4 t	8	848.7 t

Note that the vehicle masses do not include the reboost systems. From the table it is clear that the use of nuclear propulsion leads to a 25 percent saving of the IMLEO. However, a direct comparison may be misleading, for the following reasons:

- The starting orbits are assumed to be the same. Actually a 407 km orbit is likely to be too low for starting a nuclear engine, at least in terms of political acceptance. If a higher orbit is adopted, the number of launches may be increased because the vehicle is able to place a smaller mass into orbit and the reduction of the ΔV that can be obtained in this manner is too small to compensate.
- It may be questionable to exclude the need for reboost modules in the case of nuclear propulsion. It is true that the smaller number of launches reduces the waiting time in LEO, but this may be insufficient to avoid reboosting. Reboosting may be avoided by using a higher starting orbit, and this partially mitigates the disadvantage mentioned in the previous point.
- Little is said about the mass of the radiation shielding for the nuclear engine. This may be a small point because the reactor is almost 90 m from the crew, and in between is a lot of structure and propellant which will provide shielding.
- In both cases the cargo vehicles use aerobraking to achieve Mars orbit, but using NTP it would be possible to make a propulsive MOI maneuver, thereby eliminating the need for the complex and massive aeroshell for this phase. An aeroshell would be required for EDL, but this would be lighter and this would at least partially compensate for the mass of the fuel required for the propulsive maneuver.

- With nuclear propulsion, the spacecraft could enter Earth orbit at the end of the mission, although the altitude would not be very low. This would avoid high speed direct entry (and related risks), and possibly enable the crew ship to be designed to be reusable.
- Nuclear propulsion permits faster transits to and from Mars, thus reducing radiation exposure, increasing the time spent on the planet, and slightly reducing the mission duration. Hence if NTP is adopted, it should not be to reduce the IMLEO relative to chemical propulsion for a given mission, it should be to undertake fast missions and, perhaps, to allow some vehicles to be reusable (a benefit for the program as a whole that is otherwise impossible).

13.4 NEP MISSION

13.4.1 General considerations

In order to study a nuclear electric mission, it is necessary to make a guess about the specific mass of the generator. In this study a value $\alpha = 6$ kg/kW $= 0.006$ kg/W is assumed. This value is fairly optimistic, but seems reasonable for the not too distant future.

This mission is also based on launching two cargo vessels in one launch opportunity and a crewed ship in the next one. To be able to make a direct comparison, this study uses the same favorable 2035–2037 opportunities as the chemical and NTP missions. Because the spacecraft are assumed to be reusable, the less favorable 2040 opportunity is also considered.

The total number of crew carried to Mars is six. The electric thrusters are of the VASIMR type that is fed with liquid argon. Argon boils at $-189°C$, hence no specific ZBO technology is required. Assuming the overall efficiency of the thruster plus the power conditioning $\eta_t = 0.6$, the effective specific mass $\alpha_e = \alpha/\eta_t$ has a value of 10 kg/kW $= 0.010$ kg/W. The maximum specific impulse is assumed to be 8,000 seconds. All vehicles start the mission from a circular LEO at 500 km, and enter a circular LMO at that same altitude. All maneuvers are performed by the electric thrusters (the only aerodynamic maneuvers are for EDL). The cargo ships travel to Mars on a 450 day trajectory, including the time required to spiral around Earth to escape into interplanetary space and to spiral around Mars to achieve the desired orbit.

The optimal durations of the various phases of the Earth-Mars journey were computed as shown in Section C and are:

Launch opportunity	Total mission time	T_E	T_{INT}	T_M
2035	450 days	134.72 days	258.7 days	56.6 days
2037	450 days	120.7 days	280.0 days	49.3 days
2040	450 days	109.1 days	296.8 days	44.1 days

If the spacecraft have to be carried back in order to reuse them, the duration of the phases of the return journey are:

Launch opportunity	Total mission time	T_E	T_{INT}	T_M
2037	450 days	48.2 days	284.4 days	117.4 days
2040	450 days	44.6 days	300.0 days	105.4 days

The relevant dates for the outbound journey are:

Start	TMI	MOI	Arrive
January 7, 2035	May 22, 2035	February 4, 2036	April 1, 2036
March 18, 2037	July, 17, 2037	April 23, 2038	June 1, 2038
May 11, 2039	August 28, 2039	June 20, 2040	August 3, 2040

The dates for the possible inbound journeys are:

Start	TEI	EOI	Arrive
April 21, 2037	June 9, 2037	March 20, 2038	July 15, 2038
May 31, 2039	July 15, 2039	May 10, 2040	August 23, 2040

The optimal values of J (expressed in m^2s^{-3}) and γ (with $\alpha_e = 10$ kg/kW) for the outbound journey are:

Launch opportunity	J_{tot}	γ	J_1	J_2	J_3
2035	4.200	0.205	2.020	1.331	0.849
2037	5.052	0.225	2.243	1.844	0.965
2040	5.744	0.240	2.466	2.207	1.070

The optimal values of J (m^2s^{-3}) and γ for the return leg are:

Launch opportunity	J_{tot}	γ	J_1	J_2	J_3
2037	4.861	0.220	0.984	1.573	2.30
2040	5.564	0.236	1.060	1.956	2.54

In the favorable 2037 opportunity the crew ship can travel to Mars on a 210 day trajectory but it takes 225 days in the less favorable 2040 opportunity. The optimal durations of the phases of the Earth-Mars journey are:

Launch opportunity	Total mission time	T_E	T_{INT}	T_M
2037	210 days	33.9 days	162.5 days	13.5 days
2040	225 days	32.3 days	179.4 days	13.3 days

In this case, the spacecraft have to be carried back with the crew and the duration of the phases of the return journey are:

Launch opportunity	Total mission time	T_M	T_{INT}	T_E
2040	225 days	12.9 days	180.0 days	32.1 days

The relevant dates for the outbound journey are:

Start	TMI	MOI	Arrive
July 28, 2037	August 31, 2037	February 9, 2038	February 23, 2038
September 9, 2039	October 11, 2039	April 2, 2040	April 21, 2040

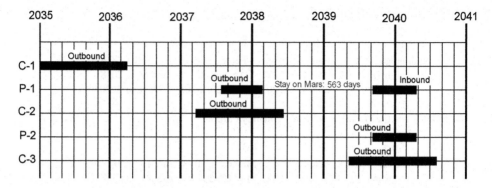

Figure 13.4 The NEP timeline of the first three cargo 'flights' (labeled C; each involving two cargo ships traveling together; disposable, no return) and the first two passenger ships (P) up to 2041.

The dates for the return leg are:

Start	TEI	EOI	Arrive
September 9, 2039	September 22, 2039	March 20, 2040	April 21, 2040

The timeline of the first three cargo 'flights' (disposable, no return) and the first two crew ships (with the second crew going out during the same 2040 opposition as their predecessors return home) is reported in Figure 13.4. Note that when a crew sets off, so does a cargo 'flight' carrying payloads for their successors. Each cargo 'flight' involves two cargo ships traveling together.

The optimal values of J (m²s⁻³) and γ (with α_e = 10 kg/kW) for the outbound journey are:

Launch opportunity	J_{tot}	γ	J_1	J_2	J_3
2037	20.478	0.453	7.332	10.01	3.13
2040	22.051	0.470	8.09	11.188	3.17

The optimal values of J (m²s⁻³) and γ for the return leg are:

Launch opportunity	J_{tot}	γ	J_1	J_2	J_3
2040	22.252	0.475	3.273	11.551	7.702

13.4.2   Cargo spacecraft

One of the two cargo vessels will carry the habitat and the other will carry the Mars Descent and Ascent Vehicle (DAV). The payload of each craft is subdivided into two smaller units of about 40 t (including the aeroshell) so that no large payload is landed with a single aeroshell. Since the landers initiate the descent from LMO, the aeroshell need not to be designed for aerocapture. Thus each lander is assumed to be 60 percent payload and 40 percent aeroshell. This is based on a NASA study in which the aeroshell mass was estimated at 50 percent of the total mass for an 80 t vehicle which had to perform aerocapture prior to the EDL phase.[3] The aeroshell for the EDL maneuver has therefore been estimated at 40 percent of the total entry mass of a 40 t vehicle.[4] This statement requires further investigations and lower values may even prove to be possible.

The total mass of the aeroshells and the payload is 80 t per cargo ship, and the payload that each carries to Mars is 48 t. The cargo ships must also carry to Mars the propellant for the return of the crew ship, which may be estimated at about 60 t (including the tanks) per cargo ship, thereby bringing the payload to 140 t. Each cargo vessel consists of two payload units and a propulsion unit that comprises a generator, the plasma thrusters, the RCS, and the propellant (liquid argon) tank.

Assuming $\gamma = 0.240$ (the least favorable value for the stated launch opportunities), a 9.08 MW generator, and a mass margin of 5 t, the mass breakdown is as follows:

Payload	Return propellant	Dry	RCS	Margins	Generator	Propellant	Total
80 t	60 t	25 t	3 t	5 t	54.5 t	71.7 t	299.2 t

Six heavy lift launchers, each carrying 100 t, are required to place into LEO all of the modules that are needed to assemble the two cargo vehicles in LEO. If two reboost modules are required, the IMLEO is increased by a non-negligible amount.

13.4.3   Crew spacecraft

The payload of the vehicle consists of the transit habitat (33 t) and the crew module plus the crew (7 t). The crew spacecraft will consist of a CEV, the transit habitat, and one propulsion unit that comprises a generator, the plasma thrusters, the RCS, and the propellant (liquid argon) tank. The nuclear generator provides ample power for the life support system and all functions associated with the crew, hence there is no need for a PVA. Nevertheless, a backup PVA may be carried for redundancy. The CEV transports the crew to the outbound spacecraft shortly prior to TMI, and then remains docked to the transit habitat until it is utilized to land on Mars. A simplified CEV is sufficient since it need only enter the Martian atmosphere from LMO (no high speed entry is required, as would occur on a mission in which the vehicle would make a direct entry into the atmosphere at the end of the mission). Upon returning to LEO, the CEV docks at a space station (or stays in independent orbit), from where the crew can return to Earth either prior to or after the quarantine.

Assuming $\gamma = 0.470$ (the worst case), a 9.30 MW generator and a mass margin of 5 t, the mass breakdown is as follows:

Dry	Payload	RCS	Margins	Generator	Propellant	Total
15 t	40 t	3 t	5 t	55.8 t	105.26 t	224.0 t

---

[3] B. Steinfeld, J. Theisinger et al., "High Mass Mars Entry, Descent and Landing, Architecture Assessment," Proc. of the AIAA Space 2009 Conf. and Exposition, AIAA 2009–6684, Pasadena, Sept. 2009.

[4] This evaluation is more optimistic than that in Figure 6.22.

The mass of propellant for the return journey is about 110 t (with a margin). To carry all the modules to assemble the crew vehicle into LEO (no reboost module is considered, due to the higher orbit and the brief wait), two heavy lift launchers would be needed, each carrying about 112 t. The first launch occurs 120 days before TMI. A human rated launcher enables the crew to board the interplanetary spacecraft a few days prior to achieving TMI, when it is already beyond the Van Allen Belts.

13.4.4   Overall mission characteristics

Each cargo vessel and the crew vehicle have IMLEO of about 300 t and 224 t, respectively. The total mass to be placed into orbit is 824 t, which requires eight heavy lift launchers plus one human rated launcher to carry the crew.

The critical components that need to be developed are:

- The nuclear generator.
- The plasma thruster. This is of the same type as that which is currently under development in order to reboost the ISS.
- The aeroshell or HIAD to perform EDL. However, this is much less critical than an aeroshell for aerocapture.
- The ISRU/ISPP facility to produce the oxidizer (and possibly also the fuel) for the MAV.

The two cargo vessels are abandoned in LMO. A total of about 25 t of propellant per ship would be needed to bring them back on a trajectory similar to that of the outbound leg. The tanks would have to be enlarged to accommodate that extra propellant, and of course more propellant would be needed to accelerate the enlarged vehicle away from Earth. This would raise the IMLEO of each cargo ship by about 45 t. However, it would be worthwhile paying this penalty to make the vehicles reusable. All vehicles would then require only to be refurbished and refueled for their next mission. In this scheme, the cargo ships arrive back in LEO 1,215 days after leaving it, thereby skipping one available launch window.

In the 2037 launch opportunity, the crew spends 563 days on Mars. Without taking into account the possibility of the astronauts boarding the interplanetary spacecraft shortly prior to escaping the Earth's sphere of influence, the total mission timing is:

Total mission time	Time spent in space	Time on Mars	Ratio t_{planet}/t_{space}
998 days	435 (210+225) days	563 days	1.29

13.5   SEP MISSION

13.5.1   General considerations

At present, it seems likely that a solar photovoltaic generator may have a lower specific mass than a nuclear generator. Reasonable values for the specific mass α and the specific area β for existing PVAs are $\alpha = 3.3$ kg/kW $= 0.0033$ kg/W and $\beta = 0.0033$ m^2/W, corresponding to an efficiency $\eta_a = 22$ percent.

This mission is based on two cargo vessels launched in a given launch opportunity, and a crew ship in the next one. All computations were made using the same opportunities as the previous examples to enable significant comparisons to be made. The first cargo mission is launched in 2035, and the crew set off in 2037. The total crew to Mars is six astronauts. The same electric thrusters of the VASIMR type fed with liquid argon, are used. Since the overall efficiency is again $\eta_t = 0.6$, the overall specific mass is assumed to have a value $\alpha_e = \alpha/\eta_t = 5.5$ kg/kW = 0.0055 kg/W. As before, the maximum specific impulse $I_{s\ max}$ is assumed to be 8,000 seconds. All spacecraft begin the mission from a circular LEO at 500 km, and enter a circular LMO at that same altitude. All maneuvers are performed by the electric thrusters (the only aerodynamic maneuvers are for EDL). The cargo vessels travel on a 450 day trajectory, including the time required to spiral around Earth to escape into interplanetary space and to spiral around Mars to achieve the desired orbit.

The optimal durations of the various phases of the Earth-Mars journey are:

Launch opportunity	Total mission time	T_E	T_{INT}	T_M
2035	450 days	111.0 days	265.2 days	73.8 days
2037	450 days	97.4 days	289.7 days	62.9 days
2040	450 days	84.7 days	310.0 days	55.3 days

For vehicles that are returned to Earth for reuse, the durations of the phases of the return journey are:

Launch opportunity	Total mission time	T_M	T_{INT}	T_E
2037	450 days	62.1 days	290.0 days	97.9 days
2040	450 days	55.5 days	310.0 days	84.5 days

The relevant dates for the outbound journey are:

Start	TMI	MOI	Arrive
February 1, 2035	May 23, 2035	February 13, 2036	April 26, 2036
April 12, 2037	July 19, 2037	May 4, 2038	July 6, 2038
June, 7, 2039	August 30, 2039	July 5, 2040	August 30, 2040

The dates for the possible return legs are:

Start	TEI	EOI	Arrive
March 29, 2037	May 30, 2037	March 16, 2038	June, 22, 2038
May, 8, 2039	July 3, 2039	May 8, 2040	July 31, 2040

The optimal values for J (m²s⁻³) and for γ (with $\alpha_e = 5.5$ kg/kW) on the outbound journey are:

Launch opportunity	J_{tot}	γ	J_1	J_2	J_3
2035	6.128	0.184	2.428	2.161	1.539
2037	7.352	0.201	2.746	2.822	1.785
2040	8.432	0.215	3.132	3.287	2.013

The optimal values of J (m^2s^{-3}) and γ for the return leg are:

Launch opportunity	J_{tot}	γ	J_1	J_2	J_3
2037	7.287	0.200	1.808	2.748	2.731
2040	8.341	0.214	2.006	3.196	3.139

The crew ship travels to Mars on a 220 day trajectory in the 2037 opportunity. A longer voyage of 235 days is possible for the less favorable 2040 opportunity. The optimal durations of the various phases of the Earth-Mars journey are:

Launch opportunity	Total mission time	T_E	T_{INT}	T_M
2037	220 days	28.9 days	172.7 days	18.4 days
2040	235 days	27.6 days	190.0 days	17.4 days

The return journey for the crew ship actually occurs during the 2040 opposition, and it lasts 235 days. The durations of the phases of the return leg are:

Launch opportunity	Total mission time	T_M	T_{INT}	T_E
2040	235 days	11.6 days	169.5 days	28.9 days

The relevant dates for the outbound journey are:

Start	TMI	MOI	Arrive
August 2, 2037	August 31, 2037	February 20, 2038	March 10, 2038
September 16, 2039	October 13, 2039	April 4, 2040	May 8, 2040

The dates for the return leg are:

Start	TEI	EOI	Arrive
August 23, 2039	September 9, 2039	March 17, 2040	April, 14, 2040

The timeline of the first three cargo 'flights' (disposable, no return) and the first two crew ships (with the second crew going out during the same 2040 opposition as their predecessors return home) is reported in Figure 13.5. Note that when a crew sets off, so does a cargo 'flight' carrying payloads for their successors. Each cargo 'flight' involves two cargo ships traveling together.

The optimal values of J (m^2s^{-3}) and γ (with $\alpha_e = 5.5$ kg/kW) for the outbound journeys are:

Launch opportunity	J_{tot}	γ	J_1	J_2	J_3
2037	27.921	0.392	8.495	13.911	5.514
2040	30.720	0.411	8.875	16.055	5.789

The optimal values of J (m^2s^{-3}) and γ for the return leg are:

Launch opportunity	J_{tot}	γ	J_1	J_2	J_3
2040	31.136	0.414	5.779	16.471	8.88

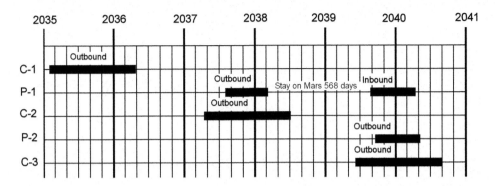

Figure 13.5 The SEP timeline of the first three cargo 'flights' (labeled C; each involving two cargo ships traveling together; disposable, no return) and the first two passenger ships (P) up to 2041.

13.5.2 Cargo spacecraft

One of the two cargo spacecraft will carry to Mars the habitat, while the other one will have the Mars Descent and Descent Vehicle (DAV). The payload of each spacecraft is subdivided into two smaller units of about 40 t (including the aeroshell) to enable the EDL to be performed without the need to land large payloads in a single unit. The total mass of the aeroshells and the payload is therefore 80 t per cargo ship. Assuming (as previously) that the lander consists of 60 percent payload and 40 percent aeroshell, each cargo ship will land 48 t on Mars.

Figure 13.6 shows the trajectory of a cargo vessel in the 2035 launch opportunity. The acceleration a, the specific impulse I_s, the propellant throughput $\dot{m}$, the power of the jet P, the thrust T and the mass m of the spacecraft are plotted as functions of time in Figure 13.7. The cargo ship needs also to transport to Mars the propellant to enable the crew ship to return. This has been estimated at about 38 t per cargo ship. Each cargo spacecraft therefore consists of two payload units plus the return propellant, and one propulsion unit made up of the generator, the plasma thrusters, the structure, the RCS, and the propellant (liquid argon) tank.

The mass breakdown for the three launch opportunities is as follows:

Launch	Payload	Return propellant	Dry	RCS	Margins	Generator	Propellant	Total
2035	80 t	38 t	25 t	3 t	5 t	33.9 t	41.6 t	226.5 t
2037	80 t	38 t	25 t	3 t	5 t	38.0 t	47.6 t	236.6 t
2040	80 t	38 t	25 t	3 t	5 t	41.4 t	52.8 t	245.2 t

The mass in LMO is 151 t, plus a maximum of 41.4 t for the generator. The power of the generator is 12.50 MW, corresponding to an array area $S_a = 41{,}500$ m^2. To use a configuration similar to that of the ISS, it would be possible to use 28 rectangular panels 74 m long and 20 m wide; i.e. each being about twice the size of those of the ISS. A returning spacecraft is likely to travel almost empty, so the generator will be rather larger than

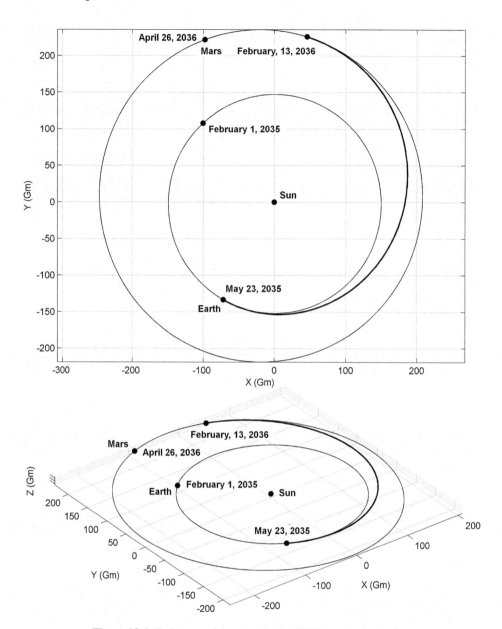

Figure 13.6 Trajectory of the cargo in the 2035 launch opportunity.

necessary for that phase of the mission. Just 7 t of propellant is required for the return journey. Recovering a cargo ship would increase the IMLEO by less than 10 t.

Four heavy lift launchers, each carrying 120 t, are required to carry to LEO all the modules needed to assemble the two cargo vehicles (no reboost module is considered here, owing to the higher orbit and the short wait in LEO). The total IMLEO is 490 t.

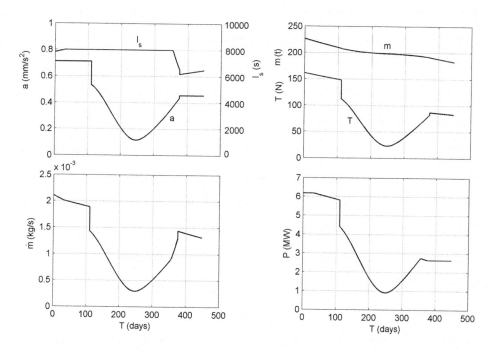

Figure 13.7 acceleration a, the specific impulse I_s, the propellant throughput $\dot{m}$ the power P, the thrust T, and the mass m of the spacecraft plotted as functions of time.

13.5.3 Crew spacecraft

The crew spacecraft will consist of a CEV, with the crew (about 7 t), the transit habitat (about 33 t), and one propulsion unit consisting of a generator, the plasma thrusters, the RCS, and the propellant (liquid argon) tank. Figure 13.8 shows the trajectory of the crew ship in the 2037 launch opportunity. Figure 13.9 plots the acceleration a, the specific impulse I_s, the propellant throughput $\dot{m}$, the power of the jet P, the thrust T, and the mass m as functions of time.

The nuclear generator provides ample power for the life support system and all functions associated with the crew, hence there is no need for a PVA. Nevertheless, a backup PVA may be carried for redundancy.

The CEV transports the crew to LEO prior to TMI, and then remains docked to the transit habitat until it lands on Mars. A simplified CEV is sufficient since it need only enter the Martian atmosphere from LMO (no high speed entry is performed, as would occur on a mission in which the vehicle would make a direct entry into the atmosphere at the end of the mission). Upon returning to LEO, the CEV docks at a space station (or stays in independent orbit), from where the crew can return to Earth either prior to or after the quarantine.

The payload here includes an inflatable habitat, in which the crew can live during both the outbound and the return legs.

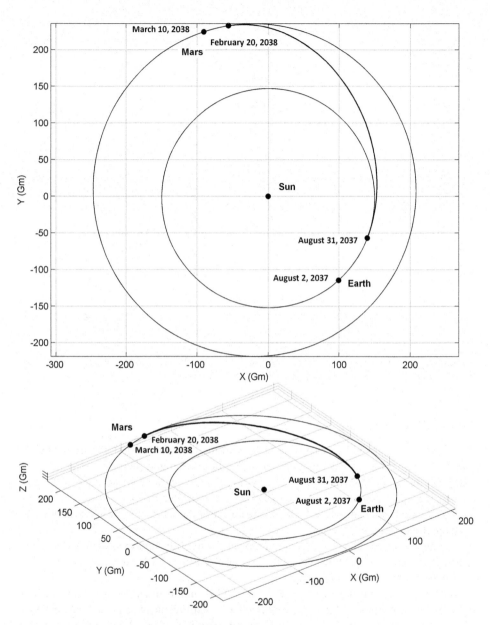

Figure 13.8 The trajectory of a crew ship for the 2037 launch opportunity.

Considering a value of γ = 0.411 (for the 2040 launch opportunity; the value for the 2037 opportunity is lower since the travel is shorter), the mass breakdown for the outbound journey is as follows:

Dry	Payload	RCS	Margins	Generator	Propellant	Total
15 t	40 t	3 t	5 t	44.0 t	74.7 t	181.7 t

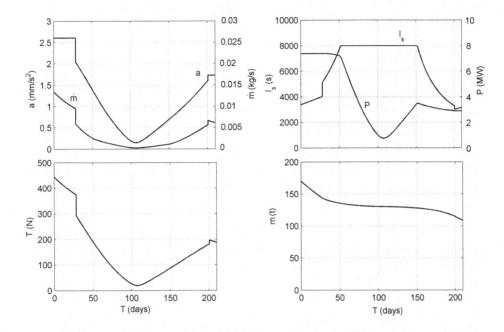

Figure 13.9 Acceleration a, the specific impulse I_s, the propellant throughput $\dot{m}$, the power P, the thrust T, and the mass m of the spacecraft plotted as functions of time.

The generator provides a power $P_a = 13.3$ MW for an array area $S_a = 44{,}000$ m²; i.e. 28 rectangular arrays 79 m long and 20 m wide, each less than twice the size of those of the ISS.

The return journey is performed during the 2040 opposition, and requires about the same propellant as the outbound journey because, although the payload is reduced, the value of γ is slightly higher. The propellant for the return journey is estimated at about 76 t.

Two heavy lift launchers, each carrying 90–95 t, will carry into LEO all the modules to assemble the crew vehicle (reboost is not considered here, owing to the higher orbit and the short wait in LEO). The first launch is planned 90 days prior to TMI, while the total mass launched to orbit is slightly larger, since it includes a crew riding a human rated launcher a few days before TMI.

13.5.4   Overall mission characteristics

The crew that sets off in the 2037 launch opportunity stays on Mars for 568 days. The return journey lasts 220 days, so the mission timing is:

Total mission time	Time spent in space	Time on Mars	Ratio $t_{\text{planet}}/t_{\text{space}}$
1,023 days	455 (220+235) days	568 days	1.25

Each cargo vessel and the crew ship have an IMLEO of 245 t and 182 t, respectively, and the total mass in LEO of 672 t requires six heavy lift launchers and one human rated launcher to carry the crew.

The critical components to be developed are:

- The plasma thruster. This is of the same type as that which is currently under development in order to reboost the ISS.
- The aeroshell or HIAD to perform EDL. However, this is much less critical than an aeroshell for aerocapture.
- The ISRU/ISPP facility to produce the oxidizer (and possibly also the fuel) for the MAV.

The two cargo spacecraft may return to LEO by their own power with the penalty of increasing the IMLEO of about 20 t.

13.6   VERY ADVANCED NEP MISSION

13.6.1   General considerations

As a further example, a very advanced NEP mission is described. Its nuclear generator has a specific mass which may be achievable in the more distant future, but to compare the results with those of the previous missions the launch opportunities are those starting in 2035. The advanced characteristics of the generator are used to reduce the mass of the cargo vessel (or, possibly to increase the cargo at the same IMLEO) and to reduce the travel time of the crew ships. The futuristic value for the specific mass of the generator is $\alpha = 0.6$ kg/kW = 0.0006 kg/W.

Like the previous missions, there will be two cargo vessels in a given launch opportunity, with a ship carrying six astronauts following in the next window. The same electric thrusters of the VASIMR type are used, fed with liquid argon. Since the overall efficiency remains $\eta_t = 0.6$, the overall specific mass is assumed as $\alpha_e = \alpha/\eta_t = 1$ kg/kW = 0.001 kg/W. As the mission is based on a more advanced technology, we assign it a slightly higher value of the maximum specific impulse $I_{s\,max}$ of 9,000 seconds. All ships begin the mission from a circular LEO at 500 km, and enter a circular LMO at that same altitude. All maneuvers are performed by the electric thrusters (the only aerodynamic maneuvers are for EDL). The cargo ships travel on a 365 day trajectory that includes the time to spiral around Earth to escape into interplanetary space and to spiral around Mars to achieve the desired orbit.

The optimal durations of the various phases of the Earth-Mars journey are:

Launch opportunity	Total mission time	T_E	T_{INT}	T_M
2035	365 days	88.3 days	241.6 days	35.1 days
2037	365 days	67.8 days	269.7 days	27.5 days
2040	365 days	56.9 days	285.0 days	23.1 days

The relevant dates for the outbound journey are:

Start	TMI	MOI	Arrive
March 7, 2035	June 3, 2035	January 31, 2036	March 6, 2036
May 23, 2037	July 29, 2037	April 25, 2038	May 23, 2038
July, 14, 2039	September 8, 2039	June 19, 2040	July 13, 2040

The optimal values of J (m²s⁻³) and γ (with $\alpha_e = 1$ kg/kW) for the outbound journey are:

Launch opportunity	J_{tot}	γ	J_1	J_2	J_3
2035	9.918	0.100	3.009	5.590	1.319
2037	12.861	0.113	3.857	7.357	1.647
2037	15.594	0.125	4.547	9.117	1.929

The crew ships travel to Mars on a 90 day trajectory during the favorable opportunity of 2037 and in 100 days in the less favorable one of 2040. The optimal durations for the various phases of the Earth-Mars journey are:

Launch opportunity	Total mission time	T_E	T_{INT}	T_M
2037	90 days	7.2 days	80.0 days	2.8 days
2040	100 days	7.2 days	90.0 days	2.8 days

The duration of the phases of the return journey are:

Launch opportunity	Total mission time	T_M	T_{INT}	T_E
2040	100 days	2.8 days	90.0 days	7.2 days

The relevant dates for the outbound journey are:

Start	TMI	MOI	Arrive
September 25, 2037	October 2, 2037	December 21, 2037	December 24, 2037
November 6, 2039	November 14, 2039	February 12, 2040	February 14, 2040

The dates for the return leg are:

Start	TEI	EOI	Arrive
November 6, 2039	November 9, 2039	February 7, 2040	February 14, 2040

The optimal values of J (m²s⁻³) and γ (with $\alpha_e = 1$ kg/kW) for the outbound journey are:

Launch opportunity	J_{tot}	γ	J_1	J_2	J_3
2037	159.28	0.399	29.86	117.272	12.14
2040	171.353	0.414	29.82	129.343	12.19

The optimal values of J (m²s⁻³) and γ for the return leg are:

Launch opportunity	J_{tot}	γ	J_1	J_2	J_3
2040	178.457	0.422	12.14	136.449	29.86

13.6.2 Cargo spacecraft

The payloads of the cargo ships are the same as in the previous examples. Assuming a value of $\gamma = 0.125$ (the least favorable value for the launch opportunities being considered) and a generator providing 36.3 MW the mass breakdown is as follows:

Payload	Return propellant	Dry	RCS	Margins	Generator	Propellant	Total
80 t	40 t	25 t	3 t	5 t	21.8 t	25.0 t	199.8 t

If no reboost module is required, then four heavy lift launchers, each carrying 100 t, are required to carry to LEO all the modules to assemble the two cargo vehicles.

13.6.3 Crew spacecraft

The payload of the vehicle comprises the transit habitat (33 t) and the crew module plus the crew (7 t). The crew spacecraft will consist of a CEV, the transit habitat, and one propulsion unit consisting of a generator, the plasma thrusters, the RCS, and the propellant (liquid argon) tank.

The CEV transports the crew to LEO prior to TMI, and then remains docked to the transit habitat until it lands on Mars. A simplified CEV is sufficient since it need only enter the Martian atmosphere from LMO (no high speed entry is performed, as would occur on a mission in which the vehicle would make a direct entry into the atmosphere at the end of the mission). Upon returning to LEO, the CEV docks at a space station (or stays in independent orbit), from where the crew can return to Earth either prior to or after the quarantine.

Assuming $\gamma = 0.461$, the mass breakdown is as follows:

Dry	Payload	RCS	Margins	Generator	Propellant	Total
15 t	40 t	3 t	5 t	44.5 t	75.9 t	183.4 t

The 74.2 MW power required of the generator is quite high, but reasonable considering that the total time in space is 3 months. The mass of propellant for the return journey is again about 75 t.

Two heavy lift launchers, each carrying about 90 t, are required to carry into LEO all the modules needed to assemble the crew vehicle (no reboost module is considered here, owing to the higher orbit and the short wait in LEO).

13.6.4 Overall mission characteristics

Each cargo vessel and the crew ship have IMLEO of about 200 t and 183.5 t, respectively, and the total mass in LEO of 583.5 t requires six heavy lift launchers plus one human rated launcher to carry the crew.

The critical components to be developed are:

- The nuclear generator, which is assumed to be much more advanced than those which are predictable in the near future.
- The plasma thruster, which is also much more advanced than those being developed for the reboost of the ISS, since a higher specific impulse (9,000 seconds) is required in order to match the advanced reactor.
- The aeroshell or HIAD for EDL, and the ISRU/ISPP facility are, however, not more advanced than those of the previous examples.

The two cargo vessels are abandoned in LMO. A total of about 15 t of propellant per ship would be needed to bring them back on a trajectory similar to that of the outbound leg. The tanks would have to be enlarged to accommodate that extra propellant, and of course more propellant would be needed to accelerate the enlarged vehicle away from Earth. This would raise the IMLEO of each cargo ship by about 20 t. However, it would be worthwhile paying this penalty to make the vehicles reusable. All vehicles would then require only to be refurbished and refueled for their next mission. In this scheme the cargo ships arrive back in LEO 1,110 days after leaving it, thereby skipping one available launch window.

In the 2037 launch opportunity, the crew spends 705 days on Mars. Without taking into account the possibility of the astronauts boarding the interplanetary spacecraft shortly prior to escaping the Earth's sphere of influence, the total mission timing is:

Total mission time	Time spent in space	Time on Mars	Ratio t_{planet}/t_{space}
872 days	190 (90+100) days	682 days	3.59

13.7   EXTREMELY FAST NEP PASSENGER SHIP

13.7.1   General considerations

As a final example, a 'SF' passenger ship is described. The specific mass of the generator is assumed to be $\alpha = 0.014$ kg/kW $= 14 \times 10^{-6}$ kg/W, a value which currently belongs more in the realm of science fiction. This example is shown to state that very fast interplanetary journeys do not require questionable breakthroughs like warp drives or propellant-less propulsion, but simply further development of present technologies. This value of α may require the use of nuclear fusion, and it will surely require new materials, but it will not require unpredictable theoretical developments.

The electric thrusters of the VASIMR type are assumed to be more developed than the current state of the art, with an improved efficiency $\eta_t = 0.7$, and a higher maximum specific impulse $I_{s\,max}$ of 15,000 seconds. The overall specific mass is $\alpha_e = \alpha/\eta_t = 0.02$ kg/kW $= 20 \times 10^{-6}$ kg/W. These values require only further development of what is now either possible or will soon be possible, not technological revolutions. As in the previous cases, the passenger ships start from a circular LEO at 500 km, and enter a circular LMO at that same altitude.

Assuming the launch opportunity of 2050 and a total travel time of 30 days, the optimal durations of the various phases of the Earth-Mars journey are:

Launch opportunity	Total mission time	T_E	T_{INT}	T_M
2050	30 days	0.5 day	29.3 days	0.2 day

The relevant dates for the outbound journey are:

Start	Arrive
July 30, 2050	August 29, 2050

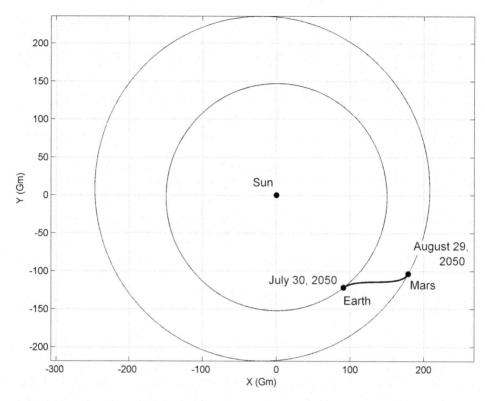

Figure 13.10 Trajectory of an extremely fast NEP spaceship for the 2050 launch opportunity that covers the Earth-Mars distance in just 30 days.

The optimal values of J (m^{2}s^{-3}) and γ (with $\alpha_e = 0.02$ kg/kW) for the outbound journey are:

Launch opportunity	J_{tot}	γ	J_1	J_2	J_3
2050	7,121.7	0.377	253.9	6,769.9	97.9

The trajectory for the 2050 launch opportunity is shown in Figure 13.10. Owing to the very high performance of the generator, this is an almost straight path. It is interesting to note that the thruster is not matching the high performance of the generator, since with an $I_{s\,max}$ of just 15,000 seconds the thruster is almost always operating in conditions of constant specific impulse. It is operating as a CEV and not a VEV system. This notwithstanding, the propellant consumption is extremely limited for such tremendous performance.

13.7.2 The spacecraft

The total payload is 40 t, including the transit habitat and the crew. The crew ship is made up of the transit habitat, and one propulsion unit comprising a generator, the plasma thrusters, the RCS, and the propellant (liquid argon) tank. Upon returning to LEO, the CEV docks at a space station (or stays in independent orbit), from where the crew can return to Earth either before or after the quarantine.

Assuming $\gamma = 0.377$, the mass breakdown is as follows:

Dry	Payload	RCS	Margins	Generator	Propellant	Total
15 t	40 t	3 t	5 t	38.2 t	61.3 t	162.5 t

At 2,728 MW the power of the generator is high, but this corresponds to the extremely impressive performance of the spacecraft. No propellant for the return journey is considered, since it is assumed either to be carried to Mars aboard a cargo ship, or to be produced on the planet. The fully assembled spacecraft can be placed into LEO by a single heavy lift launcher and then fueled for the mission. Upon returning to Earth orbit at the end of the mission, it merely requires refurbishment and refueling.

13.8 OVERALL COMPARISON

The main characteristics of the missions summarized in this chapter are summarized in Table 13.1. The first six missions are not fully comparable because the first one carries only three astronauts and much less payload to Mars, and the first three use the same trajectory and the same travel time whereas with NTP the travel time could be shortened at the expenses of the IMLEO.

Table 13.1 A comparison of the five missions described. Prop: propulsion type; n_c: number of crew; t_{tot}: total mission time; t_{space}: total time in space; t_{plan}: time on Mars; m_{cargoV}: mass of cargo vehicle(s); m_{crewV}: mass of crew vehicles; n_L: number of heavy lift launches needed to carry everything (excluding the crew) in Earth orbit.

#	Prop	n_c	t_{tot} days	t_{space} days	t_{plan} days	m_{cargoV} t	m_{crewV} t	n_L	IMLEO t
1	C	3	914	375	539	3×125	125	4	500
2	C	6	914	375	539	2×310.2	486	12	1,251.80
3	NTP	6	914	375	539	2×246.2	356.4	8	848.7
4	NEP	6	998	435 (≈385)	563	2×300.0	224	8	824
5	SEP	6	1,013	445 (≈395)	568	2×245.0	182	6	672
6	NEP	6	872	190	682	2×200.0	184	6	584
7	NEP	6	–	30 (one way)	–	–	162.5	2	–

In the case of electric propulsion, the numbers in brackets refer to the possibility that the crew reaches the spacecraft just two days before reaching escape conditions and at the end of the mission abandons it two days after being captured by the Earth's sphere of influence. In this case a reduction of the time in space of about 50 days can be obtained.

Chemical propulsion requires either a low payload or a quite large IMLEO. The case of SEP is interesting for the low IMLEO and the fairly quick trajectories to and from Mars with almost off-the-shelf technology.

Examples 6 and 7 (in the table) show that to obtain really fast interplanetary transfers it is essential to resort to nuclear power. In comparing NTP versus NEP, the studies slightly favored NEP.

The final example may seem like science fiction, but it has been added to indicate what traveling to Mars may become once lightweight and powerful nuclear generators (i.e. with low α, possibly using fusion) become practicable. The value of α assumed in this example is low, but not extreme. Once lower values are achievable, even faster and low cost transportation in deep space will enable humankind to really become a spacefaring civilization.

14

Conclusions

Humankind originated in one small region on Earth, and then slowly extended its habitat to the entire planet. In doing so, it had to develop increasingly complex technologies, both in order to reach new lands and to live in increasingly harsher environments. The last lands which were settled required fairly advanced technologies, like oceanic navigation, shelters and warm garments for living in the Arctic and technologies to obtain food in such difficult conditions.

Now it is time for humankind to further expand its sphere of influence, and the first thing is to explore newfound lands with a view to settling them. In this case, however, the technological requirements are far more demanding. Simply to reach the new lands will be a challenge beyond anything that we have faced previously, and the new environments are so harsh that living there will be even more challenging than the journey to get there.

The first extraterrestrial land, standing there and ready for us to colonize, is the Moon. The fact that we have already been there proves that the journey is not impossible. A dozen human beings lived on the lunar surface for periods ranging between several hours to a few days. Although confined by space suits, they were readily able to walk on its surface and to undertake productive work. Unfortunately, at that point our foray into deep space came to an abrupt halt, with the result that for almost half century human missions have been limited to activities in LEO.

Was this setback caused by a perceived inadequacy of our technology? Was it at least a contributory cause? At a first glance, this seems to be ruled out. The technology of the 1960s was able to transport several crews to the Moon, and the many projects for expeditions to Mars outlined in Chapter 1 are mostly based on current (or little more advanced) technologies. With current, off-the-shelf technology we have the means to visit our nearest celestial neighbors, but we cannot proceed directly to settle the Moon and Mars.

An example from history may, perhaps, be useful. There is no doubt that in the tenth century the Vikings had the technology to reach America, and Leif Eriksson succeeded in this task. Such an enterprise was repeated and a few colonists from Scandinavia established small settlements on the American coast. But these settlements were so short lived that for centuries their very existence was forgotten. Only at the end of the fifteenth century were the Americas 'discovered' by Europeans and, at that point, colonization came strikingly fast. In the space of a few decades, new colonies flourished and attracted a large

© Springer International Publishing Switzerland 2017
G. Genta, *Next Stop Mars*, Springer Praxis Books, DOI 10.1007/978-3-319-44311-9_14

number of people. Thirty years after Christopher Columbus' landing, the population of the American colonies numbered in the thousands. Forty five years after the first human landing on the Moon, however, no one is permanently living in space.

The ethnologist Ben Finney has noted that navigational techniques and the rigging of ships were sufficient to allow people to cross the oceans safely and repeatedly only after the middle of the fifteenth century. Moreover, although the appropriate technology is undeniably essential, it is also necessary to have favorable political and economic conditions.

If we want to prevent our journeys to the Moon, and perhaps later to Mars, having the same small impact on history as the Viking voyages to America had, we need to develop new technologies to be able to travel safely and sustainably through space at costs that are compatible with economic constraints at that time. When this situation is achieved, funding from government sources will no longer be the basic requirement, and raising funds via the stock market will be the modus vivendi. While space agencies perform scientific research and support major technological advancements, the operational aspects of space activities will be directly in the hands of private organizations. Perhaps something analogous to the way in which the British developed their colonies will be established – a sort of East India Company for the new settlements in the solar system.

We desperately need to improve our space technologies. In the 1980s and early 1990s there was a stagnation, if not actually a retreat. In the 1970s the Americans, and a few years later also the Russians, had a launcher in the 100 t class, research in nuclear propulsion was sufficiently advanced for prototype nuclear engines to be ground tested by both the USA and the USSR, and nuclear reactors were being launched into space. At the end of the 1990s it seemed that after a pause of more than 20 years, space exploration was about to resume. A return to the Moon as a step towards Mars appeared to be at hand. But this wave of optimism proved to be an illusion. Returning to the Moon was slipped to an undetermined future, with the phrase: "We've already been there." This dismissal shows a total lack of understanding of the true reasons for investing in space in general, and missions to the Moon in particular. It suggested we are indifferent tourist who go to a place just in order to boast that we have been there. What would we now think about John Cabot or Amerigo Vespucci if they had asked, "Why should we go to America? Columbus has already been there!"

We must return to the Moon to settle a new world, and perhaps even more, to learn how to settle new worlds and to develop the relevant technologies. And then as President Bush said upon the twentieth anniversary of the historic Apollo 11 lunar landing, "…and then a journey into tomorrow, a journey to another planet, a manned mission to Mars."

However, the first time we went to the Moon, it was a race with the USSR to show the world which nation had the best political and economic system. The context was the Cold War, and competing to be first to reach the Moon was seen as a harmless alternative to letting international tensions boil over into a 'hot' nuclear war. The remarkable thing is that when President Kennedy issued the challenge in 1961, the task had seemed utterly unbelievable.

Now things are radically different and the name of the game is cooperation and global participation. The example of the ISS illustrates that it would be possible for nations to join forces to undertake a new program of missions to the Moon to create a small outpost as an initial step towards human Mars exploration.

With ongoing technological advances, a new space economy, and an expanding global economy, the challenge of establishing a lunar base and ultimately heading for Mars will increasingly transfer from the realm of nations and space agencies into the realm of private enterprise.

A poster published by NASA depicting some of the steps leading toward human Mars exploration is shown in Figure 14.1. It shows the ideas of a single space agency (really, of

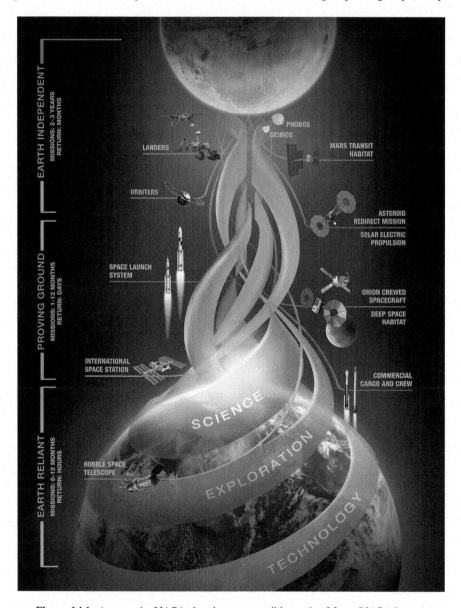

Figure 14.1 A poster by NASA showing one possible road to Mars. (NASA image)

a single agency at a certain moment of its history) and therefore is far from being complete. The first thing that is lacking is a mention of the role of the Moon in this great venture. A second point worth mentioning is that while technology, exploration, and science are clearly stated as motivators of space exploration, business is missing. However, as has been pointed out several times in this book, space business, and above all lunar business, is essential to making human Mars exploration affordable.

Essentially, what is missing in the NASA approach are other players, both public and private. It is impossible to develop all possible technologies and approaches. If a mission is to be a joint effort involving many players, it must be a synthesis of the interests and ideas of the various potential participants. To pursue too many, possibly conflicting, scenarios would only lead to failure. Informed choices will be crucial.

Things may be completely different if Mars exploration is to be undertaken by private companies. There might even be rival companies competing with one another. In this case, instead of dangerous competition between rival nations that may resort to armed conflict, it will be healthy rivalry between players trying to open new businesses and markets. In this case, it is possible the various players will compete with different technologies and pursue different goals. And as often happens in business, when huge difficulties must be overcome and the profits are sufficient to be shared around, rivals may find due cause to cooperate on certain tasks.

Whoever the partners or competitors may be in this great adventure, it is essential that those proposing a human mission to Mars realize that humankind cannot afford a false start. Affordability and safety must be paramount. A single accident, especially if it results in the loss of a mission or, worse, the loss of a crew, could prompt such a backlash as to threaten loss of program. Even relatively minor setbacks such as unexpected growth in costs, large delays, or the withdrawal of important partners, may have the same result. Even though the Apollo missions were spectacularly successful, the program was curtailed, partly owing to the cost (although there were launch vehicles and spacecraft in stock, bought and paid for), partly to public apathy (the American public was bored of watching their astronauts on the lunar surface), and partly because NASA management was becoming risk averse (the next mission might end in loss of crew).

For the sake of paradox, we could argue that, in some ways, we might now be rather better placed for human space exploration if President Kennedy had never begun a 'crash program' to reach the Moon in the 1960s. If events had followed those trends perceived as being logical at that time, then maybe we would have been in position to celebrate the new millennium by establish an outpost on the Moon as the first step in a coordinated program that had Mars as its ultimate objective.

However, we must begin where we happen to be, which means choosing the relevant technologies, then accurately balancing performance with technological readiness and safety, with an awareness that the first mission to Mars must not be merely a 'flag and footprint' exercise. Furthermore, it must not be seen as the final point of an ancient dream. It must be the opening of a new era of exploration whose purpose is to turn humankind into a spacefaring species.

Appendix A: Positions of the planets

The motions of the planets of the solar system are complex but, if mutual gravitational attractions can be neglected and the orbits are assumed to be governed only by the attraction of the Sun, they would move along elliptical orbits. This is adequate for first approximation studies, but in order to navigate through space requires detailed knowledge of their ephemerides.

A.1 FIRST APPROXIMATION (CIRCULAR ORBITS)

As Johannes Kepler discovered in the early seventeenth century, the orbits that the planets pursue around the Sun are not circular but elliptical. However, this is not strictly true because the mutual interactions of the planets prevent their orbits from being true ellipses. Furthermore, the orbits are not coplanar. The result is complexity. The positions of the planets and other solar system bodies are reported as functions of time in ephemerides. In the pre-computer age, these took the form of printed tables. Nowadays they are mathematical models running on digital computers.

The slight eccentricity and non-coplanarity of the orbits makes each interplanetary voyage a unique case. For a given launch opportunity the exact date of launch must be given in advance and the trajectory designed in an *ad hoc* manner. There are however, some regularities. In the case of Mars things repeat (although not exactly) with the synodic period of 779.94 days. There are other similarities that repeat every 7 or 8 synodic periods, and larger ones every 37 synodic periods (79 years). The synodic period of Venus is 583.92 days, so trajectories for Mars that include a Venus flyby do not repeat regularly even with the assumption of circular orbits.

The fact that the orbits of the planets are elliptical and non-coplanar cannot be neglected in the actual design of a mission. Moreover, since the orbits are not even elliptical, each mission must be studied in great detail by integrating the equations of motion numerically and by using the detailed ephemeris of the solar system. But for a preliminary study aimed at the overall design of a generic mission, it is possible to assume that the planetary orbits are both circular and coplanar, to achieve general results that apply for any launch opportunity, at least as a first approximation.

© Springer International Publishing Switzerland 2017
G. Genta, *Next Stop Mars*, Springer Praxis Books, DOI 10.1007/978-3-319-44311-9

The data for the orbits of Earth and Mars are reported in Table 3.1. Since some Mars missions include a flyby of Venus, the data for that planet are also included.

If the orbits of the planets are assumed to be circular with radius Ri, the orbital velocity of the i-th planet is

$$V_i = \sqrt{\frac{\mu_S}{R_i}} \qquad (A.1)$$

where μ_S is the gravitational parameter of the Sun, such that

$$\mu_S = GM_S \qquad (A.2)$$

where G is the universal gravitational constant ($G = 6.67384 \times 10^{-11}$ m³ kg⁻¹ s⁻²) and M_S is the mass of the Sun ($M_S = 1.99 \times 10^{30}$ kg).

The angular velocity of a planet about the Sun is thus

$$\omega_i = \sqrt{\frac{\mu_S}{R_i^3}}. \qquad (A.3)$$

The position of the planet along its orbit can be measured using the *true anomaly* angle v, measured relative to an inertial frame that is fixed in the ecliptic plane. Since, in the simplified approach, the orbits are circular and coplanar, the plane is simply the plane in which all orbits lie. The x axis is usually measured relative to the fixed direction of the Earth-Sun line at the vernal equinox. Another scheme that is suitable for circular orbits, is to use an axis passing through the position of the planets at one of the oppositions (Figure A.1).[1]

The position of the i-th planet at time t is

$$v_i = \omega_i t. \qquad (A.4)$$

Planets 1 and 2 are in opposition when

$$v_1 = v_2 + 2j\pi \qquad \text{where } j = 0,1,2,... \qquad (A.5)$$

and in conjunction when

$$v_1 = v_2 + \pi + 2j\pi \qquad \text{where } j = 0,1,2,... \qquad (A.6)$$

The planets are thus in opposition and conjunction at times

$$t_{oj} = \frac{2\pi j}{\omega_1 - \omega_2} \qquad t_{cj} = \frac{\pi(2j+1)}{\omega_1 - \omega_2} \qquad \text{where } j = 0,1,2,... \qquad (A.7)$$

---

[1] A planet is in opposition (with respect to the Sun) when the Sun, the Earth, and the planet are in line with the Earth and the planet at the same side of the Sun. When they are aligned at opposite sides of the Sun the planet is in conjunction.

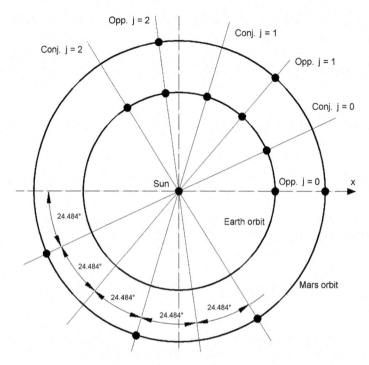

Figure A.1 Position of the planets at three sequential Earth-Mars oppositions and conjunctions. The x axis is defined with reference to the first opposition.

The positions of Mars and Earth at three sequential oppositions and conjunctions are reported in Figure A.1.

The synodic period is easily computed using the formula

$$T_s = \frac{2\pi}{|\omega_1 - \omega_2|} = \frac{1}{\left|\dfrac{1}{T_1} - \dfrac{1}{T_2}\right|} \tag{A.8}$$

where Ti is the time taken by the i-th planet to complete an entire revolution around the Sun. The synodic period of Mars with respect to Earth is 779.94 days. That of Venus with respect to Earth is 583.44 days. The synodic period of Venus with respect to Mars is 333.92 days.

The anomaly difference between a conjunction and the subsequent opposition and vice versa is 24.484° for Earth–Mars oppositions.

Using the values reported in Table 3.1 the circular orbits of Venus, Earth, and Mars are plotted in Figure A.2 as dotted lines.

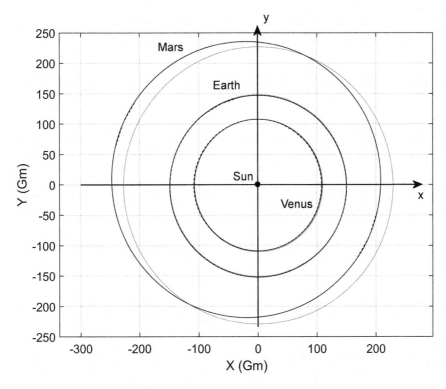

Figure A.2 A projection of the orbits of Venus, Earth, and Mars on the ecliptic plane. Full lines: JPL simplified ephemerides; dashed lines: elliptical orbits (full and dashed lines are superimposed); dotted lines: circular orbits.

A.2 SECOND APPROXIMATION (ELLIPTICAL ORBITS)

The first approximation solution cannot be used in any serious study of interplanetary missions for which at least the ellipticity of the planetary orbits must be taken into account.

To state the ephemeris of any object in the solar system it is necessary to give a space and time reference. Although the positions of the planets are often specified in terms of a spherical reference system, here the Cartesian frame O_{xyz} is used. This frame is centered in the Sun and the xy plane is that of the Earth's orbit, which is the ecliptic plane. The direction of the x axis is defined to be the line of the vernal equinox.

The planar orbits of the planets of the solar system can easily be computed once we know the distance from the Sun at perihelion d_{per}, the aphelion d_{aph}, and the longitude of perihelion ϕ_p. The semi-major axis a, the semi-minor axis b, and the distance of the focus from the center c are

$$a = \frac{d_{per} + d_{aph}}{2} \qquad c = \frac{a}{2} - d_{per} \qquad b = \sqrt{a^2 - c^2}. \qquad (A.9)$$

The equation of a planet's orbit in a reference frame where the x' axis is the line of apses (the line connecting the aphelion with the perihelion) is thus

$$\frac{\left(x'+c\right)^2}{a^2}+\frac{y'^2}{b^2}=1. \tag{A.10}$$

Given a set of values of coordinates x', the coordinates y' are

$$y'=\pm b\sqrt{1-\frac{x'^2}{a^2}}. \tag{A.11}$$

The ellipse can be easily referred to the frame whose x axis is directed toward the vernal equinox

$$\begin{Bmatrix}x'\\y'\end{Bmatrix}=\begin{bmatrix}\cos\left(\phi_p\right)&\sin\left(\phi_p\right)\\\sin\left(\phi_p\right)&\cos\left(\phi_p\right)\end{bmatrix}\begin{Bmatrix}x\\y\end{Bmatrix}. \tag{A.12}$$

The orbits of Venus, Earth, and Mars obtained in this manner using the values in Table 3.1 are plotted in Figure 3.2 as dashed lines. The orbits of Earth and Venus are clearly seen to be much less elliptical than that of Mars.

A.3 THIRD APPROXIMATION (SIMPLIFIED EPHEMERIS)

The same reference frame can be used as in the previous section, but the time reference is not so simple. Julian time was introduced in 1583 by astronomer Giuseppe Scaligero and its use is now almost universal. By this scheme, time $t = 0$ occurred at noon on January 1, 4713 BC. The Julian year is assumed to consist of precisely 365.25 days and often the time unit is the Julian century of 36,525 days. For instance, the Julian date of the moment at which this page was written (October 29, 2015 at 1:30:00 p.m.) is 2,457,325.0625 (in days).

A simplified ephemeris was devised by the JPL.[2] It is based on specifying six parameters for each planet (namely the Keplerian elements of the orbit, but things are more complex for Jupiter and the planets lying beyond and for those ten parameters are needed):

- Semimajor axis a.
- Eccentricity e.
- Inclination I.
- Mean longitude L.
- Longitude of perihelion ϖ.
- Longitude of the ascending node Ω.

---

[2] http://ssd.jpl.nasa.gov/?planet_pos.

The semimajor axis, and similarly all other parameters, are expressed by a linear equation in time

$$a = a_0 + \dot{a}t \qquad (A.13)$$

where a_0 and $\dot{a}$ are two constants and t is the Julian time in centuries.

The six constants for the Keplerian elements of the planets of interest here are

	a_0 (AU)	e_0	I_0 (°)
Venus	0.72332102	0.00676399	3.39777545
Earth	1.00000018	0.01673163	-0.00054346
Mars	1.52371243	0.09336511	1.85181869

	L_0 (°)	ϖ_0 (°)	Ω_0 (°)
Venus	181.97970850	131.76755713	76.67261496
Earth	100.46691572	102.93005885	-5.11260389
Mars	-4.56813164	-23.91744784	49.71320984

The parameters expressing the variation in time of the Keplerian elements are

	$\dot{a}$ (AU/cty)	$\dot{e}$ (1/cty)	$\dot{I}$ (°/cty)
Venus	-0.00000026	-0.00005107	0.00043494
Earth	-0.00000003	-0.00003661	-0.01337178
Mars	0.00000097	0.00009149	-0.00724757

	$\dot{L}$ (°/cty)	$\dot{\varpi}$ (°/cty)	$\dot{\Omega}$ (°/cty)
Venus	58517.81560260	0.05679648	-0.27274174
Earth	35999.37306329	0.31795260	-0.24123856
Mars	19140.29934243	0.45223625	-0.26852431

where time is expressed in Julian centuries (cty). These tables are valid for times ranging between 3000 BC and 3000 AD.

Once the six Keplerian elements are computed as functions of time t, it is possible to compute the argument of the perihelion ω and the mean anomaly M

$$\omega = \varpi - \Omega \qquad\qquad M = L - \varpi . \qquad (A.14)$$

Kepler's equation

$$M = E - e\sin(E) \qquad (A.15)$$

must now be solved numerically in E.

An iterative approach is suggested that starts at the value

$$E_0 = M + e\frac{180}{\pi}\sin(M) \qquad (A.16)$$

where the factor $180/\pi$ arises from the fact that M is expressed in degrees, while e is a number. Once the starting value E_0 is obtained, the iteration proceeds using the three equations

$$\Delta M = E_i - e \frac{180}{\pi} \sin(E_i)$$

$$\Delta E = \frac{\Delta M}{1 - e \dfrac{180}{\pi} \cos(E_i)} \tag{A.17}$$

$$E_{i+1} = E_i + \Delta E$$

for $i = 1, 2, \ldots$ until the increment ΔE becomes lower than a stated tolerance, which in the context of the approximated formulae can be one millionth of a degree.

At this point the heliocentric coordinates of the planet in its orbital plane, with x' axis passing through perihelion, can be obtained

$$\mathbf{r}' = \left\{ \begin{matrix} x' \\ y' \\ z' \end{matrix} \right\} = \left\{ \begin{matrix} a(\cos(E) - e) \\ a\sqrt{1 - e^2} \sin(E) \\ 0 \end{matrix} \right\} \tag{A.18}$$

and the coordinates in the ecliptic plane are

$$\mathbf{r} = \mathbf{R}_1 \mathbf{R}_2 \mathbf{R}_3 \mathbf{r}' \tag{A.19}$$

where

$$\mathbf{R}_1 = \begin{bmatrix} \cos(\Omega) & -\sin(\Omega) & 0 \\ \sin(\Omega) & \cos(\Omega) & 0 \\ 0 & 0 & 1 \end{bmatrix} \qquad \mathbf{R}_2 = \begin{bmatrix} 1 & 0 & 0 \\ 0 & \cos(I) & -\sin(I) \\ 0 & \sin(I) & \cos(I) \end{bmatrix} \tag{A.20}$$

$$\mathbf{R}_3 = \begin{bmatrix} \cos(\omega) & -\sin(\omega) & 0 \\ \sin(\omega) & \cos(\omega) & 0 \\ 0 & 0 & 1 \end{bmatrix} . \tag{A.21}$$

The projection of the orbits of the three planets onto the ecliptic plane are plotted in Figure A.2 as full lines. Clearly, the last two approximations give almost identical solutions, at least in terms of what can be seen on a small plot like this.

Symbols

a	semi-major axis	
b	semi-minor axis	
c	semi-distance between foci	
e	eccentricity	
t	time	
G	gravitational constant	$G = 6.67384 \times 10^{-11} \, m^3 \, kg^{-1} \, s^{-2}$
I	inclination	
L	mean longitude	
M	mean anomaly	
M_S	mass of the Sun	$M_S = 1.99 \times 10^{30} \, kg$
R	radius of Orbit of a planet	
T_s	synodic period	
T	period of revolution	
V	orbital velocity	
μ_S	gravitational parameter of the Sun	$\mu_S = 1.3281 \times 10^{20} \, m^3 \, s^{-2}$
ν	true anomaly	
ϕ_p	longitude of perihelion	
ω	angular velocity of a planet along its orbit	
ω	argument of perihelion (ephem.)	
ϖ	longitude of perihelion (ephem.)	
Ω	longitude of ascending node	

Appendix B: Impulsive trajectories

As stated in Chapter 6, the trajectories which can be achieved by chemical and nuclear thermal rockets can be obtained through the patched conic approximation. Idealized solutions obtained by assuming that planetary orbits are circular and coplanar are first described; then more realistic solutions in which the planets follow elliptical and non-coplanar orbits are considered.

B.1 LAUNCH FROM A PLANETARY SURFACE

Chemical and nuclear thermal rockets supply large thrusts usually for a short time. The trajectory of a craft powered by such an engine is usually obtained using the so-called patched conic approximation in which the thrust is assumed to be infinitely high and to last for an infinitesimal duration, producing an instantaneous change of velocity of the spacecraft ΔV. After that, the spacecraft proceeds along a conic (elliptical, parabolic or hyperbolic) trajectory; hence the term patched conic approximation. A mission is characterized by a series of velocity increments which must be applied at well-defined times.

This cannot be applied to launching a satellite into orbit, since the duration of the impulsive phase is not negligible relative to the time required to achieve orbit, and the patched conic approximation can be used only as a very rough approximation.

The computation of the speed increment needed to put a satellite into LEO is therefore quite difficult and, above all, depends on the launcher and the trajectory, which must be optimized in each case. A very rough approximation can be obtained by adding the orbital velocity to the ΔV required for reaching the orbital altitude h. In the case of a circular orbit, it follows (Eq. (B.30))

$$\Delta V = \sqrt{\frac{\mu_p}{r_p + h}} + \sqrt{\frac{2\mu_p h}{r_p\left(r_p + h\right)}} = \sqrt{\frac{\mu_p}{r_p + h}}\left(1 + \sqrt{\frac{2h}{r_p}}\right). \tag{B.1}$$

Using the values appropriate to Earth for μ_p and r_p, the two terms of the sum and the total ΔV are displayed in Figure 6.2. This value is greater than the correct one for some

© Springer International Publishing Switzerland 2017
G. Genta, *Next Stop Mars*, Springer Praxis Books, DOI 10.1007/978-3-319-44311-9

reasons, and lower than it for other reasons. It is in excess because choosing a correct turn between the vertical direction at launch and the horizontal direction in orbit gives a total ΔV less than the simple sum of the two contributions. It is less because this computation does not account for the gravity losses, which are mainly influenced by the duration of the burn, and the atmospheric drag losses, which are mainly linked with the precise velocity profile during the ascent. Their sum is in the range 1.3–1.8 km/s for a 200 km orbit.

Using this approximation, the value of ΔV to enter a LEO with a 300 km altitude is $7,739 + 2,378 = 10,117$ m/s. In a similar way, the value of ΔV to enter a low Mars orbit with the same 300 km altitude is $3,414 + 1,439 = 4,853$ m/s.

If the orbit is elliptical, with perigee and apogee altitudes h_p and h_a, it is possible (although only as a very rough approximation) to sum the velocity required to reach perigee with the orbital velocity at that point, obtaining (Eq. (B.38))

$$\Delta V = \sqrt{\frac{2\mu_p\left(r_p + h_a\right)}{\left(r_p + h_p\right)\left(2r_p + h_a + h_p\right)}} + \sqrt{\frac{2\mu_p h_p}{r_p\left(r_p + h_p\right)}} = \tag{B.2}$$

$$= \sqrt{\frac{2\mu_p}{r_p + h_p}}\left(\sqrt{\frac{r_p + h_a}{2r_p + h_a + h_p}} + \sqrt{\frac{h_p}{r_p}}\right).$$

To enter an elliptical orbit with a periareion of 300 km and an apoareion of 33,970 km about Mars, the value of ΔV, computed using the same approximation, is $4,606 + 1,439 = 6,045$ m/s.[3]

B.2 INTERPLANETARY TRAJECTORIES BETWEEN CIRCULAR COPLANAR PLANETARY ORBITS

B.2.1 The Hohmann transfer

Assuming that the planetary orbits are circular and coplanar, the interplanetary trajectory which requires the lowest energy expenditure (i.e. the lowest value of ΔV) is a semi-elliptical trajectory, tangent to the circular orbits of the departure and arrival planets. It is usually referred to as the Hohmann trajectory or the Hohmann transfer orbit.

The major axis of the Hohmann ellipse $2a_H$ is the sum of the radii of the orbits of the two planets, so in the case of an Earth–Mars transfer it is

$$2a_H = R_E + R_M. \tag{B.3}$$

The specific mechanical energy of a spacecraft on a Hohmann trajectory is

$$\varepsilon = -\frac{\mu_S}{2a_H} = -\frac{\mu_S}{R_E + R_M}. \tag{B.4}$$

---

[3] The terms periareion and apoareion are commonly used to indicate the apses of an orbit around Mars.

The period of the orbit is

$$T_H = 2\pi\sqrt{\frac{a^3}{\mu_S}} = 2\pi\sqrt{\frac{(R_E + R_M)^3}{8\mu_S}}.$$

(B.5)

The interplanetary travel time T is thus

$$T = \pi\sqrt{\frac{(R_E + R_M)^3}{8\mu_S}}.$$

(B.6)

Hence it follow that, for Mars

$$T = 258.73\,\text{days} = 8.62\,\text{months}.$$

The specific mechanical energy of the vehicle computed at the start of the interplanetary maneuver (velocity V_1, at which the craft enters the Hohmann transfer, and distance from the Sun R_E) is

$$\varepsilon = \frac{1}{2}V_1^2 - \frac{\mu_S}{R_E}.$$

(B.7)

By equating this energy with that of the Hohmann trajectory, it follows that

$$\frac{1}{2}V_1^2 - \frac{\mu_S}{R_E} = -\frac{\mu_S}{R_E + R_M}$$

(B.8)

i.e.

$$V_1 = \sqrt{2\mu_S \frac{R_M}{R_E(R_E + R_M)}}$$

(B.9)

which yields $V_1 = 32{,}740$ m/s.

When the spacecraft arrives at Mars orbit, it has a velocity V_2 which can be computed in the same manner

$$\frac{1}{2}V_2^2 - \frac{\mu_S}{R_M} = -\frac{\mu_S}{R_E + R_M}$$

(B.10)

i.e.

$$V_2 = \sqrt{2\mu_S \frac{R_E}{R_M(R_E + R_M)}}$$

(B.11)

which is too low with respect to the velocity of Mars on its orbit. In the case of the Hohmann transfer, $V_2 = 21,491$ m/s. Figure 6.3 shows a Hohmann trajectory.

Velocity V_1 is the heliocentric velocity of the spacecraft after it departs from the Earth's sphere of influence; meaning that it is no longer within the planet's gravitational well. Part of this velocity arises from the fact that it is traveling around the Sun in the same circular orbit as Earth (distance from the Sun R_E and velocity V_{o_E}). If it departed the sphere of influence with a hyperbolic excess speed $V_{\infty 1}$, with this hyperbolic excess speed assumed to be parallel to the orbital velocity (the Hohmann trajectory is tangent to the orbit of the Earth), the relationship between V_1, V_{o_E} and $V_{\infty 1}$ is simply

$$V_1 = V_{o_E} + V_{\infty 1}. \tag{B.12}$$

The hyperbolic excess speed with which it must depart the Earth's sphere of influence is therefore $V_{\infty 1} = 2,945$ m/s.

Often to express the energy that is required to enter the interplanetary trajectory we use parameter

$$C_3 = V_{\infty 1}^2 \tag{B.13}$$

and in the case of the Hohmann trajectory $C_3 = 8.673 \times 10^6$ m²/s².

When the spacecraft arrives at Mars orbit, it has a velocity V_2 and thus it enters Mars' sphere of influence with a hyperbolic excess speed

$$V_{\infty 2} = V_{o_M} - V_2 = 2,649 \, \text{m/s}. \tag{B.14}$$

Be aware also that the direction of V_2 is the same as that of V_{o_M} because the Hohmann trajectory is tangent to the orbit of Mars.

The Hohmann trajectory is precisely half of an ellipse, which amounts to saying that its angle at the Sun is $\Theta = 180°$.

The anomaly angle of the Earth at the beginning of the interplanetary transfer ν_1 and that of Mars at its end ν_2 are easily computed. A first relationship is

$$\nu_2 = \nu_1 + \Theta = 180°. \tag{B.15}$$

It is possible to state a reference frame in the orbital plane (since the orbits have been assumed to be circular and coplanar, the plane is the same for both) having its x axis in the direction of the opposition. Also time t is assumed to be $t = 0$ at the opposition. The instant t_1 at which the spacecraft leaves the Earth is thus

$$t_1 = \frac{\nu_1}{\omega_E} \tag{B.16}$$

while the instant t_2 at which the spacecraft arrives on Mars is

$$t_2 = \frac{\nu_2}{\omega_M}. \tag{B.17}$$

A relationship between those instants and the transfer duration T is easily found

$$t_2 - t_1 = \frac{v_2}{\omega_M} - \frac{v_1}{\omega_E} = T.$$

(B.18)

From these equations it follows that

$$
\begin{cases}
v_1 = \dfrac{\omega_E \left(\omega_M T - \Theta \right)}{\omega_E - \omega_M} \\[4mm]
v_2 = \dfrac{\omega_M \left(\omega_E T - \Theta \right)}{\omega_E - \omega_M}.
\end{cases}
$$

(B.19)

Hence it is easy to compute that $v_1 = -94.69°$, $v_2 = 85.31°$, $t_1 = -96.04$ days, and $t_2 = 162.7$ days. The launch on a Hohmann transfer must therefore be performed 96 days before an opposition, and the spacecraft will arrive at Mars about 163 days after the date of opposition.

B.2.2 General elliptical trajectories

As seen in the previous section, the Hohmann transfer is half of an ellipse, tangent to the two orbits of the planets, which for a first approximation are assumed to be circular and coplanar. A different trajectory is produced if at launch the velocity is higher, and perhaps not tangent to the circular planetary orbit. There are two possibilities, known as a trajectory of Type I and a trajectory of Type II (Figure 6.5). Generally, Type I trajectories are faster than Hohmann trajectories, and Type II trajectories are slower. Regardless, both require more energy than the Hohmann trajectory. However, this is true only in the simplified case of circular planetary orbits. If the orbits are elliptical and non-coplanar there are two optimal trajectories, one of Type I and one of Type II, and whether or not they require more or less energy than the theoretical Hohmann transfer (which in reality does not exist) depends on the particular launch opportunity.

General elliptical trajectories can be computed by stating the time at which the spacecraft enters the interplanetary trajectory (TMI time) and the time it crosses the orbit of Mars in the vicinity of the planet. In the case of Type II trajectories, the vehicle crosses the orbit of Mars twice, with the first one occurring at a point that is far from the planet. The positions of Earth and Mars corresponding to these two instants are easily obtained. Here the orbits are assumed to be circular and coplanar and thus these positions are expressed by two vectors $\mathbf{r}_1$ and $\mathbf{r}_2$, whose moduli are the radii of the orbits, and their directions in the common orbital planes are the anomalies ν_1 and ν_2 (Figure B.1) with respect to any direction chosen as a reference. Here, the simplest choice is to take as the x axis the one that passes through the positions of the planets at the opposition, as in Figure A.1.

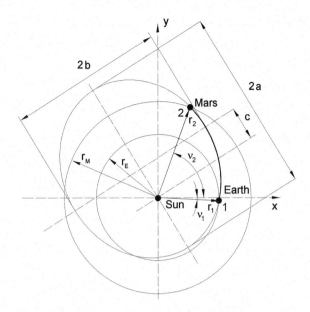

Figure B.1 An elliptical interplanetary trajectory (Type I trajectory shown).

Finding the elliptical orbit that passes through points $\mathbf{r}_1$ and $\mathbf{r}_2$ in two instants separated by time t is a well-known mathematical exercise called the Gauss problem because it was solved for the first time by Carl Friedrich Gauss in 1809 in relation to asteroid Vesta. It is absolutely general, and while here it is applied to the planar problem for two circular planetary orbits, in the following section will be used for the fully tridimensional problem and non-circular orbits.

We will use the approach based on the Newton-Raphson method for solving non-linear equations described by Shefer.[4]

To define the elliptical orbit, the velocities $\mathbf{v}_1$ and $\mathbf{v}_2$ at the starting and arrival points must be defined. Actually, these are exactly the parameters needed to compute the hyperbolic excess speeds required for the mission. Before starting the iterative procedure, two parameters named κ and σ are defined as follows

$$\kappa = 2\sqrt{r_1 r_2}\,\cos\!\left(\frac{\Delta v}{2}\right) = \sqrt{r_1 r_2}\,\left|\mathbf{U}_2 + \mathbf{U}_1\right| \tag{B.20}$$

$$\sigma = 2\sqrt{r_1 r_2}\,\sin\!\left(\frac{\Delta v}{2}\right) = \sqrt{r_1 r_2}\,\left|\mathbf{U}_2 - \mathbf{U}_1\right|$$

[4] V.A. Shefer, "New method of Orbit Determination from Two Position Vectors Based on Solving Gauss's Equations," *Solar System Research*, vol. 44, n. 3, pp. 252–266, 2010.

where Δv is the angle included between vectors $\mathbf{r}_1$ and $\mathbf{r}_2$ and

$$r_1 = |\mathbf{r}_1|, \quad r_2 = |\mathbf{r}_2|, \quad U_1 = \frac{\mathbf{r}_1}{r_1} \text{ and } U_2 = \frac{\mathbf{r}_2}{r_2} \text{ (if } 0 < \Delta v < \pi\text{)}.$$

The second formulation of Eq. (B.20) has the advantage of avoiding a singularity which may occur when Δv is close to π.

The initial and the final velocities are obtained from the formulae

$$\mathbf{v}_1 = \frac{\mathbf{r}_2 - f\mathbf{r}_1}{g} \quad \mathbf{v}_2 = \dot{f}\mathbf{r}_1 + \dot{g}\mathbf{v}_1 \tag{B.21}$$

where

$$f = 1 - \frac{2\alpha}{r_1}$$

$$g = t - X\sqrt{\frac{\alpha^3}{\mu_S}}$$

$$\dot{f} = \sqrt{\frac{\mu_S}{p}} \tan\left(\frac{\Delta v}{2}\right)\left[\frac{1 - \cos(\Delta v)}{p} - \frac{1}{r_1} - \frac{1}{r_2}\right]$$

$$\dot{g} = 1 - \frac{r_1}{p}\left[1 - \cos(\Delta v)\right], p = \frac{\sigma^2}{4\alpha}.$$

The two unknown quantities α and X that are required to compute the velocities are both functions of a third unknown x. The equation that allows a solution to the problem is

$$\alpha(x)Y^2(x) = \mu_S t^2 \tag{B.22}$$

where

$$Y(x) = \kappa + \alpha(x)X(x), \alpha(x) = \bar{r} + \kappa\left(x - \frac{1}{2}\right)$$

such that $\bar{r} = (r_1 + r_2)/2$ is the mean radius, and

$$X(x) = \frac{\text{arsin}\left(\sqrt{x}\right) - (1 - 2x)\sqrt{x(1-x)}}{2\left[\sqrt{x(1-x)}\right]^3}.$$

The last equation holds only in case of elliptical orbits. If the trajectory is hyperbolic or parabolic, then a different formulation must be used (it is not reported here since only elliptical trajectories are of interest).

To solve Eq. (B.22) using the Newton-Raphson equation, it must be rewritten in the form

$$F(x) = \alpha(x)Y^2(x) - \mu_s t^2 = 0 \tag{B.23}$$

and the equation that computes the value of x_{i+1} at the $(i+1)$-th iteration from the value x_i at the i-th iteration is

$$x_{i+1} = x_i - \frac{F(x_i)}{F'(x_i)} \tag{B.24}$$

where

$$F'(x_i) = \kappa Y^2(x) + 2\alpha(x)Y(x)\big[\kappa X(x) + \alpha(x)X'(x)\big]$$

and

$$X'(x_i) = \frac{4 - 3(1-2x)X(x)}{2x(1-x)}.$$

A starting value for the iteration may be

$$x = \frac{1}{2} - \frac{\kappa}{4\bar{r}}.$$

The algorithm converges very quickly to give the interplanetary trajectory. Once the velocities $\mathbf{v}_1$ and $\mathbf{v}_2$ have been computed, it is possible to obtain the orbit and hence the hyperbolic excess speeds at the start and at the end of the interplanetary transfer.

At the end of the Trans-Mars injection (TMI) maneuver, the spacecraft must have a speed equal to $\mathbf{v}_1$, while the Earth has its orbital velocity $\mathbf{V}_{oE}$. This means that the spacecraft must depart the Earth's sphere of influence with a speed $\mathbf{V}_{\infty 1}$

$$\mathbf{V}_{\infty 1} = \mathbf{v}_1 - \mathbf{V}_{oE}. \tag{B.25}$$

For the case of the Hohmann trajectory, the Earth's orbital velocity and the velocity on the trajectory were aligned, and then Eq. (B.12) is similar to Eq. (B.25) except that it is a relationship between moduli rather than a relationship between vectors. In this case $\mathbf{v}_1$ has a generic direction while $\mathbf{V}_{oE}$ is tangent to the circular Earth orbit. In the more general case seen later, the first vector does not lie within the Earth's orbital plane.

Once the start time and the arrival time are specified, we can compute the required hyperbolic excess speed $\mathbf{V}_{\infty 1}$ and its square

$$C_3 = V_{\infty 1}^2. \tag{B.26}$$

This is a parameter often used to assess the energy required for the interplanetary trajectory. A plot in which C_3 is reported as a function of the departure and arrival dates is known as a pork-chop plot.

A general pork-chop plot, computed using the circular orbits assumption and thus holding only as a first approximation for all launch opportunities is shown in Figure 6.6.

In a similar way, the hyperbolic excess speed at the entrance into the sphere of influence of Mars is immediately computed as

$$\mathbf{V}_{\infty 2} = \mathbf{V}_{oM} - \mathbf{v}_2 . \tag{B.27}$$

Here it is possible either to use a propulsive maneuver to enter the planet's sphere of influence or to perform an aerocapture maneuver using the aerodynamic drag in the Martian atmosphere.

B.2.3 Leaving Earth orbit

If the interplanetary orbit is entered by first reaching escape velocity and then by further increasing the velocity until the speed V_1 is obtained, the ΔV required for the TMI maneuver is

$$\Delta V_1 = V_{\infty 1} \tag{B.28}$$

and this must be added to the ΔV required to escape from the Earth's sphere of influence. As we saw for the case of the Hohmann transfer, $V_{\infty 1} = 2,945$ m/s.

However, this is the worst way to start the interplanetary trajectory. It is far more expedient to gain the speed increment from a single burn on the Earth's surface and achieve the escape condition without a pause in Earth orbit. This is usually called a 'Mars direct' strategy.

With simple computations and neglecting both aerodynamic drag and gravity losses, it follows that

$$\Delta V_1 = \sqrt{\frac{2\mu_E}{r_E} + V_{\infty 1}^2} \tag{B.29}$$

and this yields $\Delta V_1 = 11,569$ m/s for the minimum energy trajectory.

Although a direct launch from the Earth's surface is much more convenient than first achieving orbit and then starting toward Mars, it is usually not considered, both because of the need to assemble a large spacecraft in Earth orbit and for practical reasons like the expediency of checking out all of the systems in Earth orbit prior to committing to an interplanetary trajectory.

Assume that the orbit in which the spacecraft is assembled is circular, with a height h_{Eo} above the Earth's surface. The orbital velocity is

$$V_{oE} = \sqrt{\frac{\mu_E}{r_E + h_{Eo}}} . \tag{B.30}$$

The minimum speed increment required to depart from the Earth's sphere of influence ΔV_e can be computed by assuming that the energy the vehicle

$$\varepsilon = \frac{1}{2}\left(V_{oE} + \Delta V_e\right)^2 - \frac{\mu_E}{r_E + h_{Eo}} \tag{B.31}$$

must be equal to 0, which is the energy required to arrive at an infinite distance with zero speed. As a result

$$\Delta V_e = \sqrt{\frac{2\mu_E}{r_E + h_{Eo}}} - V_{oE} = \left(\sqrt{2} - 1\right)V_{oE}. \tag{B.32}$$

In the case of a 500 km orbit, the escape velocity from Earth is 3,155 m/s. If the interplanetary orbit is entered by first reaching the escape velocity and then by further increasing the velocity until the speed V_1 is obtained, the total ΔV to leave Earth is

$$\Delta V_1 = \Delta V_e + V_{\infty 1} \tag{B.33}$$

which, in the case of a minimum energy trajectory for Mars, yields $\Delta V_1 = 6{,}100$ m/s.

As already stated, this is the worst way to start the interplanetary trajectory. It is more expedient to gain the entire speed increment at the lowest possible altitude, to obtain a hyperbolic excess speed that is equal to the required speed increase to start the interplanetary trajectory. To exit the sphere of influence of the Earth with the hyperbolic excess speed $V_{\infty 1}$ the energy must be

$$\varepsilon = \frac{1}{2}\left(V_{oE} + \Delta V_1\right)^2 - \frac{\mu_E}{r_E + h_{Eo}} = \frac{1}{2}V_{\infty 1}^2. \tag{B.34}$$

The speed increment therefore must thus be

$$\Delta V_1 = \sqrt{\frac{2\mu_E}{r_E + h_{Eo}} + V_{\infty 1}^2} - V_{oE} = \sqrt{2V_{oE}^2 + V_{\infty 1}^2} - V_{oE}. \tag{B.35}$$

In the case of the minimum energy trajectory, this amounts to 3,550.5 m/s, which is not much larger than the escape velocity (3,155 m/s). The advantage of doing a single burn in LEO over two burns, one in LEO to get escape velocity, and then the other to achieve the interplanetary trajectory, is self evident. A sketch of the velocities involved in this phase of the flight, both in the Earth and Sun reference frames, is shown in Figure B.2a.

If the starting orbit is elliptical with a height h_a at apogee and h_p at perigee, then the energy of the elliptical orbit is

$$\varepsilon = -\frac{\mu_E}{2r_E + h_a + h_p}. \tag{B.36}$$

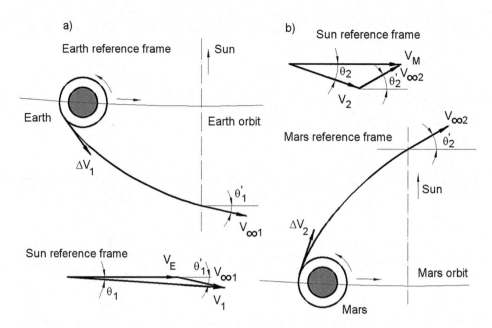

Figure B.2 Sketches of (a) the Earth departure and (b) Mars arrival phases, both in the Earth and Sun reference frames.

The velocity at the perigee Vp can be computed from the equation

$$\varepsilon = -\frac{\mu_E}{2r_E + h_a + h_p} = \frac{1}{2}V_p^2 - \frac{\mu_E}{r_E + h_p} \tag{B.37}$$

i.e.

$$V_p = \sqrt{\frac{2\mu_E(r_E + h_a)}{(r_E + h_p)(2r_E + h_a + h_p)}} \tag{B.38}$$

and the speed increment must be

$$\Delta V_1 = \sqrt{\frac{2\mu_E}{r_E + h_p} + V_{\infty 1}^2} - V_p . \tag{B.39}$$

B.2.4 Entering Mars orbit

When the spacecraft reaches Mars orbit, it must first increase its heliocentric speed to remain close to the planet, and then it must enter the sphere of influence of the planet and brake in such a manner as to enter the required orbit around it. Here too, a single ΔV at orbit insertion (Figure B.2b) is more expedient than achieving the maneuver by two separate burns.

The value of ΔV_2 required to enter directly into a circular orbit with height h_{Mo} above the surface of the planet can be computed using the same approach as for the propulsive maneuver for leaving Earth orbit, namely

$$\Delta V_2 = \sqrt{\frac{2\mu_M}{r_M + h_{Mo}} + V_{\infty 2}^2} - V_{oM} = \sqrt{2V_{oM}^2 + V_{\infty 2}^2} - V_{oM} \tag{B.40}$$

where $V_{\infty 2}$ is the hyperbolic excess speed at Mars, which can be computed from an equation similar to Eq. (B.27), and

$$V_{oM} = \sqrt{\frac{\mu_M}{r_M + h_{Mo}}}. \tag{B.41}$$

In the case of achieving Mars orbit using a propulsive maneuver, it is quite convenient to get into a fairly elongated elliptical orbit, such as with a 250 km periareion and a 33,970 km apoareion because it will have a period that is approximately the same as that required for the planet to rotate once on its axis.

The ΔV at the arrival in this case is

$$\Delta V_2 = \sqrt{\frac{2\mu_M}{r_M + h_p} + V_{\infty 2}^2} - V_p \tag{B.42}$$

where

$$V_p = \sqrt{\frac{2\mu_M (r_M + h_a)}{(r_M + h_p)(2r_M + h_a + h_p)}}. \tag{B.43}$$

Also in this case, a direct entry with a single burn when reaching the surface would be even more convenient. Since in this case the speed increments are used to brake the spacecraft and Mars possesses an atmosphere, it is possible to use aerodynamic forces instead of performing propulsive maneuvers. An aerocapture maneuver can thus supply the arrival ΔV_2 at zero cost in terms of propellant. However, it is necessary to have a heat shield and the necessary provisions, and a trade-off between the two solutions is required. This was discussed in detail in Section 6.6.

B.2.5 Optimization in terms of ΔV

The pork-chop plot of Figure 6.6 deals with the energy required to enter the interplanetary trajectory after getting out of the Earth's sphere of influence. It is thus referred to the trajectory, but it is insensitive to the starting orbit and does not take into account the braking that must be performed upon arrival at Mars.

Once the departure orbit has been stated, it is easy to compute the value of ΔV_1 for each value of C_3, producing a sort of pork-chop plot for the velocity. In the same way, after the arrival orbit about Mars has been defined, it is possible to calculate ΔV_2 for each combination of date of departure and date of arrival. This enables values of $\Delta V_{tot} = \Delta V_1 + \Delta V_2$ to be computed.

For each value of the journey time T, it is possible to express the total speed increment $\Delta V tot$ as a function of the departure time, and then to choose the departure date that minimizes it. A plot like that in Figure 6.7 can thereby be obtained. For each time T, the trajectory leading to the lowest value of $\Delta V tot$ is chosen. This will be the most economical trajectory to reach Mars in a given time.

It must be noted that this way of proceeding is possible only after the starting orbit (in the plot, a 300 km circular Earth orbit) and an arrival orbit (in the plot a 300 km circular Mars orbit) have been specified. In the same figure, a set of lines relevant to a starting circular orbit at 250 km from Earth's surface and a very elliptical arrival orbit with a 250 km periareion and a 33,970 km apoareion relative to the surface are also reported (dotted lines).

If Mars Orbit Insertion (MOI) is performed not by the propulsion system but by aerocapture and/or aerobraking, the optimization should be performed with reference to ΔV_1 only (dashed line).

From the pork-chop plot it is possible to obtain directly the departure and the arrival dates of the chosen trajectories. An example of a plot that yields this kind of information for the same trajectories studied in Figure 6.7 is shown in Figure 6.8.

B.3 ELLIPTICAL NON-COPLANAR PLANETARY ORBITS

Up to now, planetary orbits have been assumed to be circular and coplanar. These two assumptions must be dropped to obtain the desired precision (even if in the preliminary mission analysis the patched conic approach and the two bodies assumption can be retained). If the ellipticity of the orbits is considered, the computation must be referred to a calendar date rather than to a generic position of the planet in its orbit.

Once the date of TMI is specified, it is possible to obtain from the ephemeris of the starting planet its position $\mathbf{r}_1$ and its velocity $\mathbf{v}_{p1}$. And once the duration T of the interplanetary transfer is specified, from the ephemeris of the arrival planet it is possible to obtain its position $\mathbf{r}_2$ and its velocity $\mathbf{v}_{p2}$.

The same Gauss algorithm used for circular coplanar trajectories can be used here in order to obtain an interplanetary trajectory which does not lie in any of the two orbital planes. The pork-chop plots can also be obtained for any particular launch window. Those reported in Figure 6.7 are related to the launch windows of 3035, 2037, 2040 and 2042, and clearly show two optimal trajectories: one of Type I and one of Type II. The first one is faster than the second, but in some launch opportunities the latter may require less energy than the first.

It must be remembered that, when considering the starting and arrival orbits for the two planets, the inclination of the orbit must be taken into account. A maneuver may be necessary to change the plane of the spacecraft's orbit, and such actions are always costly in terms of propellant.

B.4 TRAJECTORIES WITH GRAVITY ASSIST

As stated in Chapter 6, gravity assists have been used by several robotic planetary exploration missions, but they are seldom considered for human flight owing to the long mission times involved. Nevertheless, Venus flybys have been suggested for Mars flyby and short-stay landing missions.

Once again assuming circular and coplanar orbits, consider a return from Mars that uses a Hohmann trajectory aimed at Venus. Using the same formulae already used for an Earth-Mars Hohmann ellipse, the major axis of the trajectory $2a_H$ is

$$2a_H = R_V + R_M = 336.13 \times 10^9 \text{ m.} \tag{B.44}$$

The travel time T to reach Venus is

$$T = \pi \sqrt{\frac{(R_V + R_M)^3}{8\mu_S}} = 217.39 \text{ days} = 7.24 \text{ months.} \tag{B.45}$$

The velocity to enter the Hohmann trajectory on exiting the sphere of influence of Mars is

$$V_1 = \sqrt{2\mu_S \frac{R_V}{R_M (R_V + R_M)}} \tag{B.46}$$

yielding $V_1 = 19{,}369$ m/s. The hyperbolic excess speed with which the spacecraft must exit the sphere of influence of Mars is $V_{\infty 1} = 4{,}769.8$ m/s.

When the spacecraft arrives at Venus orbit it has a velocity V_2

$$V_2 = \sqrt{2\mu_S \frac{R_M}{R_V (R_V + R_M)}} \tag{B.47}$$

which gives $V_2 = 40{,}798$ m/s. The hyperbolic excess speed with which the spacecraft enters the sphere of influence of Venus is $V_{\infty 1} = 5{,}764{,}4$ m/s.

An analysis of a trajectory which includes a Venus flyby is much more complex than that of a direct one. Although in the case of circular and coplanar orbits it is possible to define a generic direct trajectory and then find how many days prior to the planetary opposition the spacecraft must leave Earth, there is no generic trajectory that includes a flyby; each launch window is unique, even when dealing with circular planetary orbits.

The position of Venus with respect to both Earth and Mars (at the oppositions they have the same anomaly ν) at various oppositions is reported in Table 3.2. This shows that the differences between the various launch windows are large and nothing can be said in general.

The only possible approach is to refer to a certain date, for instance the date in which the spacecraft arrives at Mars. From the ephemerides of the planets, we can readily calculate the anomalies ν_{1M}, ν_{1E} and ν_{1V} of Mars, Earth and Venus respectively. If T_s and T_{r1} are the duration of the stay on Mars and the duration of the first part of the return journey (from Mars to the Venus flyby), the anomaly of Mars at the departure and that of Venus at the flyby are

$$\nu_{2M} = \nu_{1M} + \omega_M T_s \tag{B.48}$$

$$\nu_{3V} = \nu_{1V} + \omega_V \left(T_s + T_{r1}\right).$$

The anomaly of the Earth at the end of the mission is

$$\nu_{4E} = \nu_{1E} + \omega_E \left(T_s + T_{r1} + T_{r2}\right) \tag{B.49}$$

where Tr_2 is the duration of the second part of the return journey (from the Venus flyby to Earth).

This approach makes the additional assumption that the gravity assist accomplished at the flyby is instantaneous. The spacecraft leaves Mars with the heliocentric speed V_0, reaches Venus with speed V_1, leaves Venus with speed V_2, and reaches Earth with speed V_3. All these velocities are coplanar vectors, and so can be expressed by their components along x and y axes in the ecliptic plane.

The situation at the Venus flyby is shown in Figure B.3. The spacecraft approaches the planet with a heliocentric velocity $\mathbf{V}_1$. Since the planet moves at a velocity $\mathbf{V}_p$, the approach velocity is in a reference frame that is moving together with the planet and therefore

$$\mathbf{V}_{p1} = \mathbf{V}_1 - \mathbf{V}_p . \tag{B.50}$$

During the approach to the planet, the spacecraft follows a hyperbolic trajectory, hence its velocity increases. We can regard $\mathbf{V}_{p1}$ as the velocity immediately outside the sphere of influence of the planet. When the spacecraft departs the sphere of influence again, it has a velocity $\mathbf{V}_{p2}$ in the frame fixed to the planet which is equal to $\mathbf{V}_{p1}$ in modulus but has a different direction. The heliocentric speed on exiting the sphere of influence is

$$\mathbf{V}_2 = \mathbf{V}_p + \mathbf{V}_{p2} . \tag{B.51}$$

The deviation angle in the planetocentric reference frame is easily computed as

$$\delta = \text{acos}\left(\frac{\mathbf{V}_{p1}^T \mathbf{V}_{p2}}{V_{p1} V_{p2}}\right). \tag{B.52}$$

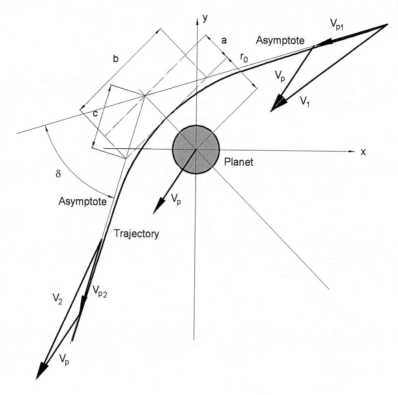

Figure B.3 A Venus flyby trajectory.

The geometric parameters defining the hyperbolic trajectory are

$$a = -\frac{\mu_V}{V_{p1}} \quad e = \frac{1}{\sin\left(\dfrac{\delta}{2}\right)} \tag{B.53}$$

$$b = a\sqrt{(e^2 - 1)} \quad c = \sqrt{a^2 + b^2}.$$

The equations above hold also in the general three-dimensional case with non-coplanar orbits. Once the vectors $\mathbf{V}_{p1}$ and $\mathbf{V}_{p2}$ have been computed, the hyperbolic trajectory develops in a plane that contains the two vectors and passes through the center of mass of the planet. In this suituation, the angle δ can be computed using Eq. (B.52).

As already stated, the velocity of the spacecraft relative to the planet is conserved during a flyby, and hence $V_{p1} = V_{p2}$. It is, however, possible to define trajectories in which $V_{p1} \neq V_{p2}$ simply by applying a thrust either in the forward direction to increase the speed, or backwards to decrease it. It is expedient to apply this thrust at the point when the spacecraft is at the minimum distance from the planet (a periapsis burn). This is often called a motorized flyby. However, this case will not be dealt with here.

B.5 FREE RETURN TRAJECTORIES

A free return trajectory starts from planet A, reaches planet B and, if no maneuver is made to approach the latter, then intersects the orbit of planet A at a point where the planet is present at that time. In the case of an Earth–Mars trajectory, the period of the free return trajectory must be a whole multiple (or a rational multiple) of one year. The trajectory must be in orbital resonance with Earth. If no maneuver is made in the vicinity of Mars, the spacecraft must thus reach a distance from the Sun R_o that is obtainable from

$$n\pi\sqrt{\frac{R_E^3}{\mu_S}} = \pi\sqrt{\frac{\left(R_E + R_o\right)^3}{8\mu_S}} \tag{B.54}$$

where if $n = 2$, 3, ... the spacecraft returns to Earth 2, 3, ... years after its launch having gone through a single heliocentric orbit; if $n = 3/2$ it returns to Earth 3 years after its launch having gone through two heliocentric orbits, etc.

From Eq. (B.54), it follows that

$$R_o = \left(2n^{2/3} - 1\right)R_E. \tag{B.55}$$

The specific mechanical energy of such trajectory is

$$\varepsilon = -\frac{\mu_S}{2n^{\frac{2}{3}}R_E}. \tag{B.56}$$

Operating as above, the speed at which the spacecraft departs its orbit around Earth is

$$V_1 = \sqrt{\frac{\mu_S}{R_E}\left(2 - \frac{1}{n^{2/3}}\right)}. \tag{B.57}$$

In the case $n = 2$, it gives $V_1 = 34{,}875$ m/s, corresponding to $V_{\infty 1} = 5{,}078$ m/s, $T = 126.7$ days and $V_{\infty 2} = 9{,}152$ m/s. Although such a trajectory is quite a fast one, it is also very expensive in terms of propellant. Free return trajectories may include gravity assist maneuvers at Mars or at Venus.

If the fact that the planetary orbits are neither circular nor coplanar is taken into account, the study of free return trajectories and of trajectories including flybys is even more complicated, and must be based on the actual ephemerides of the solar system.

B.6 PLANNING THE WHOLE MISSION

If the planetary orbits are assumed to be circular and coplanar, then the complete mission can be readily studied using Eqs. (B.19) and (B.62). The results for the missions described in Figure 6.7 are given in Figure B.4. Owing to the assumptions made, they are not linked with a specific launch opportunity but hold in general.

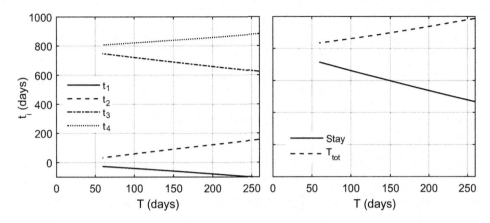

Figure B.4   Times t_1, t_2, t_3 and t_4 as functions of the transfer time T.

The return leg of the trip has been assumed to be symmetrical with respect to the outbound leg. The faster is the trip (low T), the smaller are t_1 and t_2 (in absolute terms) and the larger is t_3. In this way, a decrease of T increases the stay on Mars (also reported in the figure) and only moderately decreases the total mission time. That is, going faster reduces the time spent in space but is not a very effective way of reducing the total mission duration. For instance, if two months' travel are considered (a very expensive solution which requires values of ΔV_1 and ΔV_2 of 10,696 and 12,009 m/s respectively) the total mission time is still 827 days, but most of it (707 days) is spent on the planet instead of in space and therefore the ratio between the time spent on Mars and that spent in space is about 6 (versus a value of 0.878 for the Hohmann transfer). Anyway, the stay is always shorter than a whole synodic period. This precludes the possibility of making a short stay mission without expensive trajectories or perhaps a Venus flyby.

The anomaly angle of Mars at the start of the return journey ν_3, and that of Earth at the end of the mission ν_4 can be computed in a similar way. Since the Earth has made one more revolution about the Sun than Mars, a first relationship is

$$\nu_4 = \nu_3 + 2\pi + \Theta. \tag{B.58}$$

The instant t_3 at which the spacecraft leaves Mars is

$$t_3 = \frac{\nu_3}{\omega_M} \tag{B.59}$$

while the instant t_4 at which the spacecraft returns to Earth is

$$t_4 + nT_s = \frac{\nu_4}{\omega_E}. \tag{B.60}$$

A relationship between those instants and the transfer duration T is

$$t_4 - t_3 = \frac{v_4}{\omega_M} - \frac{v_3}{\omega_E} = T.$$
(B.61)

From these equations it follows that

$$
\begin{cases}
v_3 = \dfrac{\omega_M\left(\omega_E T - 2\pi - \Theta\right)}{\omega_M - \omega_E} \\[4ex]
v_4 = \dfrac{\omega_E\left(\omega_M T - 2\pi - \Theta\right)}{\omega_M - \omega_E}.
\end{cases}
$$
(B.62)

The optimal stay on Mars and the total duration of the mission depend on the launch opportunity that is chosen, due to the ellipticity of the planetary orbits. Two cases of a long and a short stay missions are studied below in some detail.

The main data for a number of missions having different travel times are reported in Table B.1. The table refers to missions in which no use is made of aerobraking or aerocapture. Both ΔV_1 and ΔV_2 are to be achieved by propulsive maneuvers on the outbound and return legs. Moreover, the values reported in the table refer to circular planetary orbits and to starting and arrival orbits at altitudes of 500 km, both for Earth and Mars, and thus are independent of the launch opportunity. With more realistic assumptions the launch date would affect the results.

Table B.1 Main data obtained for a number of missions having different travel times. In all cases, the orbits about Earth and Mars are circular at an altitude of 500 km.

	Hohmann			Fast		
T (days)	258.73	210	180	150	120	60
θ_1'	0	27.7	43.1	56.8	63.6	79.2
θ_1	0	2.7	4.8	7.7	10.4	24.1
ΔV_1 (m/s)	3,550.5	3,660.7	3,889.9	4,372.6	5091.7	10,696.0
θ_2	0	4.5	7.5	10.9	16.0	33.8
ΔV_2 (m/s)	2,069.9	2,325.1	2,768.6	3,487.5	4,879.2	12,008.6
ΔV_{tot} (m/s)	5,620.4	5,985.8	6,658.6	7,860.1	9,971.0	22,704.6
r_{out} (chem.)	3.53	3.83	4.45	5.83	9.35	162.51
r_{out} (NTP)	1.94	2.02	2.19	2.52	3.23	14.46
t_1 (days)	−96.0	−81.8	−70.8	−58.2	−48.4	−23.6
t_2 (days)	162.7	128.2	109.2	91.8	71.6	36.4
$\Delta V_{mission}$ (m/s)	11,240.8	11,971.6	13,317.1	15,720.1	19,941.9	45,409.2
$r_{mission}$ (chem.)	12.43	14.65	19.80	33.94	87.47	26,408.54
$r_{mission}$ (NTP)	3.75	4.09	4.79	6.36	10.44	208.99
t_3 (days)	617.2	651.7	670.8	688.1	708.4	743.5
t_4 (days)	876	861.7	850.8	838.1	828.4	803.5
Stay (days)	454.5	523.5	561.6	596.3	636.8	707.2
Total time (days)	972	943.5	921.6	896.3	876.8	827.2

B.6.1 Example of a long stay mission

Consider a mission to be performed in the 2035 launch opportunity, with a flight time of 6 months (180 days). The simplified solution of Table B.1, obtained assuming circular and coplanar orbits, yields

- Launch time: 70.8 days before opposition.
- Stay time: 561.6 days.
- Total mission time: 921.6 days.
- $\Delta V_1 = \Delta V_4 = 3,889.6$ m/s.
- $\Delta V_2 = \Delta V_3 = 2,768.3$ m/s.

The opposition in 2035 is a perihelic one and occurs on September 15, 2035. Using the values from the simplified computation, we see that the launch date is July 4, 2035.

Assuming a travel time for both the forward and return legs of 180 days and a stay on Mars of 561.6 days (a total mission time of about 922 days), the arrival date at Mars is December 31, 2035, the date of departure from Mars is July 14, 2037, and the date of arrival back at Earth is January 10, 2038. Therefore the following values are obtained for ΔV

- $\Delta V_1 = 3,648.0$ m/s.
- $\Delta V_2 = 2,108.5$ m/s.
- $\Delta V_3 = 2,775.8$ m/s.
- $\Delta V_4 = 4,135.0$ m/s.

These trajectories of the spacecraft are reported in Figure 7.3. It is clear that the ΔV values for the outbound leg are lower than those for the simplified solution, which is fairly obvious because 2035 is a perihelic opposition and Mars is in a very favorable position. However, the values of ΔV for the return journey are higher (in particular ΔV_4) owing to the fact that the return leg takes place quite far from the opposition, both in time and in space.

So we see that the optimal solution for the simplified strategy need not be optimal when taking into account elliptical and non-coplanar orbits. By changing slightly the starting dates, it follows that, for the outbound leg

- ΔV_1 decreases starting earlier. The minimum occurs starting on June 26, with $\Delta V_1 = 3,631.0$ m/s (but $\Delta V_2 = 2,161.0$ m/s).
- ΔV_2 decreases starting later. The minimum occurs starting on July 18, with $\Delta V_2 = 2,072.6$ m/s (but $\Delta V_1 = 3,791.3$ m/s).

If the maneuver upon arrival at Mars is made using aerobraking, then it is expedient to start earlier, but if all the maneuvers are propulsive it is expedient to search for the minimum value of $\Delta V_1 + \Delta V_2$, which in this case occurs on July 4 with $\Delta V_1 + \Delta V_2 = 5,756.5$ m/s.

A similar optimization can be made for the return leg by changing the length of stay on the planet

- ΔV_3 decreases reducing very slightly the duration of stay. The minimum occurs staying 560 days $\Delta V_3 = 2,775.4$ m/s (but $\Delta V_4 = 4,188.7$ m/s).
- ΔV_4 decreases increasing the duration of stay. The minimum occurs staying 605 days, with $\Delta V_4 = 3,545.5$ m/s (but $\Delta V_3 = 3,986.0$ m/s). Note that now $\Delta V_3 > \Delta V_4$.

If the maneuver upon arrival at Earth is made using aerobraking, then it is expedient to shorten the stay slightly, but if all the maneuvers are propulsive it is expedient to search for the minimum value of $\Delta V_3 + \Delta V_4$, which occurs for a stay time of 575 days, giving $\Delta V_3 = 2,909.9$ m/s and $\Delta V_4 = 3,836.6$ m/s.

B.6.2. Example of a short stay mission

Consider a short stay mission that leaves Earth in the 2033 launch opportunity. The opposition is on June 17, 2033 and the best launch date for an outbound leg of 180 days is April 17, 2033 if circular orbits are assumed. The arrival at Mars occurs on October 14, 2033.

Assume that the stay on Mars is about 40 days and that the return leg lasts 281.878 days (165 for the Mars–Venus part and 116.878 for the Venus–Earth part). The last figure was computed using a trial and error procedure to achieve the flyby, still assuming circular and coplanar orbits. Hence the total mission time is 502 days. Assuming circular orbits at an altitude of 500 km at both planets, the speed increments are

- $\Delta V_1 = 3,889.6 \text{m/s}.$
- $\Delta V_2 = 2,768.3 \text{m/s}.$
- $\Delta V_3 = 4,192.4 \text{m/s}.$
- $\Delta V_4 = 6,761.4 \text{m/s}.$

The heliocentric velocity upon arrival at Venus is 40,341.8 m/s and the heliocentric velocity upon leaving the planet is 37,735.3 m/s. It can be computed that the hyperbolic excess speeds at both arrival and departure are 12,038.5 m/s.

The deviation due to the flyby is $\delta = 14.097°$. The main parameters of the hyperbolic trajectory at Venus are $a = -2,241$ km, $b = -18,129$ and $e = 8.149$. The minimum distance from the center of the planet is $r_0 = 16,025.5$ km and the minimum altitude is $h_0 = 9,974$ km. The simplified return leg is reported in Figure B.5.

Using the values from the simplified computation, the launch and arrival dates remain April 17, 2033 and October 14, 2033 if the outbound leg is 180 days. Assuming a stay on Mars of 40 days, the departure date from Mars is November 23, 2033. Some changes must be made to the return leg duration to arrange the Venus flyby: the first part (Mars to Venus) lasts 165 days and the second part (Venus to Earth) lasts 116.878 days. This gives following values of ΔV

- $\Delta V_1 = 3,546.0 \text{m/s}.$
- $\Delta V_2 = 2,528.0 \text{m/s}.$
- $\Delta V_3 = 4,938.6 \text{m/s}.$
- $\Delta V_4 = 7,207.6 \text{m/s}.$

The hyperbolic excess speed on arrival at Venus is equal to that at departure and is 13,236.2 m/s. The Venus flyby date is May 12, 2034, and the date of arrival at Earth is September 4, 2034. The total mission time is 502 days. Somewhat smaller values of the ΔV can be obtained by slightly changing the dates of the various maneuvers.

The values of ΔV_3 and above all of ΔV_4 are much larger than those typical of long stay missions, showing that, as expected, short stay missions are much more expensive in terms

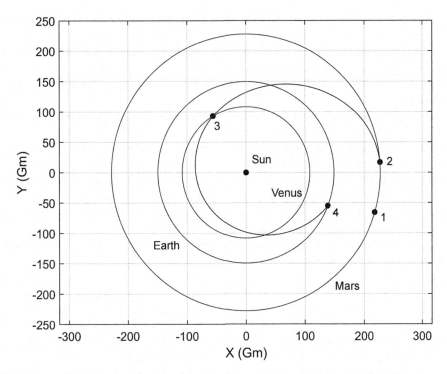

Figure B.5 Simplified return trajectory with a Venus flyby. 1: Arrival at Mars. 2: Departure from Mars. 3: Venus flyby. 4: Arrival at Earth.

of energy. If a direct atmospheric entry is made at Earth, then the high velocity of the interplanetary trajectory, compared to performing entry from Earth orbit, will pose difficulties and risks.

The deviation due to the flyby is $\delta = 11.610°$. The main parameters of the hyperbolic trajectory at Venus are $a = -26,712$ km, $b = -262,753$ and $e = 9.887$. The minimum distance from the center of Venus is $r_0 = 237,395$ km, and the minimum altitude is $h_0 = 231,343$ km. The trajectories for both the outbound and the inbound legs are reported in Figure 7.1.

It must be noted that while the trajectories of the simplified and the three-dimensional solutions are quite similar, the characteristics of the flyby are very different, and in particular, in the second case the deviation is much smaller and this translates into a larger distance from the planet.

B.7 PROPELLANT CONSUMPTION AND IMLEO

The so-called rocket equation that relates the velocity increase ΔV with the exhaust velocity v_e of the rocket engine and the initial and final masses of the spacecraft m_i and m_f is

$$\Delta V = v_e \ln\left(\frac{m_i}{m_f}\right) = v_e \ln\left(\frac{m_f + m_p}{m_f}\right). \tag{B.63}$$

The exhaust velocity depends on many factors, such as the mixture ratio, the feeding pressure, the nozzle design, etc. In a preliminary design it is reasonable to assume a realistic value, somewhat lower than the maximum achievable value in ideal conditions. The values here for the various propellants that may be used for a Mars mission are reported in Table 7.4, along with the ideal maximum value. In Table 7.5 the values for different nuclear rockets are reported.

Although in chemical rockets both the ejection mass and the energy are supplied by the propellant, in nuclear thermal rockets the propellant supplies only the former. As a consequence, in chemical rockets the maximum exhaust velocity is limited by the chemical energy that the fuel-oxidizer combination can develop. In nuclear rockets, however, the limitation comes from the maximum temperature the reactor core can withstand. The former limitation cannot be increased, while in the second case the limitation is due to the technology used and hence improvements are possible. The limitation for the nuclear rocket in Table 7.5 refers to NERVA-type solid core reactors.

The dry mass of the rocket m_r, including the tanks, the plumbing, the engine, the control devices, etc., can be expressed as a function of the propellant mass m_p as

$$m_r = A + K m_p. \tag{B.64}$$

For a first approximation, the constant term A may be neglected. Some values of the constant K, usually called the tankage factor, are reported in Table 7.4. These are very rough approximations and improvements may be expected, albeit at the cost of expensive research work.

For large rockets using storable propellants, K may be lower than 0.1. For cryogenic propellants it is larger, and here a value of 0.12 has been assumed. For nuclear rockets it is even larger and not easily predictable at our current level of technology. A value of 0.2 is assumed with the caveat that additional research is necessary.

The propellant required to achieve the velocity increment ΔV is

$$\frac{m_p}{m_i} = 1 - e^{\left(-\frac{\Delta V}{v_e}\right)}. \tag{B.65}$$

The final mass can thus be assumed to be made by the mass of the rocket m_r plus the mass of the spacecraft m_s

$$m_f = m_s + m_r = m_s + K m_p. \tag{B.66}$$

Introducing this expression into Eq. (B.63) it follows that

$$\frac{\Delta V}{v_e} = \ln \left[\frac{m_i}{m_s + K m_p} \right] \tag{B.67}$$

which means

$$\frac{m_i}{m_s + K m_p} = e^{\frac{\Delta V}{v_e}}. \tag{B.68}$$

From the two last equations it follows that

$$\frac{m_i}{m_s + K m_i \left(1 - e^{\left(-\frac{\Delta V}{v_e}\right)}\right)} = e^{\frac{\Delta V}{v_e}}.$$ (B.69)

The mass of the spacecraft is thus

$$\frac{m_s}{m_i} = e^{\left(-\frac{\Delta V}{v_e}\right)} - K \left(1 - e^{\left(-\frac{\Delta V}{v_e}\right)}\right).$$ (B.70)

The so-called 'gear ratio' (the ratio between the initial mass and the mass of the space-craft) is

$$R = \frac{m_i}{m_s} = \frac{e^{\frac{\Delta V}{v_e}}}{1 - K \left(e^{\frac{\Delta V}{v_e}} - 1\right)}.$$ (B.71)

This approach allows us to make a simple (albeit not very accurate) evaluation of the initial mass to be placed into LEO (IMLEO) in terms of the masses of the various space-craft and objects that need to be transported to Mars.

Assume a split mission with one (or more) cargo ships following a slow trajectory and one fast crew ship carrying people:

- The crewed spacecraft enters Mars orbit and then lands on Mars using an aeroshell and other devices. Its mass on the planet is m_{LAND}, which includes also the provisions for the trip and those taken to Mars (This is an approximation, because the water and the food used during the journey will no longer be on board upon entering Mars orbit or on landing, except for organic waste that is stored as radiation screening, which we shall neglect here). Refer to Figure B.6.
- The cargo ship brings to Mars orbit:

  - The outpost, which then lands on Mars using an aeroshell. Its mass on the planet is m_{OUT}.
  - The Mars Ascent Vehicle (MAV) which lands on Mars dry if the ascent propellant is to be produced on Mars, or partially dry if something is carried from Earth, or otherwise full of propellant carried from Earth. Its mass on the planet is $m_{MAV_{plan}}$.
  - The return spacecraft, which remains in Mars orbit. Its mass at the end of the return leg is m_{RET}.

Different assumptions may be made, like direct entry into Mars' atmosphere, or using the outpost habitat for landing and the outbound spacecraft for the return leg, but the modifications to the equations are straightforward.

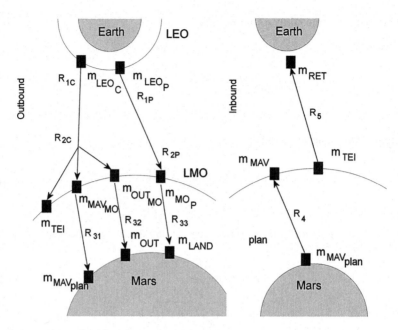

Figure B.6 A scheme detailing the masses of the various vehicles and the various 'gear ratios' used to evaluate the IMLEO.

The IMLEO required for the entire mission is the sum of the IMLEO of the cargo ship(s) $m_{\mathrm{LEO_C}}$ and of the passenger ship $m_{\mathrm{LEO_P}}$

$$\mathrm{IMLEO} = m_{\mathrm{LEO_C}} + m_{\mathrm{LEO_P}}. \tag{B.72}$$

The latter can be computed from the masses of the cargo to be carried in Mars orbit $m_{\mathrm{MO_C}}$ and of the passenger ship in Mars orbit $m_{\mathrm{MO_P}}$

$$\begin{aligned} m_{\mathrm{LEO_C}} &= R_{1C} R_{2C} m_{\mathrm{MO_C}} \\ m_{\mathrm{LEO_P}} &= R_{1P} R_{2P} m_{\mathrm{MO_P}} \end{aligned} \tag{B.73}$$

where R_{1C} and R_{1P} are the gear ratios for TMI from LEO, computed using Eq. (B.71)—the ratio for the passenger ship is usually higher because the relevant ΔV is larger in order to have a short travel time—and R_{2C} and R_{2P} relate to entering Mars orbit. In the case of a propulsive MOI they can be computed from Eq. (B.71) but if aerocapture is used it is necessary to take account of the aeroshell mass m_{aer_i}. Hence

$$R_{2_i} = \frac{m_{\mathrm{MO_i}} + m_{aer_i}}{m_{\mathrm{MO_i}}} = 1 + \frac{m_{aer_i}}{m_{\mathrm{MO_i}}} \tag{B.74}$$

where i represents either C or P for the cargo vehicle and the passenger ship respectively.

If a propulsive maneuver is combined with aerocapture, then the computed value of R_{2_i} needs to be multiplied by an expression of the type given by Eq. (B.71)

The cargo mass in Mars orbit m_{MO_C} is the sum of all masses which are sent using the cargo ship:

- The mass of the return vehicle with its propellant m_{TEI}.
- The mass of the MAV $m_{MAV_{MO}}$ (dry, if ISRU is intended, or with its propellant if there is no ISRU) in Mars orbit, i.e. including the aeroshell, parachute, and other landing systems and the propellant needed to deorbit the vehicle and perform the landing.
- The total mass of all the outpost elements $m_{OUT_{MO}}$ (habitat, ISRU, rovers, etc.) including the aeroshell and all the landing systems

$$m_{MO_C} = m_{TEI} + m_{MAV_{MO}} + m_{OUT_{MO}}. \qquad (B.75)$$

The latter two are

$$m_{MAV_{MO}} + m_{OUT_{MO}} = R_{31} m_{MAV_{plan}} + R_{32} m_{OUT} \qquad (B.76)$$

where R_{31} and R_{32}, the gear ratios for the landing phase, are expressed by equation of the type given in Eq. (B.74).

The mass of the MAV $m_{MAV_{plan}}$ to be landed on the surface of Mars is

$$m_{MAVplan} = R_4 m_{MAV} \qquad (B.77)$$

where m_{MAV} is the mass of the MAV dry, including tanks and engines, as remains after ascending into Mars orbit. R_4 is a gear ratio equal to unity if the MAV propellant is fully produced on Mars, or has a value larger than one which can be computed using an equation similar to Eq. (B.71), but in that case the tankage factor must be assumed to be equal to zero (the dry rocket has already been included in m_{MAV}) and due account made for the case where LOX and perhaps some of the methane is produced on Mars.

If the mass of the return habitat is m_{RET}, the mass in Mars orbit prior to the TEI maneuver is

$$m_{TEI} = R_5 m_{RET} \qquad (B.78)$$

where R_5 is the gear ratio at the TEI burn and can be computed using Eq. (B.71). If the returning craft enters orbit around Earth, the gear ratio R_5 takes into account the propellant for this maneuver.

The mass of the passenger ship in Mars orbit m_{MO_P} is

$$m_{MO_P} = R_{33} m_{LAND} \qquad (B.79)$$

where m_{LAND} is the mass of the lander on the surface, and R_{33} takes into account the mass of the heat shield and all devices used for landing as well as the rocket used to deorbit.

The IMLEO is therefore

$$\text{IMLEO} = R_{1C} R_{2C} R_5 m_{\text{RET}} +$$

$$+ R_{1C} R_{2C} R_{31} R_4 m_{\text{MAV}} +$$

$$+ R_{1C} R_{2C} R_{32} m_{\text{OUT}} +$$

$$+ R_{1P} R_{2P} R_{33} m_{\text{LAND}} = \tag{B.80}$$

$$= R_{\text{RET}} m_{\text{RET}} + R_{\text{MAV}} m_{\text{MAV}} + R_{\text{OUT}} m_{\text{OUT}} + R_{\text{LAND}} m_{\text{LAND}} .$$

The values of the gear ratios for the various parts of the travel, assuming chemical rockets, are reported in Table B.2.

The gear ratios were computed for generic circular planetary orbits without aiming for a particular launch opportunity. The following values of ΔV were assumed: TMI (slow trajectory) 3,600 m/s; TMI (fast trajectory) 3,900 m/s; entering circular MO from a slow trajectory 2,070 m/s; entering circular MO from a fast trajectory 2,800 m/s; entering elliptical MO from a slow trajectory 900 m/s; entering elliptical MO from a fast trajectory 1,560 m/s; propulsive landing from a circular MO 4,100 m/s; and propulsive landing from an elliptical MO 5,300 m/s. In addition, TEI was assumed to be the same as propulsively entering MO; and reaching MO from the ground was assumed to be the same as landing from orbit. The data for the propellants reported in Table 7.4 were assumed, except for the tankage factor, which has been assumed to be equal to 0.2 for cryogenic propellants that must be stored for a long time.

Table B.2 'Gear ratios' for the various parts of the interplanetary travel using chemical propulsion.

	From	To	LOX/LH2		Storable		Notes
			Cargo	Crew	Cargo	Crew	
R_1	LEO	TMT	2.673	2.927			
R_2	TMT	circ. MO	1.819	2.299	2.154	2.892	Propulsive
	TMT	ell. MO	1.285	1.558	1.383	1.770	Propulsive
	TMT	circ. MO		1.70			Aerodyn.
	TMT	ell. MO	1.30	1.50	1.30	1.50	Aerodyn.
R_3	circ. MO	Surf.	3.668		5.179		Propulsive
	eil. MO	Surf.	6.258		10.099		Propulsive
	circ. MO	Surf.		1.80			Aerodyn.
	ell. MO	Surf.		2.20			Aerodyn.
R_4	Surf.	MO		1.00			ISRU
	Surf.	circ. MO	3.668		5.179		no ISRU
	Surf.	ell. MO	6.258		10.099		no ISRU
R_5	circ. MO	TET		2.299		2.892	
	ell. MO	TET		1.558		1.770	

With the gear ratios so computed, the total gear ratios for the various spacecraft reported in Table B.3 were obtained. The table also reports the IMLEO for a split mission with a total payload of 170 t, with the following breakdown:

- $m_{RET} = 40$ t.
- $m_{MAV_{dry}} = 20$ t.
- $m_{OUT} = 70$ t.
- $m_{LAND} = 40$ t.

The eight cases are:

1. All cryogenic propellants, circular Mars orbit, propulsive MOI.
2. All cryogenic propellants, elliptical Mars orbit, propulsive MOI.
3. All cryogenic propellants, circular Mars orbit, aerobraking.
4. All cryogenic propellants, elliptical Mars orbit, aerobraking.
5. Storable propellants after TMI, circular Mars orbit, propulsive MOI.
6. Storable propellants after TMI, elliptical Mars orbit, propulsive MOI.
7. Storable propellants after TMI, circular Mars orbit, aerobraking.
8. Storable propellants after TMI, elliptical Mars orbit, aerobraking.

All cases were repeated for with ISPP and without.

The table shows the huge advantage of producing the propellant for the MAV on Mars, and also the advantage to be gained by using aerobraking for entering Mars orbit (in all cases considered, the landing is performed using aerodynamic forces). In practice, chemical propulsion cannot be used without ISPP and aerobraking. In the table, both are either used or completely neglected, as there are no intermediate solutions. Partial production of propellant or use of aerobraking only for the cargo ships and not for the crew ship have not been contemplated here, but it is easy to take this into account. The results are worse than those shown, but not so bad as to make the entire thing unfeasible.

An interesting point is that if aerobraking and ISRU are employed, the penalty from using storable propellants over LH2/LOX is not large (recalling that the tankage factor must be increased when liquid hydrogen needs to be stored for a long time). Probably, that is the most convenient solution.

Table B.3 'Gear ratios' for the various spacecraft of the missions.

Case	R_{RET}	R_{MAV}		R_{OUT}	R_{LAND}	IMLEO (t)	
		ISRU	no ISRU			ISRU	no ISRU
1	11.18	8.75	32.10	8.75	12.11	1,719	2,186
2	5.35	7.56	47.29	7.56	10.03	1,296	2,090
3	10.44	8.18	30.00	8.18	8.96	1,512	1,948
4	5.41	7.64	47.83	7.64	9.66	1,291	2,095
5	16.65	10.36	53.67	10.36	15.24	2,208	3,075
6	6.54	8.13	82.15	8.13	11.4	1,450	2,930
7	13.14	8.18	42.35	8.18	8.96	1,620	2,303
8	6.15	7.64	77.19	7.64	9.66	1,320	2,711

Case 8 seems to be the most interesting one, since it has a relatively low IMLEO without requiring long term storage of cryogenic propellants. The only technology to be developed is aerobraking of heavy spacecraft. The total IMLEO is 1,320 t for a total payload of 170 t, giving an overall gear ratio of 7.76.

One point which was not touched upon, but may increase the IMLEO by a non-negligible amount, is the need to reboost the elements carried into LEO, since the assumed orbit has an altitude of only 300 km.

The same analysis can be repeated for NTP. The relevant values of the gear ratios for the phases to be performed using nuclear propulsion are reported in Table B.4 The values of R_3 and R_4 are the same as those indicated for chemical propulsion. Owing to the uncertainty of the value of the tankage factor, the computations have been performed for both $K = 0.2$ and $K = 0.4$.

The total gear ratios for the nuclear spacecraft for four cases are shown in Table B.5. In all cases the MOI maneuver is propulsive and the descent to the planet uses aerodynamic braking.

The four cases are as follows:

1. Circular Mars orbit, cryogenic propellants are used on the MAV (if no ISRU is used).
2. Elliptical Mars orbit, cryogenic propellants are used on the MAV (if no ISRU is used).
3. Circular Mars orbit, storable propellants are used on the MAV (if no ISRU is used).
4. Elliptical Mars orbit, storable propellants are used on the MAV (if no ISRU is used).

Table B.4 'Gear ratios' for the various parts of the interplanetary travel using nuclear propulsion.

	From	To	K=0.2		K=0.4	
			Cargo	Crew	Cargo	Crew
R_1	LEO	TMT	1.731	1.818	1.971	2.105
R_2	TMT	circ. MO	1.360	1.523	1.447	1.669
	TMT	ell. MO	1.141	1.259	1.168	1.316
R_5	circ. MO	TET	1.360	1.523	1.447	1.669
	ell. MO	TET	1.141	1.259	1.168	1.316

Table B.5 'Gear ratios' for the various spacecraft of the missions using nuclear propulsion.

	Case	R_{RET}	R_{MAV}		R_{OUT}	R_{LAND}	IMLEO (t)	
			ISRU	no ISRU			ISRU	no ISRU
$K=0.2$	1	3.59	4.24	15.55	4.24	4.99	724	951
	2	2.49	4.34	27.18	4.34	5.03	692	1,149
	3	3.59	4.24	21.95	4.24	4.99	724	1,079
	4	2.49	4.34	43.87	4.34	5.03	692	1,482
$K=0.4$	1	4.76	5.13	18.84	5.13	6.32	906	1,180
	2	3.03	5.07	31.70	5.07	6.09	821	1,353
	3	4.76	5.13	26.59	5.13	6.32	906	1,335
	4	3.03	5.07	51.16	5.07	6.09	821	1,743

Table B.6 'Gear ratios' for the various spacecraft of the missions using nuclear propulsion, corrected to take into account the high starting orbit.

	Case	R_{RET}	R_{MAV} ISRU	no ISRU	R_{OUT}	R_{LAND}	IMLEO (t) ISRU	no ISRU
$K=0.2$	1	4.41	6.41	19.12	4.24	6.13	999	1,253
	2	3.06	6.57	33.44	4.34	6.19	961	1,499
	3	4.41	6.41	27.00	4.24	6.13	999	1,411
	4	3.06	6.57	53.96	4.34	6.19	961	1,909
$K=0.4$	1	5.86	7.77	23.17	5.13	7.78	1,245	1,553
	2	3.73	7.66	38.99	5.07	7.49	1,138	1,765
	3	5.86	7.77	32.71	5.13	7.78	1,245	1,743
	4	3.73	7.66	62.92	5.07	7.49	1,138	2,244

The IMLEO is much lower and the penalty from not using ISPP is less severe. The most favorable condition uses an elliptical Mars orbit; if the MAV uses propellant produced on Mars then an IMLEO of 692 t allows 170 t to be landed on Mars, giving an overall gear ratio of 4.07. Doubling the tankage factor from 0.2 to 0.4 increases the IMLEO to 821 t for an overall gear ratio of 4.83.

However, the results obtained for the chemical and nuclear propulsion cannot be compared because NTP requires a starting orbit at a higher altitude, and thus more launches are required at equal IMLEO. It is impossible to compensate exactly for this effect, since each launcher has its own characteristics, but a simple evaluation can be done by comparing the performance in 300 km LEO and in a 1,200 km orbit—usually considered suitable for NTP—for a given launcher. For instance, the payload into LEO for the current version of the SpaceX Falcon 9 is about 1.23 times that for an orbit at 1,200 km. The values given in Table B.5 have therefore been multiplied by a factor of 1.23 to yield Table B.6. The results so obtained can be compared to those in Table B.2 to show that NTP still has advantages over chemical propulsion, even taking into account the different starting orbits.

This advantage may be used in part to reduce the travel time of the crew ship, and in part to place the return ship into Earth orbit so that it can be recovered, refurbished and reused. Another advantage for the nuclear ship is that it becomes pointless to use two travel habitats, one for the outbound leg (which lands on Mars) and one for the return. A single habitatt may do both legs, and the crew can dock to a lander that has been taken to Mars orbit by the cargo ship. A saving of several tons in Mars orbit would obviously be very welcome.

Symbols

a	semi-major axis	
g	gravitational acceleration	
m	mass of the spacecraft	
$\mathbf{r}$	position vector of the spacecraft	
r_E	radius of Earth	$r_E = 6,371\,\mathrm{km}$

r_M	radius of Mars	$r_M = 3,396.2\,\text{km}$
t	time	
v_e	exhaust velocity	
C_3	parameter $C_3 = V_{\infty1}^2$	
E	specific mechanical energy	
G	gravitational constant	$G = 6.67384 \times 10^{-11}\,\text{m}^3\,\text{kg}^{-1}\,\text{s}^{-2}$
M_E	mass of Earth	$M_E = 5.9742 \times 10^{24}\,\text{kg}$
M_M	mass of the Mars	$M_M = 6.4185 \times 10^{23}\,\text{kg}$
M_S	mass of the Sun	$M_S = 1.99 \times 10^{30}\,\text{kg}$
R	gear ratio	
R_E	radius of Earth Orbit	$R_E = 149.6 \times 10^9\,\text{m}$
R_M	radius of Mars Orbit	$R_M = 227.9 \times 10^9\,\text{m}$
T	period, transit time	
V	velocity	
V_{oE}	orbital velocity of Earth	$V_E = 29,795\,\text{m}/\text{s}$
V_{oM}	orbital velocity of Mars	$V_M = 24,140\,\text{m}/\text{s}$
V_{∞}	hyperbolic excess speed	
X, Y	Cartesian coordinates	
μ_E	gravitational parameter of Earth	$\mu_E = 3.9871 \times 10^{14}\,\text{m}^3\,\text{s}^{-2}$
μ_M	gravitational parameter of Mars	$\mu_M = 4.2836 \times 10^{13}\,\text{m}^3\,\text{s}^{-2}$
μ_S	gravitational parameter of the Sun	$\mu_S = 1.3281 \times 10^{20}\,\text{m}^3\,\text{s}^{-2}$
ν	true anomaly	
ΔV	speed increment	

Appendix C: Low thrust trajectories

Electric propulsion is characterized by values for specific impulse that are much higher than for thermal (chemical or nuclear) rockets but thrusts which are usually much lower. This is often countered by using the thruster for long periods of time, and possibly even for the entire journey. The design of the trajectory must therefore be integrated with that of the thrust profile. It is possible to obtain very good performance in this way, although at the expense of much greater design complexity. Usually an approach similar to the patched conic approximation is used, with the difference that now the trajectory is no longer made of arcs of ellipses, parabolas, or hyperbolas, but of lines which look like spirals and have to be calculated numerically. As previously, we shall begin by describing idealized solutions obtained by assuming that planetary orbits are circular and coplanar, and then consider more realistic solutions in which the orbits of the planets are elliptical and non-coplanar.

C.1 INTRODUCTION

In contrast to chemical and nuclear thermal rockets, electric thrusters for NEP and SEP systems deliver a low thrust, but this can be applied for long periods of time and possibly even for the entire journey. An electric propulsion system is characterized by a high specific impulse, so a fast interplanetary trajectory can be obtained with low propellant consumption. On the one hand, low thrust propulsion makes possible a wide variety of trajectories, giving the mission designer a lot of freedom, but it also makes selecting and optimizing the trajectory (and indeed the whole mission profile) much more complicated.

Although impulsive trajectories can be obtained in closed form (at least if the two-body assumption is used) no analytical solution for a continuous-thrust trajectory is possible, except for the simplest cases. Their optimization requires using numerical procedures. There are two main approaches for optimizing a trajectory. With direct methods the problem is solved by means of gradient-based procedures, and with indirect methods it becomes a boundary value problem which is solved using shooting procedures. Both require the generation of an initial set of optimization parameters to start the procedure, and this is often the most expensive computational phase.

© Springer International Publishing Switzerland 2017

G. Genta, *Next Stop Mars*, Springer Praxis Books, DOI 10.1007/978-3-319-44311-9

Whether direct or indirect methods are best is a controversial topic. In this section, use will be made of indirect methods, following the method for optimizing low thrust rocket missions developed by Irving and recalled by Keaton.[5] They separate the spacecraft optimization from the thrust program optimization and then connect the two parts of the analysis by using the mass-to-power ratio (or specific mass) α of the power generator that feeds the thruster.

The thrust the rocket produces is

$$\mathbf{T} = \dot{m}\mathbf{v}_e = -q\mathbf{v}_e \tag{C.1}$$

where $q = -\dot{m}$ is the propellant throughput and $\mathbf{v}_e$ is the ejection velocity of the engine. If the throughput is known, the mass $m(t)$ is a known function of time. If q is constant then

$$m(t) = m_i - qt \tag{C.2}$$

where m_i is the initial mass. If the rocket operates for a time t_f, the final mass is

$$m_f = m_i - \int_0^{t_f} q \, dt = m_i - m_p \tag{C.3}$$

where m_p is the propellant mass.

The spacecraft must be provided with a power generator that provides the electrical power to the thruster. If the latter has a mass m_w, the final mass can be considered as

$$m_f = m_l + m_w + m_s \tag{C.4}$$

where m_l and m_s are respectively the payload and the structural mass. By assuming that the mass of the generator is proportional to its power by way of the specific mass α (which for now is assumed to be a constant) and the structural mass is proportional to the mass of the propellant through the tankage factor K, then

$$m_w = \alpha P, \quad m_s = Km_p \tag{C.5}$$

and it follows that

$$m_f = m_l + \alpha P + Km_p. \tag{C.6}$$

The payload mass is thus

$$m_l = m_i - \alpha P - (K+1)m_p. \tag{C.7}$$

The power of the jet is

$$P_j = \frac{1}{2}qv_e^2 = \frac{T}{2}v_e \tag{C.8}$$

---

[5] P.W. Keaton, *Low Thrust Rocket Trajectories*, LA-10625-MS, Los Alamos, 2002.

and by introducing the efficiency of the thruster η_t (which includes also the regulator and the whole propulsive chain) the power that the generator must produce is

$$P = \frac{1}{2\eta_t} q v_e^2 = \frac{T}{2\eta_t} v_e. \tag{C.9}$$

The payload mass is thus

$$m_l = m_i - \alpha \frac{1}{2\eta_t} q v_e^2 - (K+1) \int_0^{t_f} q \, dt = \tag{C.10}$$

$$= m_i - \frac{1}{2} \alpha_e T v_e - (K+1) \int_0^{t_f} \frac{T}{v_e} \, dt$$

where the effective specific mass α_e

$$\alpha_e = \frac{\alpha}{\eta_t}$$

takes into account also the efficiency of the thruster η_t.

It is clear that any increase of the specific impulse (or the ejection velocity) will provide a decrease in the propellant mass (for a given thrust profile) but an increase in the required power, and likewise in the generator mass. To optimize the payload mass for a given mission, the optimal specific impulse must be identified.

C.2 CONSTANT EJECTION VELOCITY

As described in Section 6.5, it is possible to distinguish between constant ejection velocity (CEV) and variable ejection velocity (VEV) systems.

Consider a rocket of the first type which is accelerating from standstill in three-dimensional space, free of a gravitational field, under the effect of a constant thrust **T**. Because both v_e and **T** are constant, the throughput q and the power P are also constant. The motion occurs in a straight line (say along the x axis) and therefore the equation motion in the inertial reference frame is simply

$$m(t)\ddot{x} = T \tag{C.11}$$

and therefore

$$\ddot{x} = \frac{T}{m_i - qt}. \tag{C.12}$$

By integrating this equation, the velocity attained in the generic time t is

$$V(t) = \frac{T}{q} \ln\left(\frac{m_i}{m_i - qt}\right) = v_e \ln\left(\frac{m_i}{m_i - qt}\right). \tag{C.13}$$

The total ΔV thereby achieved is

$$\Delta V = v_e \ln\left(\frac{m_i}{m_f}\right) = v_e \ln\left(\frac{m_i}{m_i - m_p}\right) \tag{C.14}$$

which is nothing other than the famous rocket equation (B.63).
The ratio between the propellant mass and the initial mass is

$$\frac{m_p}{m_i} = 1 - e^{-\frac{\Delta V}{v_e}}. \tag{C.15}$$

From Eq. (C.10), the ratio between the payload and the initial mass is

$$\frac{m_l}{m_i} = 1 - \frac{1}{2}\alpha_e \frac{v_e^2}{t_t}\left(1 - e^{-\frac{\Delta V}{v_e}}\right) - (K+1)\left(1 - e^{-\frac{\Delta V}{v_e}}\right) \tag{C.16}$$

where the last two terms are m_w/m_i and $(m_s + m_p)/m_i$ respectively.
The ratio

$$V_c = \sqrt{\frac{2t_f}{\alpha_e}} \tag{C.17}$$

is dimensionally a velocity, usually referred to as the characteristic velocity or Langmuir velocity,[6] and it enables us to write the equations in non-dimensional form as

$$\frac{m_l}{m_i} = 1 - \frac{v_e^2}{V_c^2}\left(1 - e^{-\frac{\Delta V}{v_e}}\right) - (K+1)\left(1 - e^{-\frac{\Delta V}{v_e}}\right). \tag{C.18}$$

The search for the specific impulse (or the ejection velocity) that achieves the maximum payload is straightforward once we state the values of ΔV and the time t_f required to achieve that effect, plus the α_e and K parameters of the vehicle. For instance, in Figure C.1a $\Delta V = 10$ km/s, $t_f = 10^7$ seconds (3.86 months), $K = 0.05$, and $\alpha_e = 1, 2, 5, 10, 20, 50$ and 100 kg/kW. The ratio between the initial mass and the payload mass m_i/m_l is reported as a function of the ejection velocity for the various values of the effective specific mass. The optimal value of the ejection velocity, which is the value that minimizes m_i/m_l, increases as α_e decreases.

---

[6] D.B. Langmuir, "Low-Thrust Flight. Constant exhaust velocity in Field-Free Space," in [2] and [5].

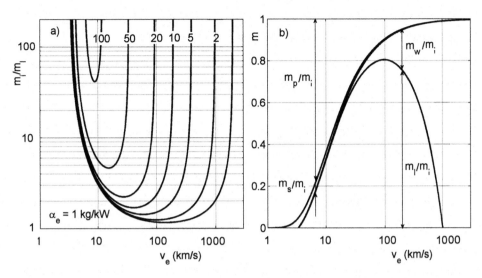

Figure C.1 Performance of a constant specific impulse system during an acceleration in free space with the following data: $\Delta V = 10$ km/s, $t_f = 10^7$ seconds (3.86 months), $K = 0.05$. (a) The ratio between the initial mass and the payload as a function of the ejection velocity for $\alpha_e = 1, 2, 5, 10, 20,$ 50 and 100 kg/kW. (b) Non-dimensional masses of the payload, the propellant, the structure, and the generator as functions of the ejection velocity. Case with $\alpha_e = 2$ kg/kW.

Figure C.1b reports the non-dimensional masses of the payload, the propellant, the structure, and the generator as functions of the ejection velocity in the case with $\alpha_e = 2$ kg/ kW. In this case the optimum value of the ejection velocity is about 100 km/s.

C.3 VARIABLE EJECTION VELOCITY

C.3.1 Spacecraft mass distribution

If it is possible to vary the ejection velocity during acceleration, it is evident that the higher the ejection velocity the lower is the propellant throughput (i.e. the propellant consumption) to supply a given thrust, but the higher is the power required. The best way to use the generator is to have it supply continually the maximum power that it can, and then vary the ejection velocity in order to regulate the thrust. The system therefore works on the parabola labeled ' $P =$ constant' in Figure 6.12.

Consider again an acceleration in a portion of space that is free from gravitational fields. Equation (C.11) can be written in the form

$$\ddot{\mathbf{x}} = \frac{\mathbf{T}(t)}{m(t)} = \mathbf{a}(t) \qquad (C.19)$$

where $\mathbf{a}(t)$ is the acceleration due to the thrust. Since in general it is not stated that the direction of the thrust is constant, $\mathbf{x}$, $\mathbf{T}$, and $\mathbf{a}$ are generic vectors.

From Eqs. (C.8) and (C.19) it follows that

$$\frac{\mathbf{a}^2}{2P} = -\frac{\dot{m}}{m^2} = \frac{d}{dt}\left(\frac{1}{m}\right).$$ (C.20)

This differential equation can be integrated to obtain

$$\frac{1}{m(t)} = \frac{1}{m_i} + \int_0^t \frac{\left[\mathbf{a}(u)\right]^2}{2P(u)} du.$$ (C.21)

As already stated, the power is assumed to be constant, so the final mass is

$$\frac{1}{m_f} = \frac{1}{m_i} + \frac{1}{2P} \int_0^{t_f} \left[\mathbf{a}(t)\right]^2 dt$$ (C.22)

and therefore

$$\frac{m_i}{m_f} = 1 + \frac{m_i}{m_w}\frac{\alpha_e}{2} \int_0^{t_f} \left[\mathbf{a}(t)\right]^2 dt.$$ (C.23)

Equation (C.23) allows us to deduce that for any value of P (or of m_w), the final mass, and hence the sum $m_l + m_s$, is maximum when a law $\mathbf{a}(t)$ that minimized the integral

$$J = \frac{1}{2}\int_0^{t_f} a^2 dt$$ (C.24)

is chosen.

It must be noted that in the case of CEV thrusters, the integral to be minimized is

$$J' = \int_0^{t_f} a \, dt.$$ (C.25)

By introducing the non-dimensional parameter

$$\gamma^2 = \frac{\alpha_e}{2}\int_0^{t_f} \left[\mathbf{a}(u)\right]^2 du = \alpha_e J$$ (C.26)

it is possible to write

$$\frac{m_l + m_s}{m_i} = \frac{m_f}{m_i} - \frac{m_w}{m_i} = \frac{1}{1 + \dfrac{m_i}{m_w}\gamma^2} - \frac{m_w}{m_i} = \tag{C.27}$$

$$= \frac{m_w}{m_i}\left(\frac{1}{\dfrac{m_w}{m_i} + \gamma^2} - 1\right).$$

Clearly the condition $\gamma < 1$ must be satisfied, as otherwise a negative value of $m_l + m_s$ would be obtained.

Usually the optimization is done by searching for a mass distribution that allows us to maximize the ratio

$$\frac{\left(m_l + m_s\right)}{m_i}. \tag{C.28}$$

in the manner described by Irving.[7]

From Eq. (C.27) it follows that

$$\left[\frac{d}{d\left(m_w / m_i\right)}\right]\frac{m_l + m_s}{m_i} = -\frac{\left(\dfrac{m_w}{m_i}\right)^2 + 2\gamma^2\,\dfrac{m_w}{m_i} + \gamma^4 - \gamma^2}{\left(\dfrac{m_w}{m_i} + \gamma^2\right)^2}. \tag{C.29}$$

By equating the derivative to zero, the following optimal value of the generator mass is obtained

$$\frac{m_w}{m_i} = \gamma\left(1 - \gamma\right). \tag{C.30}$$

It then follows that

$$\frac{m_l + m_s}{m_i} = \left(1 - \gamma\right)^2 \tag{C.31}$$

[7] J.H. Irving, "Low-Thrust Flight. Variable exhaust velocity in Gravitational Fields," in [2]; P.W. Keaton, *Low Thrust Rocket Trajectories*, LA-10625-MS, Los Alamos, 2002.

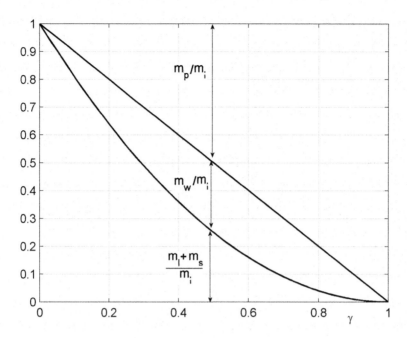

Figure C.2 Optimal mass distribution as a function of γ.

and

$$\frac{m_p}{m_i} = \gamma \,. \tag{C.32}$$

The optimal mass distribution is shown in Figure C.2.

The ejection velocity can be easily computed as a function of time. Since

$$\left| v_e \right| = \frac{2P}{|T|} = \frac{2P}{|a| m} \tag{C.33}$$

and remembering that P is constant, it follows from Eq. (C.21) that

$$\left| v_e \right| = \frac{2P}{|a|} \left(\frac{1}{m_i} + \int_0^t \frac{\left[\mathbf{a}(u) \right]^2}{2P} du \right) \tag{C.34}$$

and

$$\left|v_e\right| = \frac{1}{\left|a\right|} \left(\frac{2P}{m_i} + \int_0^t \left[\mathbf{a}(u)\right]^2 du \right).$$

(C.35)

By introducing the specific mass α_e it follows that

$$\left|v_e\right| = \frac{1}{\left|a\right|} \left(\frac{2m_w}{\alpha_e m_i} + \int_0^t \left[\mathbf{a}(u)\right]^2 du \right)$$

(C.36)

and thus

$$\left|v_e\right| = \frac{1}{\left|a\right|} \left[\frac{2}{\alpha_e} \gamma \left(1-\gamma\right) + \int_0^t \left[\mathbf{a}(u)\right]^2 du \right].$$

(C.37)

The propellant throughput can be easily computed as

$$\frac{\dot{m}}{m_i} = \frac{2}{\alpha_e v_e^2} \gamma \left(1-\gamma\right).$$

(C.38)

C.3.2 Straight motion in field-free space

Optimizing the law $a(t)$ in the case of a straight line acceleration in space free of gravitational fields is easy. Since the problem is one-dimensional, x, T, and a are now scalars. Using the standard calculus of variations, an extremum of J occurs when

$$\delta J = \frac{1}{2} \int_0^{t_f} a \delta a \, dt = 0.$$

(C.39)

Since

$$a = \ddot{x}$$

(C.40)

it follows that

$$\delta a = \frac{d^2 \delta x}{dt^2}.$$

(C.41)

By integrating-by-parts, it follows that

$$2\delta J = \int_0^{t_f} a \delta a \, dt = \int_0^{t_f} \frac{d^2 a}{dt^2} \delta x \, dt + \left[a \frac{d\delta x}{dt} \right]_0^{t_f} - \left[\frac{da}{dt} \delta x \right]_0^{t_f}.$$

(C.42)

The last two terms vanish when the initial position and velocity are stated. If some of the boundary conditions are not specified, additional conditions, namely transversality conditions, must be stated. The condition that allows us to minimize J is thus

$$\frac{d^2 a}{dt^2} = 0. \tag{C.43}$$

By integrating, it follows that

$$\begin{cases} a = c_1 t + c_2 \\ V = \dfrac{1}{2} c_1 t^2 + c_2 t + c_3 \\ x = \dfrac{1}{6} c_1 t^3 + \dfrac{1}{2} c_2 t^2 + c_3 t + c_4 \end{cases} \tag{C.44}$$

and

$$J = \frac{t_f}{2} \left(\frac{c_1^2}{3} t_f^2 + c_1 c_2 t_f + c_2^2 \right). \tag{C.45}$$

C.3.3 Acceleration from standstill

Assuming that the spacecraft starts from standstill at time $t = 0$ and accelerates until it reaches a speed ΔV at time $t = t_f$, at time $t = 0$, x, and V vanish, and it follows that $c_3 = c_4 = 0$. The final position is not stated and thus the transversality condition

$$\frac{da}{dt} = 0 \text{ for } t = t_f \tag{C.46}$$

must be stated, so that the last term of Eq. (C.42) vanishes. From this condition, it can also be seen that $c_1 = 0$.

To reach the speed ΔV at time t_f it follows that

$$c_2 = \frac{\Delta V}{t_f} \tag{C.47}$$

and

$$a = \frac{\Delta V}{t_f} \quad V = \frac{\Delta V}{t_f} t \quad x = \frac{1}{2} \frac{\Delta V}{t_f} t^2 \quad J = \frac{\Delta V^2}{2 t_f}. \tag{C.48}$$

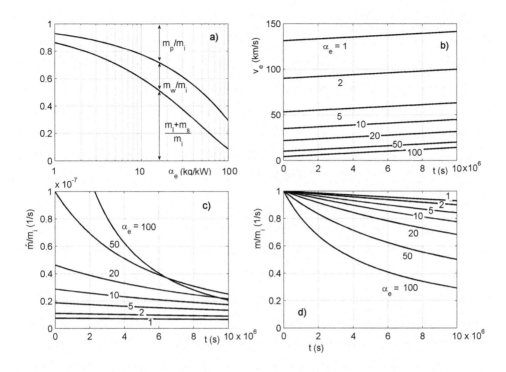

Figure C.3 A vehicle performing an acceleration with $\Delta V = 10$ km/s in $t_f = 10^7$ s. (a) Optimal mass distribution as a function of α_e. (b), (c) and (d) Time histories of the ejection velocity, the propellant throughput and the vehicle mass for different values of α_e.

Parameter γ is therefore

$$\gamma = \sqrt{\alpha_e J} = \frac{\Delta V}{\sqrt{2t_f / \alpha_e}} = \frac{\Delta V}{V_c} \qquad (C.49)$$

where V_c is the characteristic velocity.

Consider the same example as in Figure C.1, an acceleration with $\Delta V = 10$ km/s that is performed in $t_f = 10^7$ seconds. It follows that $J = 5$ m^2/s^3. The optimal mass distribution is reported as a function of α_e in Figure C.3a. The ejection velocity is reported as a function of time in Figure C.3b for different values of α_e. From this plot, it is apparent that the variation of the exhaust velocity which optimizes the payload mass is not very great. For instance, if $\alpha_e = 2$ kg/kW the ejection velocity changes slowly from about 90 km/s (specific impulse of about 9,000 seconds) to about 100 km/s. From Figure C.3b it is clear that the optimum value of the ejection velocity for the same value of α_e was slightly less than 100 km/s. The two results are very close because, in this case, the variation of the ejection velocity is not great and hence the CEV and VEV cases almost coincide.

The time histories of the propellant throughput and of the vehicle mass are reported in Figure C.3c, d.

C.3.4 Point-to-point motion

Consider a spacecraft starting from rest at point 1 with $x = 0$, and reaching point 2 with $x = x_f$ at time t_f and stopping there. The initial and final conditions are $x = 0$ and $V = 0$ at $t = 0$, and $x = x_f$ and $V = 0$ at $t = t_f$. It therefore follows that

$$c_1 = -\frac{12x_f}{t_f^3}, \quad c_2 = \frac{6x_f}{t_f^2}, \quad c_3 = 0, \quad c_4 = 0. \tag{C.50}$$

The position, velocity and acceleration laws are

$$\begin{cases} a = \dfrac{6x_f}{t_f^2}\left(1 - 2\dfrac{t}{t_f}\right) & V = \dfrac{6x_f}{t_f^2}t\left(1 - \dfrac{t^2}{t_f^2}\right) \\[2ex] x = \dfrac{3t^2 x_f}{t_f^2}\left(1 - \dfrac{2t}{3t_f}\right) & J = \dfrac{6x_f^2}{t_f^3} \end{cases}. \tag{C.51}$$

The acceleration time history for the case with $x_f = 3 \times 10^7$ km and $t_f = 10^7$ seconds is reported in Figure C.4a. The ejection velocity, the propellant throughput, and the mass are reported as functions of time for different values of α_e in Figure C.4b–d. This case is just

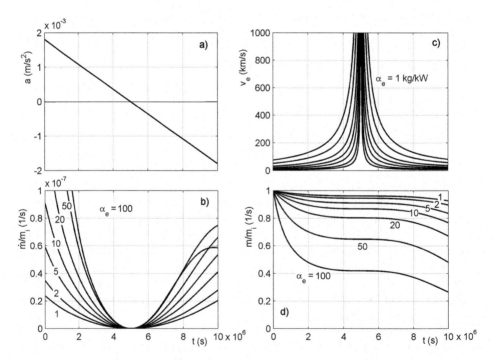

Figure C.4 The acceleration-deceleration cycle. (a) Time history of the acceleration. (b), (c) and (d) Time histories of the ejection velocity, the propellant throughput, and the mass for different values of α_e.

a theoretical one because the acceleration becomes very small in the central part of the trajectory, and at time $t = t_f / 2$ it is precisely zero. To obtain this by regulating the ejection velocity means that very large values of the latter must be achieved, and at $t = t_f / 2$ the ejection velocity must reach an infinite value, which is impossible. In an actual case, the thruster can be switched off when a very low value of thrust is requested and a phase of coasting is inserted. This does not much change the law of motion, since the thrust is almost negligible.

C.3.5 Motion in a gravitational field

The equation of motion for a generic spacecraft that is moving in space under the effect of a gravitational field whose potential is U, with a thrust $\mathbf{T}(t)$, can be obtained from Eq. (C.19) by adding a gravitational term

$$\mathbf{r} + \nabla U = \mathbf{a}. \qquad (C.52)$$

Vector $\mathbf{r}(t)$ defines the position of the spacecraft in an inertial reference frame and, as indicated by Eq. (C.19), vector $\mathbf{a}$ is the thrust divided by the mass.

In the present formulation of the equation of motion, the thrust and the mass of the spacecraft do not appear explicitly. The law $\mathbf{a}(t)$ is obtained, and then the thrust is regulated to obey this law, taking into account the fact that the mass varies in time due to the consumption of propellant. This is consistent with the VEV approach.

Generally speaking, the gravitational field influencing the spacecraft, and its potential U, are due to a number of celestial bodies that are in motion relative to each other. But here, the two-body approach is made use of. That is, at any given moment the spacecraft is assumed to be under the influence of just one celestial body. The trajectory is subdivided into patches, each in the sphere of influence of some celestial body. Obviously, these patches are not conics (hyperbolas, parabolas, or ellipses) but trajectories which cannot be described in closed form. In particular, an Earth–Mars transfer will be initiated as a spiral-like trajectory about Earth with a large number of turns. It starts off in an orbit around Earth (possibly LEO) until it escapes from the Earth's sphere of influence. Then there is a spiral trajectory around the Sun that usually involves less than a full revolution. And finally, upon entering the sphere of influence of Mars there is a spiral of many turns that ends when the desired orbit is achieved.

Resorting to the two-body assumption, the gravitational field is due to a single spherical celestial body located in the origin of the reference frame. Its potential is

$$U = -\frac{\mu}{r} \qquad (C.53)$$

where $\mu = GM$, is the standard gravitational parameter of the body whose mass is M.

Assuming a planar motion under the effect of a thrust contained in the plane defined by vectors **r** and **ṙ** , the equation of motion (C.52) can be written in polar coordinates (r, θ) as

$$\begin{cases} \ddot{r} = -\dfrac{\mu}{r^2} + r\left(\dfrac{d\theta}{dt}\right)^2 + a_r \\[4mm] \ddot{\theta} = -\dfrac{2}{r}\dfrac{dr}{dt}\dfrac{d\theta}{dt} + \dfrac{1}{r}a_\theta \end{cases}$$

(C.54)

where a_r and a_θ are the radial and tangential components of **a**, whose modulus is

$$a(t) = \frac{T(t)}{m(t)} = \sqrt{a_r^2 + a_\theta^2} .$$

(C.55)

As seen in Section C.3.1, in this case it is possible to demonstrate that the trajectory can be optimized from the viewpoint of the final mass of the spacecraft by assuming (as a cost function to be minimized) the integral defined in Eq. (C.24)

$$J = \frac{1}{2}\int_t^{t_2} a^2(t)\,dt$$

where t_1 and t_2 are the initial and final times along the trajectory. Eqs. (B.5) and (C.30) and the following ones allow us to compute the optimal mass distribution once the value of J is obtained. To optimize the thrust profile, Eqs. (C.39), (C.40), (C.41), and (C.42) must be modified in order to take account of the gravitational potential and the fact that the problem is now three-dimensional and **a** is a vector whose components are a_i for $i = 1, 2, 3$

$$\delta J = \frac{1}{2}\int_0^{t_f}\sum_{i=1}^{3} a_i \delta a_i \, dt = 0$$

(C.56)

$$a_i = \ddot{x}_i + \frac{\partial U}{\partial x_i}$$

(C.57)

$$\delta a = \frac{d^2 \delta x_i}{dt^2} + \sum_{j=1}^{3} \frac{\partial^2 U}{\partial x_j \partial x_i}$$

(C.58)

$$\int_{t_1}^{t_2} a_i \delta a_i \, dt = \int_{t_1}^{t_2}\left[\frac{d^2 a_i}{dt^2} + \sum_{j=1}^{3} a_i \frac{\partial^2 U}{\partial x_j \partial x_i}\right]\delta x \, dt + \left[a_i \frac{d\delta x_i}{dt}\right]_{t_1}^{t_2} - \left[\frac{da_i}{dt}\delta x_i\right]_{t_1}^{t_2} .$$

(C.59)

With simple considerations, it is possible to show that the equation allowing us to compute the law for $\mathbf{a}(t)$ is

$$\frac{d^2 a_i}{dt^2} + \sum_{j=1}^{3} a_j \frac{\partial^2 U}{\partial x_j \partial x_i} = 0 \tag{C.60}$$

and hence

$$\ddot{\mathbf{a}} + \mathbf{a}\nabla(\nabla U) = 0. \tag{C.61}$$

Equations (C.52) and (C.61), along with the boundary conditions

$$\begin{Bmatrix} \mathbf{r}(t_1) = \mathbf{r}_1 \\ \dot{\mathbf{r}}(t_1) = \mathbf{v}_1 \end{Bmatrix} \quad \begin{Bmatrix} \mathbf{r}(t_2) = \mathbf{r}_2 \\ \dot{\mathbf{r}}(t_2) = \mathbf{v}_2 \end{Bmatrix} \tag{C.62}$$

allow us to obtain the trajectory and the acceleration program to travel from point $\mathbf{r}_1$ and velocity $\mathbf{v}_1$ at time t_1 to point $\mathbf{r}_2$ and velocity $\mathbf{v}_2$ at time t_2.

Equations (C.52) and (C.61) may be written in Cartesian coordinates to produce a set of six second order differential equations

$$\begin{cases} \ddot{x} = -\dfrac{\mu x}{\sqrt{\left(x^2 + y^2 + z^2\right)^3}} + a_x \\[4mm] \ddot{y} = -\dfrac{\mu y}{\sqrt{\left(x^2 + y^2 + z^2\right)^3}} + a_y \\[4mm] \ddot{z} = -\dfrac{\mu z}{\sqrt{\left(x^2 + y^2 + z^2\right)^3}} + a_z \\[4mm] \ddot{a}_x = -\mu \dfrac{a_x\left(-2x^2 + y^2 + z^2\right) - 3a_y xy - 3a_z xz}{\sqrt{\left(x^2 + y^2 + z^2\right)^5}} \\[4mm] \ddot{a}_y = -\mu \dfrac{-3a_x xy + a_y\left(x^2 - 2y^2 + z^2\right) - 3a_z yz}{\sqrt{\left(x^2 + y^2 + z^2\right)^5}} \\[4mm] \ddot{a}_z = -\mu \dfrac{-3a_x xz - 3a_y yz + a_z\left(x^2 + y^2 - 2z^2\right)}{\sqrt{\left(x^2 + y^2 + z^2\right)^5}} \end{cases} \tag{C.63}$$

with initial and final conditions that specify the positions x, y, and z and the velocities $\dot{x}$, $\dot{y}$, and $\dot{z}$ at times t_1 and t_2.

By introducing six auxiliary variables, namely the velocities v_x, v_y, and v_z, and the derivatives of a_x, a_y, and a_z—namely v_{ax}, v_{ay}, v_{az}—we can generate a set of 12 first order differential equations. Their numerical integration is not difficult in itself, so long as a starting solution that is not very far from the final result is available. As this is the main difficulty in obtaining a satisfactory solution, the first point dealt with here is to state a first approximation solution. This requires 12 laws $\dot{x}(t)$, $\dot{y}(t)$, $\dot{z}(t)$, $\dot{a}_x(t)$, $\dot{a}_y(t)$, $\dot{a}_z(t)$, $x(t)$, $y(t)$, $z(t)$, $a_x(t)$, $a_y(t)$, and $a_z(t)$ which reasonably satisfy Eq. (C.63) and the related boundary conditions.

This formulation of the problem permits us to study the generic three-dimensional motion without further assumptions. If the planetary orbits are assumed to be coplanar, it becomes a two-dimensional problem and the number of second order differential equations reduce to four (equivalent to eight first order equations).

As already stated, in the case of low, continuous thrust systems, the optimization of the trajectory and the thrust profile is strictly linked with the optimization of the spacecraft. Their linkage is a single parameter, the specific mass of the power generator α_e. In the present approach it is assumed that the power generator always works at full power, with a constant specific mass α_e, and the thrust level is regulated by suitably altering the specific impulse of the thruster I_s. Such devices are currently being developed and, hopefully, will soon be tested in space. The most advanced is the VASIMR (Variable Specific Impulse Magnetoplasma Rocket) that is to be tested on the ISS.[8]

Clearly, the actual system may be unable to match the requirements of the theoretically computed trajectory; i.e. it may be unable to achieve the very high values of the specific impulse required by the optimized trajectory, but this can be obviated by switching off the thruster when an I_s in excess of the maximum practicable value is required, a condition which happens about halfway in an interplanetary transfer (see Section C.9). The effect of this maneuver is small, since in such conditions both the thrust and the propellant consumption are very low. It might even be beneficial, since it introduces a coasting phase at midcourse in which, in crewed spacecraft, checking and maintenance of the propulsion system can be performed by the astronauts.

Another option is to reduce the power P in order to regulate the thruster at constant v_e once it has attained its maximum value, thereby operating the engine with a mixture of VEV and CEV.

C.4 LEAVING LEO

As already pointed out, an interplanetary trajectory to Mars is divided into a sequence of steps, starting with a phase in the sphere of influence of Earth, a heliocentric phase in interplanetary space, and finally a phase in the sphere of influence of Mars.

In the first phase the spacecraft uses the thrust of its engine to progressively enlarge its orbit around Earth until it achieves escape conditions. Since this is a low thrust maneuver, the trajectory and the thrust profile can be computed numerically using Eq. (C.63).

---

[8] http://www.adastrarocket.com/aarc/VASIMR; http://www.nasa.gov/home/hqnews/2008/dec/HQ_08-332_VASMIR_engine.html.

However, if the acceleration due to the thrust is much smaller than the gravitational acceleration, an approximated closed form solution can be obtained. If the ratio a/g_o between the acceleration due to the thrust and the gravitational acceleration at the orbit altitude is of the order of 10^{-4} or less, then in order to escape the gravitational well of the planet, starting from a circular orbit whose radius is R_1, the thruster must impart a ΔV equal to

$$\Delta V \approx \sqrt{\frac{\mu}{R_1}}. \tag{C.64}$$

Since the required ΔV for escaping from orbit using an impulsive maneuver is given by Eq. (B.32), we see that

$$\Delta V \approx \left(\sqrt{2}-1\right)\sqrt{\frac{\mu}{R_1}}$$

and therefore the low thrust maneuver requires a ΔV that is greater by a factor of

$$\frac{1}{\sqrt{2}-1} \approx 2.4.$$

However, such low values of acceleration are too low to start a Mars mission because they would imply an escape time of the order of 100 days or more.

Even for higher accelerations it is possible to use approximated solutions, such as that reported by Keaton.[9] The maneuver is assumed to be performed so that the circumferential velocity is always very close to the orbital velocity at each altitude , so that the gravitational acceleration is always compensated for by the acceleration caused by the curvature of the trajectory, and the spacecraft moves as if it were in a field-free space (Section C.3.3).

Hence the thrust acceleration must be kept constant, and the direction of the thrust must be between the tangent to the trajectory and the direction perpendicular to the radius. The ΔV required for escaping from the Earth's sphere of influence, starting from a circular orbit with radius R_1 is

$$\Delta V = \sqrt{\frac{\mu}{R_1}}\left(1-\sqrt[4]{\frac{\sqrt{2}R_1^2 a}{\mu}}\right) \tag{C.65}$$

where, as noted, the acceleration a is kept constant.

The escape time t_e is linked to the acceleration and the ΔV by the relationship

$$t_e = \frac{\Delta V}{a} \tag{C.66}$$

---

[9] P.W. Keaton, *Low Thrust Rocket Trajectories*, LA-10625-MS, Los Alamos, 2002.

and thus

$$J = \Delta V^2 \frac{1}{2t_e}. \tag{C.67}$$

It is therefore possible to obtain two relationships that link the acceleration a, the time required to escape, and parameter J, i.e. the propellant consumption.

In Keaton's paper, this approximation is suggested for values of a of up to 0.005 g_o, and this yields escape times from Earth of slightly more than a day.

If a propulsive maneuver is used to achieve a circular orbit upon returning from Mars, a propulsive phase that is symmetric to the escape phase described above can be used.

C.5 ENTERING MARS ORBIT

If the orbit around Mars is circular, then the arrival and departure maneuvers are similar to those for the case of LEO. The choice of a low circular Mars orbit is quite reasonable if the propellant for the MAV is not produced on Mars. However, if ISPP is used on Mars, it may be expedient to enter a highly elliptical orbit on arrival on Mars so that the subsequent manoeuver to depart the sphere of influence of Mars does not require a large quantity of propellant. Clearly, the choice depends mostly on the value of α_e for the available generator. If the specific mass is low then it is better to use the low thrust system more because it has a low propellant consumption, but if the specific mass is high it is better to produce more propellant on Mars and to initiate the interplanetary phase from a highly elliptical orbit.

The ΔV for MOI is

$$\Delta V = \sqrt{\frac{2\mu_M}{r_M + h_p}} \left[1 - \sqrt{\frac{r_M + h_a}{2r_M + h_a + h_p}} \right] \tag{C.68}$$

and the same applies to subsequently departing the sphere of influence of Mars.

If an orbit with a 250 km periareion, a 33,970 km apoareion, and a period of 1 sol is used, then it follows that $\Delta V = 220.5$ m/s. This value is so low that we can employ Eqs. (C.66) and (C.67) to compute t_e and J without large errors. However, Eq. (C.67) was obtained on the basis of an impulsive perigee burn, and therefore may be a rough approximation when applied to a thrusting maneuver lasting several hours. For instance, if $a = 0.005$ m/s^2 then it follows that $t_e = 44,100$ seconds (12.25 hours) and $J = 0.55$ m^2/s^3.

C.6 INTERPLANETARY CRUISE

C.6.1 Optimization of the thrust profile

The design of the interplanetary cruise cannot be performed in closed form, even if (as will be evident below) there are approximated closed form solutions that may be used to start an iterative computation. The solution can be obtained through Eq. (C.63) plus the relevant boundary conditions.

The specific mass of the power generator is usually assumed to be constant for the entire duration of the voyage.[10] This is true for NEP, at least as an approximation. In fact, the specific mass is defined with respect to the power of the propellant jet, which is the power of the generator multiplied by the efficiency of the power converter and the thruster. The efficiency is not exactly constant, because it varies when the specific impulse of the thruster varies.

The situation is more complex in the case of SEP where obviously the power generated decreases, at least as a first approximation, with the square of the distance from the Sun. The assumption of constant specific mass is substituted by the assumption that the specific mass is a known function of the distance between the spacecraft and the Sun, which in the simplest case is

$$\alpha_e = \alpha_0 \left(\frac{|\mathbf{r}|}{R_E} \right)^2 \tag{C.69}$$

where R_E is the radius of the Earth's orbit (1 AU) and α_0 is the effective specific mass of the generator at 1 AU.

SEP has been extensively used for interplanetary probes (albeit in most cases they used trajectories which included gravity assisted maneuvers) and has been proposed for manned Mars mission, at least to carry cargo.

It must be noted that when SEP was used to reach destinations in the outer solar system, this was for flyby missions in which no propulsive maneuver was intended on reaching the destination, and often one or more gravity assist maneuvers were used to gain speed.

In a maneuver to achieve escape velocity or to enter into a planetary orbit, the distance from the Sun can be assumed to be constant. In this situation the equations above remain valid so long as Eq. (C.69) is used to express the effective specific mass. Hence Eq. (C.67) becomes

$$J = \left(\Delta V \frac{R_P}{R_E} \right)^2 \frac{1}{2t_e} \tag{C.70}$$

where R_P is the radius of the orbit of the planet around the Sun, introduced to take into account that the specific mass of the generator increases with increasing distance from the Sun. A further point should be made. During the spiral trajectory within the planet's sphere of influence, the spacecraft may pass several times through the shadow of the planet and so the periods when the thruster is switched off must be taken into account. However, a detailed numerical integration requires knowing all of the details of the starting orbit.

On the contrary, during the interplanetary cruise, the distance from the Sun varies considerably, and in the case of SEP the variation of α_e must be accounted for. If α_e can be assumed to be expressed by Eq. (C.69), the ratio between the mass of the propellant used from time t_1 and t_2 and the initial mass is no longer expressed by Eq. (C.32), where the expression of γ is the usual one, but is

$$\frac{m_p}{m_i} = \sqrt{\frac{1}{2} \int_{t_1}^{t_2} \alpha_0 \left(|\mathbf{a}| \frac{|\mathbf{r}|}{R_E} \right)^2 dt} \,. \tag{C.71}$$

---

[10] G. Genta, and P.F. Maffione, "Sub-optimal Low-thrust Trajectories for Human Mars Exploration," Atti dell'Accademia delle Scienze di Torino, to be published.

By introducing the function of time

$$\mathbf{q} = \mathbf{a}\left(\frac{|\mathbf{r}|}{R_E}\right) \tag{C.72}$$

the integral to be minimized in the optimization procedure is

$$J = \frac{1}{2}\int_{t_1}^{t_2} |\mathbf{q}|^2 (t)\, dt = \frac{1}{2}\int_{t_1}^{t_2} \left(|\mathbf{a}|\frac{|\mathbf{r}|}{R_E}\right)^2 dt. \tag{C.73}$$

In the case of SEP, equations (C.52) and (C.61) become

$$\begin{cases} \ddot{\mathbf{r}} + \nabla U = \mathbf{q}\dfrac{R_E}{|\mathbf{r}|} \\[2mm] \ddot{\mathbf{q}} + \mathbf{q}\nabla(\nabla U) = 0 \end{cases} \tag{C.74}$$

The unknowns are now the position of the vehicle and the vector $\mathbf{q}$, which is in the same direction as the thrust but has a modulus that depends also on the position (or better, on the heliocentric distance). The boundary conditions are the same as in the previous case.

Equation (C.74) can be written in Cartesian coordinates to yield a set of six second order equations or, in the state space, a set of 12 first order equations

$$\begin{cases} \dot{v}_x = -\dfrac{\mu x}{\sqrt{\left(x^2 + y^2 + z^2\right)^3}} + q_x\dfrac{r_E}{\sqrt{x^2 + y^2 + z^2}} \\[4mm] \dot{v}_y = -\dfrac{\mu y}{\sqrt{\left(x^2 + y^2 + z^2\right)^3}} + q_y\dfrac{r_E}{\sqrt{x^2 + y^2 + z^2}} \\[4mm] \dot{v}_z = -\dfrac{\mu z}{\sqrt{\left(x^2 + y^2 + z^2\right)^3}} + q_z\dfrac{r_E}{\sqrt{x^2 + y^2 + z^2}} \\[4mm] \dot{v}_{qx} = -\mu\dfrac{q_x\left(-2x^2 + y^2 + z^2\right) - 3q_y xy - 3q_z xz}{\sqrt{\left(x^2 + y^2 + z^2\right)^5}} \\[4mm] \dot{v}_{qy} = -\mu\dfrac{-3q_x xy + q_y\left(x^2 - 2y^2 + z^2\right) - 3q_z yz}{\sqrt{\left(x^2 + y^2 + z^2\right)^5}} \\[4mm] \dot{v}_{qz} = -\mu\dfrac{-3q_x xz - 3q_y yz + q_z\left(x^2 + y^2 - 2z^2\right)}{\sqrt{\left(x^2 + y^2 + z^2\right)^5}} \end{cases} \qquad \begin{cases} \dot{x} = v_x \\[1mm] \dot{y} = v_y \\[1mm] \dot{z} = v_z \\[1mm] \dot{q}_x = v_{qx} \\[1mm] \dot{q}_y = v_{qy} \\[1mm] \dot{q}_z = v_{qz} \end{cases} \tag{C.75}$$

The initial and final conditions specify the positions x, y, and z, and the velocities $\dot{x}$, $\dot{y}$, and $\dot{z}$ at the initial time ($t = 0$) and the final time ($t = T$).

Although the general three-dimensional problem can readily be solved for an interplanetary transfer between elliptical, non-coplanar orbits, we will start here by dealing with the planar case of circular and coplanar planetary orbits, for which a set of eight first order differential equations is obtained.

C.6.2 Simplified laws $a(t)$

To numerically integrate Eq. (C.75), it is necessary to start from a simplified solution. This step is quite critical for two reasons: the numerical procedure may experience difficulties in obtaining convergence on a solution, and it may converge on a local minimum instead of obtaining an absolute minimum.

C.6.2.1 Approximation 0

Keaton's aforementioned paper provides two highly approximated laws $a(t)$ for two opposite limiting cases.

Define Case 1 as when a is much larger than the gravitational acceleration μ/r^2. This is the case of a fairly fast interplanetary trajectory between circular orbits around the departure and destination planets, such as Earth-Mars trajectories that take less than 4 months, assuming the planetary orbits to be circular.

The optimized acceleration profile $a(t)$ is

$$a(t) = \frac{6L}{t_f^2}\left(1 - \frac{2t}{t_f}\right) \tag{C.76}$$

where $L = R_M - R_E$ is the difference between the radii of the orbits of the two planets and t_f is the total duration of the interplanetary transfer. The thrust is directed radially and pushes outwards in the first half of the trajectory and inwards in the second half, as was seen in Section C.3.4. Clearly, this solution is just an approximation, as can be observed by integrating numerically the equation of motion (C.52) using this thrust profile because in the solution the spacecraft does not precisely achieve the destination orbit.

Parameter J has the value

$$J = \frac{6L^2}{t_f^3}. \tag{C.77}$$

Define Case 2 as when a is much smaller than the gravitational acceleration μ/r^2. This is the case of a slow interplanetary trajectory (e.g., Earth–Mars trajectories requiring more than one year) or of spiral trajectories either to raise the vehicle's orbit around a planet or to escape from its gravitational well.

The optimized acceleration profile $a(t)$ in this case is

$$a(t) = \frac{\sqrt{\mu}}{t_f}\left(\frac{1}{\sqrt{r_1}} - \frac{1}{\sqrt{r_0}}\right) = \text{constant.} \qquad (C.78)$$

The direction of the thrust is tangential to the trajectory, but this solution is also an approximation, as will become evident by integrating numerically in time using Eq. (C.52).

The value of J is

$$J = \frac{\mu}{t_f^2}\left(\frac{1}{\sqrt{r_1}} - \frac{1}{\sqrt{r_0}}\right)^2. \qquad (C.79)$$

As it will be shown later, interplanetary trajectories that will become possible in the near future are too slow for us to apply the solution labeled Case 1, yet too fast for Case 2.

C.6.2.2 Quasi-optimal solution (approximation 1)

A second approximation solution can be obtained by initially assuming that the gravitational acceleration is negligible, as in Case 1. Rewriting Eqs. (C.52) and (C.61) in Cartesian coordinates and indicating as $\mathbf{a}'$ the value of $\mathbf{a}$ for this initial approximation, they become

$$\begin{cases} \ddot{X} = a'_X \\ \ddot{Y} = a'_Y \\ \dot{a}'_X = 0 \\ \dot{a}'_Y = 0. \end{cases} \qquad (C.80)$$

These equations hold only for $\alpha = \text{constant}$, as is the case for NEP. For SEP, the integration of the relevant equations in closed form is not possible and thus no attempt to find a simplified solution will be done. When optimizing trajectories for SEP, the computation will be started using the NEP approximated solution. In particular, in the case of Earth-Mars trajectories the variation of the distance from the Sun is moderate and the solution that is obtained is close enough to the correct one to allow convergence. The boundary conditions are $X(0) = X_0$, $Y(0) = Y_0$, $\dot{X}(0) = v_{x0}$, $\dot{Y}(0) = v_{y0}$, $X(t_f) = X_1$, $Y(t_f) = Y_1$, $\dot{X}(t_f) = v_{x1}$, and $\dot{Y}(t_f) = v_{y1}$. They can be applied for both circular or elliptical orbits so long as they are coplanar.

The equations can be solved in closed form

$$\begin{cases} \dot{a}_X = C_{11}t + C_{12} \\ \dot{a}_Y = C_{21}t + C_{22} \\ X = \dfrac{1}{6}C_{11}t^3 + \dfrac{1}{2}C_{12}t^2 + C_{13}t + C_{14} \\ Y = \dfrac{1}{6}C_{21}t^3 + \dfrac{1}{2}C_{22}t^2 + C_{23}t + C_{24} \end{cases} \tag{C.81}$$

The values of the constants can be obtained from the boundary conditions

$$\begin{cases} C_{11} = \dfrac{6}{T^3}\left[2(x_0 - x_1) + T(v_{x0} + v_{x1})\right] \\ C_{12} = \dfrac{2}{T^2}\left[-3(x_0 - x_1) - T(2v_{x0} + v_{x1})\right] \\ C_{13} = v_{x0} \\ C_{14} = x_0 \end{cases} \tag{C.82}$$

and

$$\begin{cases} C_{21} = \dfrac{6}{T^3}\left[2(y_0 - y_1) + T(v_{y0} + v_{y1})\right] \\ C_{22} = \dfrac{2}{T^2}\left[-3(y_0 - y_1) - T(2v_{y0} + v_{y1})\right] \\ C_{23} = v_{x0} \\ C_{24} = x_0 \end{cases} \tag{C.83}$$

Equation (C.81) defines a trajectory and a thrust profile, so in order to reintroduce the gravitational acceleration it is sufficient to add the latter to the acceleration that is due to the thrust. This yields a new thrust profile $\mathbf{a}(t)$

$$\begin{cases} a_X = C_{11}t + C_{12} + \dfrac{\mu X}{\left(\sqrt{X^2 + Y^2}\right)^3} \\ a_Y = C_{21}t + C_{22} + \dfrac{\mu Y}{\left(\sqrt{X^2 + Y^2}\right)^3} \end{cases} \tag{C.84}$$

Thus both the trajectory and the thrust profile are obtained in closed form, and they are exact in the sense that applying the computed thrust profile will yield the computed trajectory that complies exactly with the boundary conditions. However, the thrust profile obtained in this manner is not optimal, as can be verified by introducing it into Eq. (C.61). In Cartesian coordinates they become

$$\begin{cases} \ddot{a}_x + a_x \dfrac{\partial^2 U}{\partial X^2} + a_x \dfrac{\partial^2 U}{\partial X \partial Y} = 0 \\[4mm] \ddot{a}_y + a_x \dfrac{\partial^2 U}{\partial X \partial Y} + a_x \dfrac{\partial^2 U}{\partial Y^2} = 0. \end{cases} \tag{C.85}$$

It is easy to show that this solution is quasi-optimal for fast trajectories, and becomes increasingly poor as the transfer time increases.

The same solution can be obtained by introducing a constant value of the gravitational acceleration (constant in absolute value and direction) and repeating the relevant computations. It is easy to see that because the second derivatives of the potential appear in Eq. (C.61) it is immaterial whether the latter is zero or constant: its second derivatives are anyway equal to zero.

C.6.2.3 Quasi-optimal solution (approximation 2)

A different approach to obtaining closed form quasi-optimal solutions can be obtained by taking the first element of Eq. (C.54)

$$\ddot{r} = -\frac{\mu}{r^2} + r \left(\frac{d\theta}{dt} \right)^2 + a_r \tag{C.86}$$

and assuming that

$$r \left(\frac{d\theta}{dt} \right)^2 = \frac{\mu}{r^2} \tag{C.87}$$

i.e., that the centrifugal acceleration balances the gravitational acceleration. This allows us to compute the angular velocity $\dot{\theta}$

$$\dot{\theta} = \sqrt{\frac{\mu}{r^3}} \tag{C.88}$$

and then the angular acceleration

$$\ddot{\theta} = -\frac{3}{2} \dot{r} \sqrt{\frac{\mu}{r^5}}. \tag{C.89}$$

Notice that Eq. (C.88) must hold also in the initial and final instants, and therefore this approach is valid only for circular departure and arrival orbits.

The radial equation of motion reduces to

$$\ddot{r} = a_r .$$ (C.90)

In this condition the optimal acceleration profile is that of Case 1, and we can write

$$\begin{cases} r = \dfrac{1}{6} C_1 t^3 + \dfrac{1}{2} C_2 t^2 + C_3 t + C_4 \\ a_r = C_1 t + C_2 \end{cases} .$$ (C.91)

The second Eq. (C.54) thus becomes

$$-\frac{3}{2} \dot{r} \sqrt{\frac{\mu}{r^5}} = -\frac{2}{r} \dot{r} \sqrt{\frac{\mu}{r^3}} + \frac{1}{r} a_\theta$$ (C.92)

and hence

$$a_\theta = \frac{1}{2} \dot{r} \sqrt{\frac{\mu}{r^3}} .$$ (C.93)

Constants C_i can be obtained from the boundary conditions, which for a transfer between circular orbits produce

$$\begin{cases} C_1 = -\dfrac{12L}{T^3} & C_2 = \dfrac{6L}{T^2} \\ C_3 = 0 & C_4 = r_0 \end{cases}$$ (C.94)

where T is the travel time. Functions $r(t)$, $a_r(t)$, $\dot{\theta}(t)$ and $a_\theta(t)$ can therefore be computed easily in closed form.

The above equations hold only for circular, coplanar orbits, but it is easy to obtain the positions and the accelerations relative to the Cartesian xyz reference frame and then to modify them by introducing the actual three-dimensional initial and final positions. Position $z(t)$ and its corresponding velocity can be approximated by linearly interpolating the initial and final positions and velocities. Acceleration in the z direction a_z and the corresponding derivative $\dot{a}_z$ can be assumed to be negligible in this approximated solution.

There is no difficulty in computing $\theta(t)$ and parameter J by simple numerical integration. Again, the trajectory is exact (in the sense that the same results are arrived at by applying the computed thrust profile and then numerically integrating the trajectory), but is only near-optimal since the thrust profile does not satisfy Eq. (C.61) and therefore isn't optimal. However, it is much closer to the optimal thrust profile, even when the previous solution fails in the case of slow and intermediate trajectories.

Consider an Earth–Mars trajectory performed using NEP. In Figure C.5 the value of $\sqrt{J}$ computed using the various approximations is plotted as a function of the travel time T

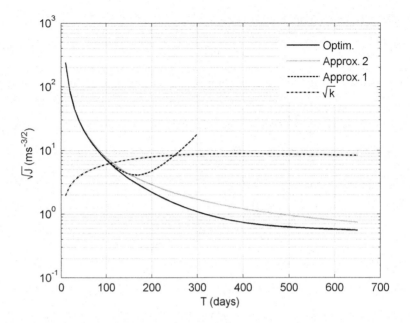

Figure C.5 Value of $\sqrt{J}$ computed using various approximations as a function of the travel time T.

spanning the range 10–650 days.[11] The plot also reports the results obtained by numerically integrating Eq. (C.63).

From the plot it is evident that:

- Approximation 1 yields fairly good results if $T < 150$ days, then becomes ever less efficient.
- Approximation 2 yields fairly good results across the entire range of T.
- The optimized solution obtained by starting from Approximation 2 follows a pattern similar to the latter but, as expected, yields a lower value of J.

Also the value of

$$\sqrt{k} = \sqrt{\frac{1}{2}\int_0^T \left(\frac{\mu}{r^r}\right)^2 dt}$$

(C.95)

is reported in the figure. This is similar to J, but is computed using the gravitational acceleration μ/r^2 instead of the acceleration due to the thrust. Where $J \gg k$ the thrust affects the trajectory much more than the attraction by the Sun and Approximation 0, Case 1 (and

---

[11] R_{33} is dimensionally a specific power which is measured in SI units as $W/kg = m^2/s^3$. In the following, $\sqrt{J}$ will be measured in $ms^{-3/2}$.

also Approximation 1) holds well. But in situations where $J \ll k$, the thrust affects the trajectory much less than the attraction by the Sun and it is possible to use Approximation 0, Case 2. Actually, in the case of exceedingly short or long journeys k is either much smaller or much larger than J, while in all other cases their values are of the same order of magnitude, showing that both cases of Approximation 0 are inapplicable. At any rate, Approximation 2 has proved to be a good starting point for the optimization procedure.

As a final consideration, the optimization procedure was performed by developing a relationship between the angle formed by the two planets around the Sun, and the travel time that is derived from Approximation 2. Clearly it is possible to assume that these two parameters can change independently and to say that the relationship here defined imparts an unnecessary constraint on the optimization. To investigate this point, a number of computations were performed by keeping time T constant and slightly varying the angle between the planets, with the results showing that the computed value of J was very close to the absolute minimum. Since the exact value of this angle will be set by the launch and arrival dates when studying a given mission in detail, this point will not be investigated further and the value of the angle obtained from Approximation 2 will be retained.

C.6.3 Earth-Mars transfer

Solving Eq. (C.63) for both NEP and SEP with different travel times produced the values of $\sqrt{J}$ reported in Figure 6.13. The plot expresses the lowest possible values of $\sqrt{J}$ for the above mentioned interplanetary transfer, and does not depend on the spacecraft characteristics. To study in greater detail one of these possible transfers we assume a 160 day transfer. The trajectories are plotted in Figure C.6. They are clearly quite close to each other. Figure C.7 reports the time histories of the acceleration $a(t)$, the angle between the thrust and the radius $\theta_T(t)$, the radius $r(t)$, and the polar angle $\theta(t)$.

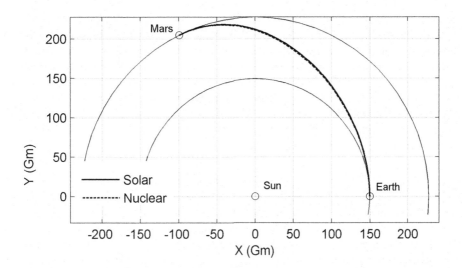

Figure C.6 Optimal trajectories for a 160 day Earth–Mars transfer using NEP and SEP.

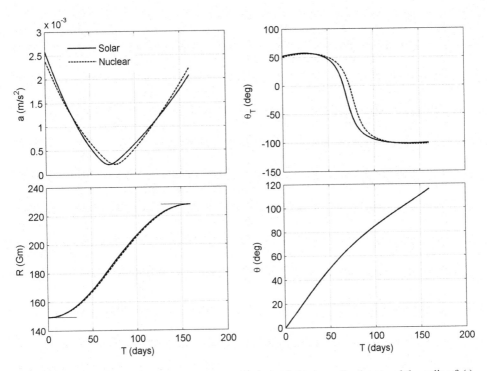

Figure C.7 Time histories of the acceleration $a(t)$, the angle between the thrust and the radius $\theta_T(t)$, the radius $r(t)$, and the polar angle $\theta(t)$.

It is evident that the thrust profile (or better, the acceleration profile) is much more different in the two cases than the trajectory. A first guess would be that the spacecraft tends to remain for a longer time closer to the Earth's orbit in the case of solar propulsion because, by doing so, it better exploits the energy from the Sun, but this is incorrect. The result of the optimization procedure is that the thruster works with a higher power when closer to the Sun, producing a higher thrust (this is not strictly required, because a large power can be employed to produce a small thrust simply by increasing the specific impulse, but that is a result of the optimization process) and it pushes less at larger distances. This has the consequence of raising the heliocentric orbit more rapidly, which is the opposite effect to that intuitively assumed.

The values of J are $J = 15.58$ m²/s³ for SEP and $J = 10.54$ m²/s³ for NEP. In order to make the trajectory possible the condition

$$\gamma = \sqrt{\alpha_0 J} < 1$$

must be met, therefore the maximum value of the specific mass of the power generator that allows a 160 day transfer is 64.2 kg/kW in the case of SEP and 94.9 kg/kW in the case of NEP. These values are only theoretical ones, since a value of $\gamma = 1$ means that the spacecraft is all generator and propellant, with no margins for the structure and the payload. To allow the interplanetary transfer, the specific mass must be smaller than the computed values.

C.7 WHOLE INTERPLANETARY JOURNEY

The interplanetary voyage consists of at least three phases. Firstly the acceleration around Earth to reach escape velocity, then the interplanetary transfer, and finally the braking into orbit around the destination planet. The optimization of the spacecraft must take into account all aspects of the entire mission. If the return journey is performed without resupplying at the destination planet, six phases must be considered for optimization. In the following, however, only the outbound journey is considered because in order to take into account the whole two-way journey many design choices must be made: what is the mass to be left on the planet, what must be taken back, etc.

Consider a one-way interplanetary journey made of the above mentioned three phases. To optimize the entire voyage

$$\gamma_{tot} = \sqrt{\gamma_1^2 + \gamma_2^2 + \gamma_3^2} = \sqrt{\alpha_0 \left(J_1 + J_2 + J_3 \right)} \tag{C.96}$$

must be minimized. If the voyage starts from Earth and ends at a destination planet, the integral J_1 is the same for SEP as it is for NEP and it must be computed by using Eqs. (C.65), (C.66), and (C.67), J_2 must take into account that the specific mass of the generator increases (if, as in the case of Mars, the direction of travel is beyond the heliocentric radius of the Earth's orbit) while the journey proceeds, while in J_3 the specific mass is constant and its value is that at the destination planet. However, strictly speaking this is not true, since the distance of Mars from the Sun is variable due to its elliptical orbit. If the orbit about the planet is a LMO, then J_3 is computed like J_1. If a highly elliptical Mars orbit is used, J_3 can be approximated by using Eq. (C.68).

Equations (C.65), (C.66), and (C.67) allow us to compute the law $J(t_e)$ in an implicit form, by stating a number of values of a and computing both J and t_e. This law depends only on the parameters of the planet and on the starting orbit, even if, as in the case of SEP, this depends also on the planet's heliocentric distance.

To optimize the entire trajectory the value of

$$J_{tot} = J_1 + J_2 + J_3 \tag{C.97}$$

must be minimized. The goal of this optimization is to find the minimum value of J_{tot} as a function of the total transfer time $T_{tot} = T_1 + T_2 + T_3$.

As a first step, the durations of the two planetocentric phases that allow us to obtain the minimum of J_{tot} at fixed T_{tot} must be obtained. Because function $J_{tot}(T_1, T_3)$ can be obtained numerically for each value of T_{tot}, the minimum value is readily found.

As an example, Figure C.8 shows the surface $J_{tot}(T_1, T_3)$ for a 180 day Earth–Mars transfer using a NEP spacecraft that starts and ends in a circular orbit at an altitude of 500 km above each planet. The plot was obtained by starting with a maximum value of the acceleration equal to 0.15 m/s² (for both the first and third phases, points with minimum values of T_1 and T_3) and then decreasing both accelerations until the duration of the relevant phases is equal to one third of the total travel time (60 days in the figure).

The contours of the surface are reported in Figure C.8b. A descent line that starts from the point of maximum acceleration (minimum times T_1 and T_3) is then computed. This leads to the optimal values of the durations of the first and third phases. In the case of this surface, the relevant

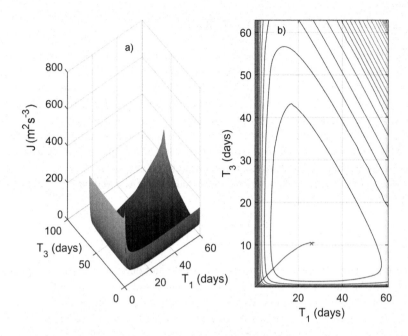

Figure C.8 Surface $J_{tot}(T_1, T_3)$ for a 180 day Earth-Mars transfer using NEP. (a) Surface. (b) Contour lines, and descent line that starts from the point corresponding to the minimum values of T_1 and T_3 and leads to the minimum of the surface.

values are $T_1 = 26.1$ days and $T_3 = 10.3$ days. It is clear that the shape of the surface allows this procedure to converge to the absolute minimum and that the bottom of the 'cup' is almost flat, which means that even non-negligible displacements of T_1 and T_3 from the optimal values yield quite small displacements of the value of J_{tot} from its minimum. The optimal conditions are quite insensitive to small variation of the durations of the first and third phases. Increasing the value of T_1 (at equal total time) may reduce, almost at no cost, the duration of the journey for the astronauts if they board the spacecraft as it completes the first phase of the journey.

The final results for an interplanetary transfer between circular orbits at an altitude of 500 km above the surfaces of Earth and Mars are plotted in Figure 6.17, where the minimum value of $\sqrt{J_{tot}}$ is reported as a function of T_{tot}, together with the values of T_1 and T_3 which allow us to reach it. By comparing the results obtained for SEP and NEP, it is clear that in the former case the optimal trajectory means spending a longer time around Mars, where the available power is smaller and thus the braking maneuver must be more gradual, and a shorter time close to Earth, where the larger available power allows the initial orbit to be raised more rapidly.

Note that this approach makes the following approximations:

- Planetary orbits have been assumed to be circular.
- Interruptions of the thrust when the spacecraft is in the shadow of a planet were neglected.
- The thruster has been assumed to be ideal, since there is no limitation to its specific impulse.

The first and the third simplifications will be eliminated in the ensuing sections.

C.8 OPTIMIZATION OF THE SPACECRAFT

Once J_{tot} has been computed, the non-dimensional parameter γ

$$\gamma = \sqrt{\alpha_e J_{tot}}$$ (C.98)

can be computed. In the case of SEP, α_0 must be substituted for α_e. The optimal mass distribution is obtained from Eq. (C.30) and so

$$\frac{m_p}{m_i} = \gamma, \quad \frac{m_w}{m_i} = \gamma(1-\gamma), \quad \frac{m_l + m_s}{m_i} = (1-\gamma)^2.$$

Operating in a way which is similiar to what we did for high thrust rockets, we can assume that

$$m_s = A + K m_p$$ (C.99)

or, in an even simpler way

$$m_s = K m_p$$ (C.100)

where K is the tankage factor.

It follows that

$$m_i = m_p (1 + K) + m_w + m_l$$ (C.101)

and thus it is possible to define a 'gear ratio'

$$R = \frac{m_i}{m_l} = \frac{1}{1 - \gamma(1 + K) - \gamma(1 - \gamma)}.$$ (C.102)

It is clear that γ must be smaller than 1, as otherwise the power generator will be too heavy (α too high) to perform the mission.

Once α_e is stated, it is possible to compute the required power

$$\frac{P}{m_i} = \frac{m_w}{\alpha_e m_i} = \frac{\gamma(1-\gamma)}{\alpha_e}$$ (C.103)

and this enables the ejection velocity v_e (or the specific impulse I_s) to be computed using Eq. (C.37).

Consider for example, a 120 day interplanetary trajectory from the Earth's heliocentric orbit to the orbit of Mars with circular, coplanar planetary orbits. The value $\sqrt{J} = 5.28$ ms$^{-3/2}$ is obtained through numerical integration. Assuming that $\alpha_e = 0.01$ kg/W $= 10$ kg/kW (which is a reasonable value for the near future) it follows that

$$\gamma = \frac{m_p}{m_i} = 0.528, \quad \frac{m_w}{m_i} = 0.249, \quad \frac{m_i + m_s}{m_i} = 0.222.$$

The optimum specific impulse at the starting instant is

$$\left(I_s\right)_i = 1,160\,\text{s}. \tag{C.104}$$

The propellant throughput $\dot{m}$ can be easily computed as

$$\dot{m} = -\frac{ma}{gI_s} = -\frac{m^2 a^2}{2P}. \tag{C.105}$$

As a consequence, the mass of the spacecraft can be computed as a function of time by using the differential equation

$$\frac{dm}{dt} = -\frac{m^2 a^2}{2P} - \frac{m^2\left(a_x^2 + a_y^2 + a_z^2\right)}{2P}. \tag{C.106}$$

The integration is easily performed by introducing the law $a(t)$ obtained from the optimization of the trajectory in Eq. (C.106). In the case of NEP the power P is constant, whereas for SEP it decreases with the square of the heliocentric distance.

Figure C.9 plots the specific impulse and the mass throughput relative to the initial mass $\dot{m}/m_i$ as functions of time.

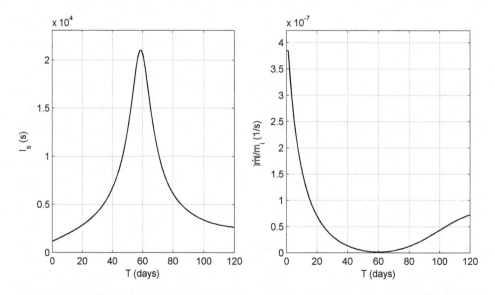

Figure C.9 Specific impulse and propellant throughput as functions of time.

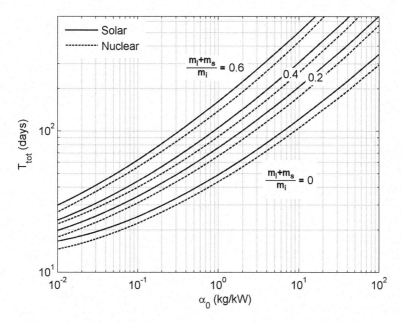

Figure C.10 A plot linking together the travel time T_{tot}, the payload fraction of the spacecraft (or better, $(m_l + m_s)/m_i$), and the specific mass α_e (α_0 in the case of SEP) of the power generator of the spacecraft. The plot refers to Earth-Mars transfers with circular, coplanar planetary orbits and the initial and final spacecraft orbits at altitudes of 500 km above the surfaces of the planets, performed using both SEP and NEP.

Once the minimum of $\sqrt{J_{tot}}$ has been obtained, it is easy to find a relationship linking together the travel time T_{tot}, the payload fraction of the spacecraft (or better, $(m_l + m_s)/m_i$), and the specific mass α_e of the power generator of the spacecraft. Such a plot is reported in Figure C.10 for both SEP and NEP.
Remembering that

$$\frac{m_l + m_s}{m_i} = \left(1 - \sqrt{J_{tot}\alpha_e}\right)^2 \tag{C.107}$$

it is possible to compute the value of J that allows us to take to Mars a payload fraction $(m_l + m_s)/m_i$ with a given specific generator mass

$$\sqrt{J_{tot}} = \frac{1}{\sqrt{\alpha_e}}\left(1 - \sqrt{\frac{m_l + m_s}{m_i}}\right). \tag{C.108}$$

The travel time can be computed simply by inverting the relationship $\sqrt{J_{tot}}(T_{tot})$ obtained above. The lower curve, labeled $(m_l + m_s)/m_i = 0$ must be considered as an absolute minimum time for the Earth–Mars transfer, but this cannot be achieved because the vehicle would consist wholly of propellant and power generator, with no allowance for either the payload or the structure. The other lines relate to more realistic situations.

C.9 ACTUAL CASE: FINITE SPECIFIC IMPULSE

As already stated, the optimization was performed by assuming that the power is kept constant (or will decrease with the square of the heliocentric distance in the case of SEP) and that the specific impulse is increased in order to reduce the thrust. As Figure C.9 shows, this may require attaining values of specific impulse that are considerably above what can be achieved by current technology, or indeed, by thrusters that we can reasonably predict.

If values in excess of those allowed by the available technology are required, there is no other chance than to choose a non-optimal thrust profile, reducing the performance of the system. The strategies which may be implemented are:

1. Switching off the thruster when the desired specific impulse exceeds the feasible one (this strategy has already been mentioned).
2. Maintaining the acceleration at the required value by reducing the power, so that the same acceleration is achieved with the maximum possible specific impulse.
3. Keeping the specific impulse constant at the maximum possible specific impulse , without reducing the power.

Strategy 2 has no influence on the trajectory because the spacecraft has, instant by instant, the same acceleration as in the optimized case. But it causes an increase in the propellant consumption because for some portion of the journey the thruster works at a specific impulse that is less than optimal and then, to provide the same thrust, it requires a larger propellant throughput.

Strategy 1 fails to reach the required target, since the total energy imparted by the thruster is less than that computed in the optimization process. The total fuel consumption is only slightly smaller, since the thruster is switched off where the thrust is low and the specific impulse is high, so the throughput is very low. To reach the required target it is possible to multiply the thrust by a constant larger than 1, which can be computed through an iterative procedure.

Strategy 3 overshoots the required target because the total energy imparted by the thruster is larger than that computed in the optimization process, and the total fuel consumption is much larger. To reach the target it is possible to multiply the thrust by a constant smaller than 1, which can be computed through an iterative procedure.

Consider for instance, the Earth–Mars 120 day transfer between circular orbits that was described in the previous section and apply Strategy 2. Assuming again that $\alpha_e = 0.01$ kg/W $= 10$ kg/kW and that the thruster cannot supply a specific impulse larger than 6,000 seconds (less than one third of that for the optimal trajectory), the differential equation that allows us to compute the mass of the vehicle is

$$
\begin{cases}
\dfrac{dm}{dt} = -\dfrac{m^2\left(a_x^2 + a_y^2 + a_z^2\right)}{2P} & \text{if } I_s \leq I_{s\max} \\[4mm]
\dfrac{dm}{dt} = -\dfrac{m\sqrt{a_x^2 + a_y^2 + a_z^2}}{gI_{s\max}} & \text{if } I_s > I_{s\max}.
\end{cases}
\tag{C.109}
$$

The final mass ratio is thus $m_f / m_i = 0.4661$, yielding

$$\frac{m_p}{m_i} = 0.534, \quad \frac{m_l + m_s}{m_i} = 0.217$$

and in this case the limitation placed upon the specific impulse produces a very small increase of the propellant mass and a reduction of the payload. Figure C.11 plots the specific impulse, the propellant throughput, thrust, and power.

The effect of the limitation of the specific impulse can be expressed by introducing a value of J computed as

$$J = \frac{1}{\alpha_e} \left(\frac{m_p}{m_i} \right)^2 . \tag{C.110}$$

In the present case, its value is $J = 28.5088$ against the value $J = 27.9212$ computed without a limitation on the specific impulse. By introducing a value of J that is corrected to take into account this limitation in the optimization procedure for the whole interplanetary transfer, it is possible to adjust the planetocentric phases to this new condition. It is expected that the

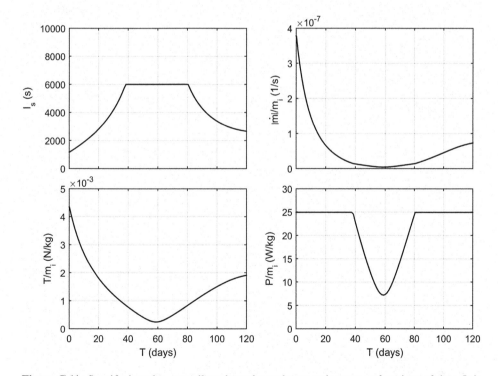

Figure C.11 Specific impulse, propellant throughput, thrust, and power as functions of time. It is the same case as Figure C.9, but with the specific impulse limited to a value of 6,000 seconds.

reduction in performance on the entire journey will be smaller than that for the interplanetary phase, because the specific impulse during these parts of the journey is much smaller.

It may seem surprising that such a limitation has so little an effect on the fuel consumption, but the one considered is a peculiar case. The increase of J depends on the value of α_e (it can be expected to be larger when α_e is smaller, since with more available power the optimization tends to a higher specific impulse) and on the duration of the journey (the effect increases with journey time, since going slowly involves a lower thrust and therefore a higher specific impulse).

The increase of $\sqrt{J}$ (or better, of $\sqrt{J} / \sqrt{J_0}$, where J_0 is the value of J with no limitations placed on the specific impulse) can be plotted for a number of values of T and α_e as a function of the maximum specific impulse, as is done in Figure C.12.

From the plot it is clear that the increase of fuel consumption may be large for slow transfers and for very lightweight power generators. Advanced cargo ships suffer in particular from this problem, whilst relatively backward passenger ships (i.e. with a heavy generator) have no problem at all. As a conclusion it may be asserted that it is useless to employ a very advanced (i.e. lightweight) generator, if an equally advanced (i.e. a high specific impulse) thruster is not used. The advantage of slow transfers may also be mitigated by this effect. It is to be expected that the reduction of the power with increasing heliocentric distance may slightly reduce this effect for SEP.

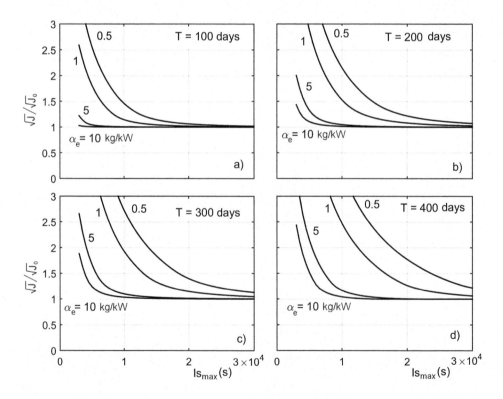

Figure C.12 $\sqrt{J} / \sqrt{J_0}$ as a function of the maximum value of the specific impulse, for a number of values of T and α_e.

C.10 IMLEO

A first approximation computation of the initial mass in LEO can be made from the mass of the various spacecraft and items to be carried to Mars.

As was done for impulsive propulsion, let us assume a split mission with one (or more) cargo ships following a slow trajectory and one crew ship carrying people:

- The crewed spacecraft enters Mars orbit, where it remains while the crew transfers to a lander that arrived with the cargo ship. It then acts as a return spacecraft, and at the end of the entire mission it enters orbit around Earth. The launch mass into Earth orbit is m_{VEH} plus the power generator, plus the other structures incorporated in the tankage factor. A detailed design of the spacecraft must be made in order to clarify whether this includes the structures which held the propellant during the outbound journey, and also to determine the value of the tankage factor.
- The cargo ship brings to Mars orbit:

  - The lander, which lands on Mars using an aeroshell and other devices. Its mass on the planet is m_{PLAN}.
  - The outpost, which then lands on Mars using an aeroshell. Its mass on the planet is m_{OUT}.
  - The Mars Ascent Vehicle (MAV) which lands on Mars dry if the propellant for ascent is produced on Mars, or partially dry if some is carried from Earth, or even together with the propellant carried from Earth. Its mass on the planet is $m_{MAV_{plan}}$.

Operating the mission this way, the propellant for the return journey is carried by the crew spacecraft on a fast trajectory, but it is more expedient to carry it by cargo ship. This option is not considered here, to avoid transferring the propellant from one vehicle to the other and also the complexity of keeping a large quantity of a propellant like argon (which is not as critical as liquid hydrogen, but anyway must be stored at low temperature) for a long time in Mars orbit.

Consider a mission with a cargo spacecraft reaching Mars in 400 days and a crew ship traveling in 210 days (although possibly the crew would board their ship about 30 days after it left LEO and so would travel for only 180 days). Assuming circular planetary orbits and no limitations on specific impulse, the values of J are $J_c = 5.90$ and $J_p = 21.10$ for the cargo and crew ships respectively with a NEP system, and $J_c = 8.61$ and $J_p = 30.56$ with a SEP system. For a NEP vehicle, assume $\alpha_e = 8$ and $K = 0.2$, aerobraking for landing all payloads on the surface, and the propellant for the MAV ascent is produced completely on Mars. This produces the following gear ratios: outbound $R_{1c} = 1.757$ and $R_{1p} = 3.775$, landing $R_{3c} = R_{3p} = 1.800$, and return $R_s = 3.775$. The overall ratios are $R_{RET} = 14.25$, $R_{MAV} = 3.16$, $R_{LAND} = 3.16$, and $R_{OUT} = 3.16$. Assuming a total mass of 150 t for all vehicles ($m_{VEH} = 30$ t, $m_{LAND} = 30$ t, $m_{MAV_{dry}} = 20$ t, and $m_{OUT} = 70$ t) the IMLEO of the cargo vehicle is 379 t and the crew ship is 427 t, giving a total IMLEO of 807 t. The overall gear ratio is therefore 5.38. However, a small decrease of α_e, or using the cargo ship to bring to Mars the propellant for the return journey, will greatly improve the situation.

Consider now a SEP vehicle where we assume $\alpha_e = 5.5$ and $K = 0.2$ and retain the same values for all the other parameters. The following gear ratios are obtained: outbound $R_{1c} = 1.759$ and $R_{1p} = 3.758$, landing $R_3c = R_{3p} = 1.800$, and return $R_5 = 3.758$. The overall gear ratios are $R_{RET} = 14.12$, $R_{MAV} = 3.17$, $R_{LAND} = 3.17$, and $R_{OUT} = 3.17$. Assuming a total of 150 t for all vehicles with the same breakdown as above, the IMLEO of the cargo vehicle is 380 t and the crew ship is 424 t, giving a total IMLEO of 804 t. The overall gear ratio is 5.36.

The two cases are almost identical, because the use of SEP leads to a lower value of J but to a higher value of the specific mass of the generator, and these two features almost exactly balance each other.

It must be noted that, also in this case, a small decrease of α_e or the use of the cargo ship to bring to Mars the propellant for the return journey would much improve the situation.

Symbols

a	ratio between the thrust and the mass of the spacecraft
h	height of the orbit
m	mass of the spacecraft
q	propellant throughput
r, θ	polar coordinates
r	vector defining the position of the spacecraft
t	time
t_f	final time
u	control input vector
v_e	exhaust velocity
xyz	Cartesian coordinates
x	state vector
I_s	specific impulse
J	cost function
K	tankage factor
P	power
R	gear ratio
R_E	radius of Earth orbit
T	thrust
U	potential of the gravitational field
α	mass/power ratio of the power generator
α_e	effective mass/power ratio of the power generator
γ	optimization parameter
η_t	thruster efficiency
μ	gravitational parameter
ΔV	velocity increment

Appendix D: Locomotion on Mars

The mobility on the planet is granted by vehicles which, in most cases, either move on the surface or fly in the atmosphere. The basic theory of wheeled ground vehicles and of atmospheric vehicles is summarized in this section.

D.1 MOBILITY ON WHEELS

D.1.1 Elastic wheels

Although wheels are best suited to prepared ground, their simple mechanical design and control also makes them a good choice for off-road locomotion, particularly on dry ground.

If a rigid wheel (which can be thought as a short cylinder) rolls on a rigid flat ground, the contact occurs along a line and the pressure that it exerts on the ground is infinitely large; an impossible result. The high contact pressure at the wheel-ground interface causes deformations of both bodies that act to substantially reduce the contact pressure.

The vehicle-soil contact must therefore be considered as the interaction between compliant bodies. Even if their compliance is low, the relevant phenomena can be understood only by taking it into account. Three cases can be considered:

1. A stiff wheel rolling on compliant ground.
2. A compliant, possibly elastic, wheel rolling on stiff ground.
3. A compliant wheel rolling on a compliant ground.

As a general rule, the resistance to motion that is encountered by the wheel results from the energy dissipation in both the wheel and the ground. Since the wheel can be designed to minimize dissipations and there is little that can be done to alter the properties of the ground, it is expedient that all deformations are concentrated in the wheels (Case 2).

Case 2 is typical of automotive technology, where low stiffness (mostly pneumatic) wheels are used on prepared roads that are covered with tarmac or concrete and are therefore quite stiff. Case 3 occurs in off-road locomotion, where the ground is compliant and the wheels are made

G. Genta, *Next Stop Mars*, Springer Praxis Books, DOI 10.1007/978-3-319-44311-9

as compliant as possible. Case 1 is considered the worst situation because rigid wheels cause much permanent deformation in the ground and so the resistance to locomotion is large.

As already stated, the wheel must be compliant to prevent the ground from deforming excessively. This is usually achieved by building the wheel in two parts, where a stiff (usually metallic) hub is inset into a tire that supplies the required elasticity.

When a wheel with a pneumatic or an elastic tire rolls on a prepared surface like concrete or tarmac, the deformation of the wheel is large and the ground can be regarded as a rigid surface. When the same wheel rolls on a natural surface, both of the objects in contact must be considered as compliant. On some robotic rovers, wheels with no compliant tire are used. In this case the deformations are localized only in the ground.

In standard vehicles on Earth, both on and off road, the rigid structure of the wheel (made by the disc and the rim) is thus surrounded by a compliant element comprising the tire and the tube. The latter can be absent in tubeless tires, in which the tire fits airtight on the rim and the carcass directly contains the air. The tire is a complex structure that has several layers of rubberized fabric, with a large number of cords running in the direction of the warp and only a few in the direction of the weft. The number of plies, their orientation, the formulation of the rubber, and the material of the cords are widely variable. These are the parameters that give each tire is particular characteristics.

In addition to its main function of distributing the vertical load on a sufficiently large area, the tire has the secondary function of ensuring adequate compliance to absorb the irregularities of the road. It is essential that the compliance in different directions is suitably distributed. The tire must be compliant in the vertical direction, while being stiffer in circumferential and lateral directions. This second function of the compliance of a tire becomes increasingly important as the speed of travel increases. For the speeds at which all present robotic rovers operate, rigid wheels are adequate in this respect.

When designing the Lunar Roving Vehicle (LRV) for the Apollo missions, it was apparent that some sort of elastic tires would be required, but pneumatic or solid rubber tires were discarded, mainly in order to reduce the mass. The designers opted for metal elastic tires resembling those that were widely tested at the end of the nineteenth century, when alternatives to pneumatic wheels were sought. Tires made by an open steel wire mesh, with a number of titanium alloy plates acting as treads in the ground contact zone, were then built. Inside the tire there was a second smaller more rigid frame that served as a stop to avoid excessive deformation under high impact loads (Fig. 9.9a).

It remains to be decided whether specifically designed pneumatic tires can be utilized on Mars, but most designs for both robotic and human-carrying rovers envisage using elastic, non-pneumatic wheels. Non-pneumatic wheels have been recently developed for automotive (in particular military) applications, like those shown in Figs. 9.5 and 9.10. Michelin patented a non-pneumatic wheel that is called a 'tweel' (tire + wheel).

D.1.2 Resistance to motion

Consider a vehicle moving at constant speed on a straight trajectory on level ground. The forces that must be overcome to maintain that speed are the rolling resistance and, if there is an atmosphere, aerodynamic drag. The former is usually the most important form of drag at low speed; the second is important only at high speed. In the case of the very thin atmosphere of Mars, the aerodynamic drag is usually assumed to be negligible.

If the ground is not level, the component of weight acting in a direction parallel to the velocity V (the grade force) must be added to the resistance to motion. It may become far more important than all other forms of drag, even for moderate values of the grade.

Neglecting aerodynamic forces, the total resistance to motion, or road load, can be written in the form

$$R = mg\left[\cos(\alpha)\left(f_0 + KV^2\right) + \sin(\alpha)\right] \tag{D.1}$$

where f_0 is the rolling coefficient at low speed, K is a coefficient expressing the increase of rolling resistance with the speed, and α is the slope of the ground (measured positive going uphill).

The road load is usually expressed as

$$R = A + BV^2 \tag{D.2}$$

where

$$A = mg\left[f_0\cos(\alpha) + \sin(\alpha)\right]$$

and

$$B = mgK\cos(\alpha).$$

The power needed to move at constant speed V is simply obtained by multiplying the road load given by Eq. (D.2) by the value of the velocity

$$P_r = VR = AV + BV^3. \tag{D.3}$$

At low speed the power required for motion is

$$P_r = mgV\left[f_0\cos(\alpha) + \sin(\alpha)\right] \approx mgV\left(f_0 + i\right) \tag{D.4}$$

where the last expression holds if $i = \tan(\alpha)$, the slope of the ground, is small.

D.1.3 Forces transmitted in longitudinal and transversal direction

Motion at constant speed is possible only if the power available at the wheels at least equals the required power given by Eq. (D.3). This means that the engine must supply a sufficient power (taking into account also the losses in the transmission) and that the road-wheel contact be able to transmit it.

In fact, the wheel has two functions. In addition to supporting the vehicle, it must produce forces in the wheel-road contact plane. The latter are usually subdivided into *longitudinal forces* (to accelerate or slow the vehicle) and *lateral* or *cornering forces* (to control the trajectory). Although in the normal use of a wheeled vehicle there is no macroscopic slippage of the wheels, longitudinal forces are generated only in the presence of a longitudinal slip, which occurs when the wheel rotates at a speed that is slightly faster (driving forces) or slightly slower (braking forces) than the speed in 'pure rolling.' In the same way, to produce lateral forces the wheel mid-plane must make a (small) angle with the direction of motion. The deformation of the wheel, and sometimes the deformation of the ground, allow the wheel to work in these longitudinal and lateral slip conditions without producing a global slip in the contact area.

The trajectory control of wheeled vehicles is usually performed as follows. Either the driver or the automatic trajectory control system operates the steering, causing some wheels to work with a sideslip and generate lateral forces. These forces cause a change of attitude of the vehicle and then a sideslip of all wheels, so that the resulting forces bend the trajectory. However, the linearity of the behavior of the tires, at least for small sideslip angles, the very low value of the lateral slip, and the short time delay with which the wheels respond to changes in the sideslip angles, give the driver the impression that the wheels are in pure rolling and the trajectory appears to be determined by the directions of the midplanes of the wheels.

When high values of the sideslip angles are reached, the average driver has the impression of losing control of the vehicle; much more so if this occurs abruptly. This impression is verified by the fact that in normal road conditions the sideslip angles become large only when approaching the limit lateral forces.

This way of controlling a wheeled vehicle is typical of motor vehicles used on hard surfaced roads, but there is another possibility. The torque about the z axis causing the vehicle to rotate and to assume the required attitude can be produced by differential traction on the wheels of the same axle, instead of by steering some of the wheels. In this case, which is usually referred to as *slip steering*, no steering wheel in the classical sense may be present.

This way of controlling the trajectory may be implemented together with the more usual means, as in the case of many VDC (Vehicle Dynamics Control) systems where the driver chooses the trajectory by steering the front wheels while the device maintains the vehicle in the required trajectory by differentially braking the left and right wheels according to suitable control algorithms. In other cases this may be the main means of controlling the trajectory, such as in some wheeled earth-moving machines and, above all, in tracked vehicles. Many three-wheeled small robots work in this way too, but the third wheel is either an omnidirectional wheel or a swiveling wheel.

The steering function can thus be implemented in the following ways:

- Slip steering (Fig. D.1a).[12] There is no particular steering mechanism, but the driving wheels can produce differential traction on the two sides of the vehicle. This can be achieved by using independent motors in the wheels in such a way that two motors actuate the wheels of the two sides independently, or by using a differential gear. The brakes can also be used in differential mode to achieve some slip steering.

---

[12] Sometimes the term *skid steering* is used instead of slip steering. It should be avoided, since this way of steering is due to longitudinal and lateral slip of the wheels, but only in extreme cases does it produce actual skidding (i.e. global slipping) of the wheels on the ground.

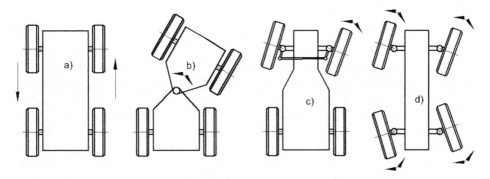

Figure D.1 (**a**) Slip steering. (**b**) Articulated steering. (**c**) Coordinated steering. (**d**) Independent steering.

> Very small trajectory curvature radius can be achieved by rotating backwards the wheels on one side, and on-the-spot turning is possible by using equal and opposite rotation speed.

- Articulated steering (Fig. D.1b). One or more whole axles, each carrying the wheels of both sides, are articulated and can steer under the control of one or more actuators. The body can be made of two or more subunits, each carrying one or more axles, or the wheels can be carried by bogies, pivoted under the body. As an alternative, instead of using actuators to rotate one of the parts of the vehicle relative to the other, it is possible to resort to differential longitudinal forces, like in slip steering. In this case, however, lower sideslip angles of the wheels usually occur.
- Coordinated steering (Fig. D.1c). The two wheels of the same axle steer about two different steering axes (kingpin axes), but their steering angles are coordinated by a mechanical linkage. If more than one axle is steering, the steering angles of the various axles can be independent or (seldom) coordinated by another mechanical linkage. Most automotive vehicles have a steering of this kind.
- Independent steering (Fig. D.1d). Each wheel of one or more axles can steer independently about a steering (kingpin) axis, under the action of an actuator. If large steering angles can be achieved and the control system is sufficiently flexible, then any kind of trajectory is possible, including turning on the spot, moving sideways, etc. A steering of this kind was employed on the Apollo LRV and in most robotic rovers used on Mars.

D.1.4 Power transmission

In most cases, the motor is connected to the wheels through a transmission system that incorporates a number of gear wheels and possibly joints and various other mechanical or electromechanical devices. Taking into account the efficiency η_t of the transmission, the power that the motor must supply is

$$P_m = \frac{P_r}{\eta_t}.$$ (D.5)

The most common type of motors considered for planetary vehicles and rovers are electric motors. A very interesting layout is that of using a motor in each wheel, without the use of transmission shafts and differential gears, which are needed when a motor drives the two wheels of an axle or, even more, all the wheels of a vehicle. However, in most cases it is not possible to connect directly the wheel to the motor shaft because the speed at which the wheels must turn is much lower than that at which the motor is at its best. A reduction gear is thus usually interposed between the motor and the wheels. If the speed range of the vehicle is large, it may be impossible to have the motor working in good conditions through the whole speed range, and a variable ratio transmission, either continuously variable (CVT) or with a number of different ratios, is required. Low speed, high torque, electric motors (known as torque motors) are a very interesting alternative to using reduction gears.

The low gravity on Mars reduces the power requirements for locomotion, and the ratio between the power required to move a given vehicle on Earth and that required by the same vehicle at the same speed and on similar ground on Mars is

$$\frac{P_m}{P_e} = \frac{g_m}{g_e} = 0.38.$$ (D.6)

Since this figure has been obtained by neglecting aerodynamic drag, it holds at low speed. At high speed the power required on Mars is much lower than that required on Earth, owing to the much lower value of aerodynamic drag on the Red Planet.

D.2 MARS' ATMOSPHERE

A planetary atmosphere is usually assumed to be a mixture of perfect gases, so the pressure p and the density ρ are linked by the relationship

$$\frac{p}{\rho} = R^*T$$ (D.7)

where T is the absolute temperature and R^* is a constant that characterizes the given gas or mixture of gases. It is the ratio between the universal gas constant $R = 8,314$ J/(molK) and the average molecular mass of the gas M

$$R^* = \frac{R}{M}.$$ (D.8)

The average molecular mass for Earth's atmosphere is 29 kg/mol and therefore $R^* = 287$ m^2/s^2 K. On Mars, the atmosphere is made mainly by carbon dioxide, whose molecular mass is 44 kg/mol. If the presence of argon and nitrogen are also accounted for, the molecular mass reduces to about 43.1 kg/mol and hence $R^* = 193$ m^2/s^2 K.

The density is thus

$$\rho = \frac{p}{R^*T} \tag{D.9}$$

and depends much on the temperature.

The temperature and pressure in the Martian atmosphere are much more variable than those of the terrestrial atmosphere, so any efforts to define a 'Mars Standard Atmosphere' are bound to yield only a rough average which is less significant than the one for Earth. NASA published on a website the model obtained by Jonathon Donadee and Dave Hiltner from the data obtained by the Mars Global Surveyor in April 1996.[13] The model divides the atmosphere into two zones: a lower one up to an altitude of 7000 m and a higher one above that altitude.

The temperature has a linear behavior

$$\begin{cases} T = -31 - 0.000998h & \text{for } h \le 7{,}000 \\ T = -23.4 - 0.00222h & \text{for } h > 7{,}000 \end{cases} \tag{D.10}$$

(in °C), while the pressure is exponential

$$p = 0.699e^{-0.00009h} \tag{D.11}$$

measured in kPa.

This model has at least 2 weak points, however. First, the linear temperature law cannot hold at high altitude because the temperature would go below absolute zero, which is impossible. Second, it is not specified whether the altitude is measured from the ground or from the conventional zero level. In Fig. 3.5 the latter interpretation is given and the altitude is limited to 100 km.

The characteristics of the Martian atmosphere are very variable with time, and this makes the design of aerodynamic or aerostatic devices for use there considerably more difficult than in the case of Earth.

Consider the limiting case of a place at a very low elevation with a pressure of 11 millibars (1100 Pa) and a temperature of -120 °C. The density is 0.038 kg/m³ which is about 3.1 % of the Earth's atmospheric pressure at sea level in standard conditions (1.225 kg/m³). The dynamic pressure $\frac{1}{2}\rho V^2$ of a wind blowing at 400 km/h in such an 'unfavorable' place (i.e. unfavorable from the point of view of the forces imparted by the wind, but very favorable in terms of protection from cosmic radiation) is the same as that of a wind blowing at about 70 km/h on Earth.

The speed of sound in air is

$$V_s = \sqrt{\frac{\gamma RT}{\mathcal{M}}} \tag{D.12}$$

---

[13] https://www.grc.nasa.gov/www/k-12/airplane/atmosmrm.html.

where γ is the adiabatic index, whose value is $7/5 = 1.4$ for diatomic molecules and $5/3$ for monatomic gases. On Earth, at sea level in a standard atmosphere its value is $V_s = 340$ m/s. On Mars, $\gamma = 1.29$ and therefore at a temperature of -50 °C the speed of sound is 235.6 m/s (848 km/h).

Since Mars has an atmosphere, albeit a thin one, it is possible to use aerostatic or aerodynamic forces to support a vehicle or a rover.

D.3 FLUIDOSTATIC SUPPORT

The force which a fluid exerts on any body that is immersed in it, is the so-called Archimedes force

$$F = \rho g V \tag{D.13}$$

where ρ is the density of the fluid and V is the volume of the object, or better, of the displaced fluid. The force is directed vertically upwards and Eq. (D.13) holds only if the gravitational acceleration is constant throughout the zone which is occupied by the body. If the body is fully immersed in the fluid, the force is applied in the geometric center of the volume of displaced fluid, or the center of buoyancy. The aerostat is stable if the center of buoyancy is located above the center of mass. Equation (D.13) implies that whilst in liquids it is possible to obtain large forces even with relatively small volumes because of their high density, in gases aerostatic support must displace large volumes of fluid.

Consider an aerostat filled with a gas with molecular mass M_1 flying in an atmosphere made by gases with an average molecular mass M_2. Assuming that the pressures of the gas inside and outside the aerostat are equal and also that the temperatures are equal, the lift is the difference between the aerostatic force and the weight of the gas

$$F = gV\left(\rho_o - \rho_i\right) = \frac{pgV}{RT}\left(\mathcal{M}_o - \mathcal{M}_i\right) \tag{D.14}$$

where subscripts o and i refer to the gases outside and inside the aerostat.

To lift a mass m, the aerostat must exert a force mg. Its volume must thus be

$$V = \frac{mRT}{p\left(\mathcal{M}_o - \mathcal{M}_i\right)}.$$

It may seem strange that aerostats have been proposed for planets with very thin atmospheres, but the low temperature and the atmospheric composition may produce a density high enough to sustain a vehicle. This is because the colder the planet, and the higher the molecular mass and the pressure of the atmosphere, then the greater is the lift that is produced.

Consider, for instance, a balloon filled with hydrogen or helium which can lift a payload of 100 kg (including the structure of the balloon) in typical Martian conditions of an atmospheric pressure of 600 Pa, a temperature of -50 °C (223 K) and $g = 3.77$ m/s^2. Assuming that the balloon is in equilibrium of pressure and temperature with the surrounding atmosphere, the volume is easily computed as 7721 m^3 if the balloon is filled with helium (molecular mass 4) or 7354 m³ if filled with hydrogen. If the balloon is spherical, its diameter is 24.5 m for helium and 24.1 m for hydrogen.

To make a comparison, the ratio between the volume of a balloon lifting a given mass on Mars and a balloon that lifts the same mass on Earth is

$$\frac{V_m}{V_e} = \frac{T_m}{T_e} \frac{p_e}{p_m} \frac{\left(\mathcal{M}_e - \mathcal{M}_i\right)}{\left(\mathcal{M}_m - \mathcal{M}_i\right)} = 81.5.$$

Since the size of the aerostat is proportional to the cube root of the volume, the aerostat for Mars is about 4.3 times as large as that for conditions on Earth.

D.4 FLUID-DYNAMICS SUPPORT

D.4.1 Aerodynamic forces

On Earth, both hydrodynamic and aerodynamic forces are used for transportation. Generally speaking, fluid dynamic forces exerted on an object that is moving in a fluid are proportional to the square of the relative speed, the density of the fluid, and the square of the linear dimension of the object

$$F = \frac{1}{2}\rho V^2 S C_f \tag{D.15}$$

where coefficient 1/2 is included just for historical reasons, surface S is a reference surface, and C_f is a coefficient depending on the shape of the body and its position with respect to the direction of the relative velocity. However, C_f depends also on two non-dimensional parameters, the Reynolds number and the Mach number

$$\mathcal{R}_e = \frac{VL}{\nu}, \quad \mathcal{M}_a = \frac{V}{V_s}$$

where ν is the kinematic viscosity of the fluid, L is a reference length, and V_s is the speed of sound in the fluid. The first one is a parameter showing the relative importance of the viscous and inertial effects in determining the aerodynamic forces. If the value of the Reynolds number is low, aerodynamic forces are primarily caused by viscosity, and if it is high they are mainly caused by the inertia of the fluid. The Mach number shows the importance of fluid compressibility. The reference surface S and length L are arbitrary, to the point that in some cases a surface is used that does not physically exist, like in the case of airships where S is the power 2/3 of the displacement. It is, however, clear that the numerical values of the coefficients depend on the choice of S and L, and these must be clearly stated.

An analysis of the aerodynamic forces is performed using a reference frame G_{xyz} that is fixed to the body and moves with it (Fig. D.2). It is located in the center of mass G, in such a way that the x axis has the direction of the resultant air velocity vector V_r of the object with respect to the atmosphere. The z axis is within the symmetry plane of the body (if that

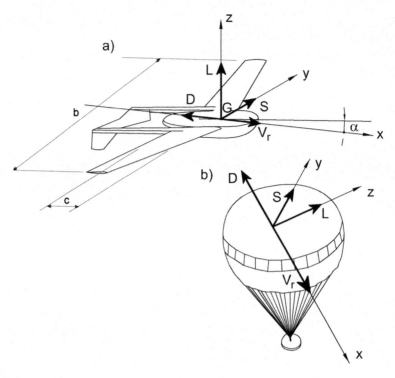

Figure D.2 Aerodynamic forces, reference frame, and components. (**a**) Aircraft (the sketch shows the UAV that NASA once planned to launch to Mars in 2003). (**b**) Parachute.

exists), perpendicular to the x axis. The y axis is perpendicular to the other two. Angle α, known as the *angle of attack*, is the angle between the x axis and a reference direction in the symmetry plane, usually orientated so that the lift vanishes when $\alpha = 0$. Another angle is often defined, the *sideslip angle* δ, between the x axis and the symmetry plane; if there is no symmetry plane, it is defined so that the side force vanishes when $\delta = 0$.

If the aerodynamic force is decomposed along the axes of frame *xyz*, the components are referred to as *drag D*, *side force S*, and *lift L*. In Fig. D.2 the drag is shown pointing backwards, corresponding to the physical situation; however, it would have been more consistent with sign conventions to plot the drag pointing forward and state that its value is negative.

The expressions for the components of the aerodynamic force are the same Eq. (D.15), but instead of using the generic force coefficient C_f we introduce the drag coefficient C_D, the lift coefficient C_L, and the side force coefficient C_S. As a first approximation, for small angles α and δ, the lift can be taken to be proportional to α and the side force to δ.

The reference situation is often that which occurs in a wind tunnel, with the object stationary and the air rushing against it; the velocity of the air relative to the body is then usually displayed, instead of the velocity of the body.

When a body falls in a planetary atmosphere under the action of inertia forces and aerodynamic drag, its ability to overcome the drag is expressed by the *ballistic coefficient*

$$\beta = \frac{m}{SC_D}$$

where S is the cross-sectional area of the body.

The ballistic coefficient of a re-entry vehicle should be as low as possible in order to slow down. A low value of the ballistic coefficient causes the spacecraft to decelerate at high altitude, and thus results in a lower peak of both temperature and deceleration. Decelerating early provides more time for the entire deceleration maneuver, and this is highly desirable when landing on a planet that has a thin atmosphere, such as Mars. To reduce the value of β, the thermal shield must be as large as possible, but this usually conflicts with the need to accommodate the aeroshell within the payload shroud of the launch vehicle. As an example, the Viking lander had a fairly low ballistic coefficient $\beta = 64$ kg/m² whereas previous Mars landers had values in the range 63–94 kg/m².

Because the mass of an object grows with the cube of its size, and the area with the square of its size, it is evident that it will be difficult to make large landers with low ballistic coefficients. Using inflatable aeroshells may be the only way to address this problem. Very large Hypersonic Inflatable Aerodynamic Decelerators (HIADs) are presently under study (Fig. 6.20).

D.4.2 Aircraft

In the case of airplane wings, S is the wing surface and the length L used in computing the Reynolds number is the mean chord c (i.e. the average width of the wing; Fig. D.2). The wing area is $S = bc$, where b is the wingspan.

The boundary layer is the thin layer of air in contact with the body where the viscous effects can be thought to be concentrated. Its thickness increases as the fluid it contains loses energy owing to viscosity, and slows down. If the fluid outside the boundary layer increases its velocity, then a negative pressure gradient along the separation line between the external flow and the boundary layer is created, and this decrease of pressure tends to help the flow within the boundary layer, countering its slowing down. In contrast, if the outer flow slows down, the pressure gradient is positive and the airflow in the boundary layer is hampered. In any case, a separation of the flow and the formation of a wake eventually occurs, but this is facilitated by a positive pressure gradient.

The efficiency of a wing is

$$E = \frac{L}{D} = \frac{C_L}{C_D}. \tag{D.16}$$

Figure D.3a reports the lift and drag coefficients and the efficiency of a wing as functions of the angle of attack. The curves are for a given wing, but are typical.

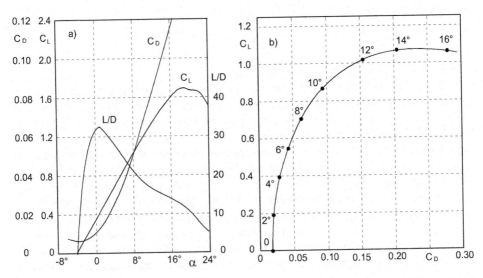

Figure D.3 (a) Lift and drag coefficients and efficiency of a wing as functions of the angle of attack. (b) The polar diagram of an airplane.

The curves $C_L(\alpha)$ and $C_D(\alpha)$ are influenced by many characteristics of the wing, like the airfoil and the planform. In particular, the drop of the lift after it reaches its maximum value (a phenomenon called *stall* of the wing) can be very abrupt.

It is possible to increase the lift coefficient (at the expense of increasing the drag coefficient) by using suitable moving surfaces located at the trailing edge (flaps) or at the leading edge (slats) which change the characteristics of the airfoil. Such high lift devices are used by all modern aircraft during take-off and landing. They might be even more important in aircraft intended for a planet with a low density atmosphere.

The variation of the aerodynamic characteristics of a body in terms of the angle of attack can be plotted in the polar diagram (Fig. D.3a) and the lift coefficient can be plotted in terms of the drag coefficient (Fig. D.3b).

The most common types of aircraft have fixed wings (airplanes) and rotary wings (autogyros and helicopters). Both have been considered for planetary exploration robots and vehicles.

The speed at which a fixed wing aircraft must fly to sustain itself can be computed by equating the weight with the aerodynamic lift

$$mg = \frac{1}{2}\rho V^2 S C_L. \tag{D.17}$$

This yields the flying speed

$$V = \sqrt{\frac{2mg}{\rho S C_L}}. \tag{D.18}$$

The minimum take-off speed can thus be computed by introducing the maximum value of the lift coefficient into Eq. (D.18). At higher speeds the airplane flies with a lower lift coefficient.

The drag at the flight speed is

$$D = \frac{1}{2}\rho V^2 S C_D = mg\frac{C_D}{C_L} = \frac{mg}{E} \qquad (D.19)$$

which is equal to the weight divided by the aerodynamic efficiency. The reciprocal of the efficiency is therefore a sort of a friction coefficient, in the sense that it is a number which, when multiplied by the weight, gives the force that opposes to motion.

The attitude of the aircraft that minimizes the drag is characterized by the maximum aerodynamic efficiency.

The power required for flight is the product of the drag and the speed

$$P = DV = mg\frac{C_D}{C_L}\sqrt{\frac{2mg}{\rho S C_L}} = \sqrt{\frac{2m^3 g^3 C_D^2}{\rho S C_L^3}}. \qquad (D.20)$$

For an aircraft with a given mass and a given wing area, the ratio between the take-off speed on Mars and that on Earth is

$$\frac{V_m}{V_e} = \sqrt{\frac{g_m}{g_e}\frac{T_m}{T_e}\frac{p_e}{p_m}\frac{\mathcal{M}_e}{\mathcal{M}_m}} = 5.7. \qquad (D.21)$$

The ratio between the power required to fly on Mars and that required on Earth is

$$\frac{P_m}{P_e} = \sqrt{\frac{g_m^3}{g_e^3}\frac{T_m}{T_e}\frac{p_e}{p_m}\frac{\mathcal{M}_e}{\mathcal{M}_m}} = 2.2. \qquad (D.22)$$

To diminish the speed and the power required, it is possible to increase the wing area. If the area is multiplied by a factor of 4.7, then the power is the same on Mars as on Earth and the ratio of the speeds reduces to 2.6. These values hold if the mass of the aircraft is assumed to be the same; a thing that can be achieved by taking account of the fact that the lower weight reduces the structural stresses.

The attitude (i.e. the angle of attack) of the aircraft that minimizes the power required for motion is that which maximizes

$$\sqrt{\frac{C_L^3}{C_D^2}} = \sqrt{C_L}\,E$$

This attitude also permits the maximum flight duration for a given quantity of energy stored on board.

A glider is an unpowered aircraft. The time that a glider can fly by losing an altitude Δz (a value that must be small enough to enable the density ρ to be considered constant) can be computed by equating the loss of potential energy with the energy required for flying for a time t

$$mg\Delta z = Pt \tag{D.23}$$

and hence

$$t = \frac{mg\Delta z}{P} = \Delta z \sqrt{\frac{\rho SC_L^3}{2mgC_D^2}}. \tag{D.24}$$

The attitude that maximizes the flight time is the same as the one that minimizes the power required for flight. If a glider is released in the high atmosphere of a planet, then the time it will take to reach the surface can be easily computed by integrating Eq. (D.24), taking into account the fact that the density of the atmosphere changes with the altitude.

D.4.3 Parachutes

Other aerodynamic devices used in planetary atmospheres are parachutes. The drag coefficient of a domed parachute is about 1.5, but there are other types of parachutes which possess different shapes. Inflatable devices, usually referred to as ballutes, have been proposed. These combine the ways that parachutes and balloons function.

For example, by assuming the same data for the Martian atmosphere as used previously, we can compute the minimum diameter of a parachute that can land 100 kg with a vertical speed of 5 m/s. It descends at constant speed when the drag is equal to the weight

$$mg = \frac{1}{2}\rho V^2 SC_D. \tag{D.25}$$

The surface area of the parachute is thus

$$S = \frac{2mg}{\rho V^2 C_D} = 1,416\,\mathrm{m}^2$$

corresponding to a diameter of 42.4 m. The assumed speed of 5 m/s is the asymptotic speed reached in steady state conditions, but during the descent the speed will be higher because the atmospheric density declines as the altitude increases.

The descent speed is linked to the ballistic coefficient by the relationship

$$V = \sqrt{\frac{2g}{\rho}\beta}.$$

D.4.4 Propellers and helicopter rotors

A propeller that is rotating at an angular velocity Ω produces a thrust which can be approximated by the equation

$$F = 2k\rho S\Omega^2 \tag{D.26}$$

where k is a constant which, in hovering, depends only upon the geometry of the propeller and on the Mach number of the tip of the blades. S is the area of the propeller disc. If a multicopter (Fig. 9.13) with n rotors has a mass m, the equilibrium equation at take-off is

$$mg = 2nk\rho S\Omega^2 . \tag{D.27}$$

In the case of a low atmospheric density which compels an increase in the angular velocity of the propellers, it may be convenient to use a larger number of smaller propellers to avoid working at a too high Mach number. The total power (all rotors) needed to produce the thrust is

$$P = n\frac{F^{3/2}}{\sqrt{2\rho S}} = n\frac{(mg)^{3/2}}{(n)^{3/2}\sqrt{2\rho S}} \tag{D.28}$$

and hence

$$P = \frac{(mg)^{3/2}}{\sqrt{2n\rho S}} . \tag{D.29}$$

As in the case of the velocity of an airplane, the ratio between the rotor speed of a multicopter of a given mass and a given rotor area on Mars and on Earth is

$$\frac{\Omega_m}{\Omega_e} = \sqrt{\frac{g_m}{g_e}\frac{T_m}{T_e}\frac{p_e}{p_m}\frac{\mathcal{M}_e}{\mathcal{M}_m}} = 5.7 . \tag{D.30}$$

The ratio between the power required to fly on Mars and that required on Earth is

$$\frac{P_m}{P_e} = \sqrt{\frac{g_m^3}{g_e^3}\frac{T_m}{T_e}\frac{p_e}{p_m}\frac{\mathcal{M}_e}{\mathcal{M}_m}} = 2.2 . \tag{D.31}$$

The power can be decreased by using larger propellers. As an example, if the diameter of the rotors is increased by a factor of 2.17, then the power remains the same and the speed of the propeller is increased by a factor of 2.6. However, with increasing propeller size and angular velocity, the Mach number at the tips of the blades increases and the performance of the propeller decreases.

Symbols

f_0	Rolling coefficient at low speed
g	Gravitational acceleration
i	Road slope
m	Mass
p	Pressure
C_i	i-th aerodynamic coefficient
D	Aerodynamic drag
E	Aerodynamic efficiency
F	Force
K	Coefficient
L	Aerodynamic lift
M	Molecular mass
M_a	Mach number
P	Power
R	Road load, universal gas constant
R_e	Reynolds number
S	Surface, aerodynamic side force
T	Temperature (absolute)
V	Velocity, volume
V_s	Speed of sound
α	Slope angle, angle of attack
β	Ballistic coefficient
γ	Adiabatic index
δ	Sideslip angle
η_t	Transmission efficiency
ρ	Density

References

[1] W. Von Braun, *The Mars Project*, University of Illinois Press, 1952 (2nd Ed. 1991).

[2] H. Seifert (editor), *Space Technology*, Wiley, New York, 1959.

[3] E. Stuhlinger E., J.C. King, "Concept for a manned mars expedition with electrically propelled vehicles," AIAA, 1963.

[4] R.R. Bate, D.D. Mueller, J.E. White, *Fundamentals of Astrodynamics*, Dover, New York, 1971.

[5] J.P. Marec, *Optimal Space Trajectories*, Elsevier, New York, 1979.

[6] B. Finney, *The Prince and the Eunuch, in Interstellar Migration and the Human Experience*, University of California Press, Berkeley, 1985.

[7] IAA, *The International Exploration of Mars, 4th Cosmic Study of the IAA*, April 1993.

[8] IAA *Human Exploration of Mars: The Reference Mission of the NASA Mars Exploration Study Team*, NASA SP-6107 - NASA Johnson Space Center, 1997.

[9] R. Zubrin, C.P. McKay, "Terraforming Mars," in S. Schmidt, R. Zubrin, *Islands in the sky*, Wiley, New York, 1996.

[10] R. Zubrin, D.A. Baker, "Mars Direct: A Proposal for the Rapid Exploration and Colonization of the Red Planet," in S. Schmidt, R. Zubrin, *Islands in the sky*, Wiley, New York, 1996.

[11] R. Zubrin, *The Case for Mars*, Touchstone, New York, 1997.

[12] B.G. Drake ed., *Reference Mission Version 3.0 Addendum to the Human Exploration of Mars*, EX13-98-036, NASA Johnson Space Center, 1998.

[13] B. Jakosky, *The Search for Life on Other Planets*, Cambridge University Press, Cambridge, 1998.

[14] J. Wiley, W.J Larson, L.K. Pranke, *Human Spaceflight: Mission Analysis and Design*, McGraw-Hill, New York, 1999.

[15] L. Bergreen, *The Quest for Mars*, Harper Collins, London, 2000.

[16] K.R. Lang, *The Sun from Space*, Springer, New York, 2000.

[17] W.K. Hartmann, *A Traveler's Guide to Mars*, Springer, New York, 2000.

[18] G. Genta, Rycroft M., *Space, The Final Frontier?*, Cambridge University Press, Cambridge, 2003.

© Springer International Publishing Switzerland 2017

G. Genta, *Next Stop Mars*, Springer Praxis Books, DOI 10.1007/978-3-319-44311-9

[19] P.A. Clancy Brack and G. Hornick *Looking for Life, Searching the Solar System*, Cambridge University Press, Cambridge 2006.

[20] W. Von Braun, *Project Mars: A Technical Tale*, 2006.

[21] N. Kanas, D. Manzey, *Space Psychology and Psychiatry*, 2nd Edition. El Segundo, California: Microcosm Press; and Dordrecht, The Netherlands: Springer, 2008.

[22] D. Rapp, *Human Missions to Mars*, Springer-Praxis, Chichester, 2008.

[23] J. Dewar, *The Nuclear Rocket*, Apogee Books, Burlington, Canada, 2009.

[24] B.G. Drake ed., *Mars Architecture Steering Group, Human Exploration of Mars, Design Reference Architecture 5.0* (and addendums), NASA Johnson Space Center, 2009.

[25] G.Genta, *Introduction to the Mechanics of Space Robots*, Springer, New York, 2011.

[26] ISECG, *The Global Exploration Roadmap*, ISECG Technical Report, August 2013.

[27] P.A. Swan, D.I. Raitt C.W. Swan, R.E. Penny, J.M. Knapman (editors) *Space Elevators, an Assessment of the Technological Feasibility and the Way Forward*, IAA, Paris, 2013.

[28] Committee on Human Spaceflight Pathway to Exploration: *Rationales and Approaches for a U.S. Program of Human Space Exploration*, The National Academies Press, Washington D.C., 2014.

[29] C. Miller et al., *Economic Assessment and System Analysis of an Evolvable Lunar Architecture*, NexGen Space, 2015.

[30] A. Dula, Z. Zhenjun (editors) *Space Mineral Resources, a Global Assessment of the Challenges and Opportunities*, IAA, Paris, 2015.

[31] S. Hubbars, J. Logsdon, C. Dreier, J. Callahan, *Humans Orbiting Mars*, The Planetary Society, Pasadena, 2015.

[32] G. Genta, J.M. Salotti (editors), *Global Human Mars System Missions Exploration Goals, Requirements and Technologies*, IAA, Paris, 2016.

About the author

Giancarlo Genta is professor of Construction of Machines at the Politecnico (Technical University) of Torino, in Italy. Since 1987 he has taught courses in Astronautic Propulsion, Construction of Aircraft Engines, Space Robotics and several courses in motor vehicle technology. He is a member of the Academy of Sciences of Torino and of the International Academy of Astronautics.

His research activity has been focused on static and dynamic structural analysis, dynamics of rotating machinery, and of controlled systems and space robots. Since 2012 he has chaired study group SG 3.16 of the International Academy of Astronautics on human Mars exploration.

He has authored 24 books, some of which are used as text books in Italian and American Universities, has published 90 papers in Italian, American, and English journals, and has presented 263 papers to symposia. He has also written two popular science books, one on space exploration and the other on the search for extra-terrestrial intelligence, and two science fiction novels in the Science and Fiction series published by Springer.

© Springer International Publishing Switzerland 2017
G. Genta, *Next Stop Mars*, Springer Praxis Books, DOI 10.1007/978-3-319-44311-9

Index

© Springer International Publishing Switzerland 2017
G. Genta, *Next Stop Mars*, Springer Praxis Books, DOI 10.1007/978-3-319-44311-9

CPSIA information can be obtained
at www.ICGtesting.com
Printed in the USA
LVOW01s1432280217

525681LV00003B/12/P